Handbook of Smart Manufacturing

This handbook covers smart manufacturing development, processing, modifications, and applications. It provides a complete understanding of the recent advancements in smart manufacturing through its various enabling manufacturing technologies, and how industries and organizations can find the needed information on how to implement smart manufacturing towards sustainability of manufacturing practices.

Handbook of Smart Manufacturing: Forecasting the Future of Industry 4.0 covers all related advances in manufacturing such as the integration of reverse engineering with smart manufacturing, industrial internet of things (IIoT), and artificial intelligence approaches, including artificial neural network, Markov decision process, and heuristics methodology. It offers smart manufacturing methods like 4D printing, micro-manufacturing, and processing of smart materials to assist the biomedical industries in the fabrication of human prostheses and implants. The handbook goes on to discuss how to accurately predict the requirements, identify errors, and make innovations for the manufacturing process more manageable by implementing various advanced technologies and solutions into the traditional manufacturing process. Strategies and algorithms used to incorporate smart manufacturing into different sectors are also highlighted within the handbook.

This handbook is an invaluable resource for stakeholders, industries, professionals, technocrats, academics, research scholars, senior graduate students, and human healthcare professionals.

Handbook of Smart Manufacturing

Forecasting the Future of Industry 4.0

Edited by
Ajay, Hari Singh, Parveen,
and Bandar AlMangour

CRC Press
Taylor & Francis Group
Boca Raton London New York

CRC Press is an imprint of the
Taylor & Francis Group, an **informa** business

Designed cover image: Shutterstock

First edition published 2023
by CRC Press
6000 Broken Sound Parkway NW, Suite 300, Boca Raton, FL 33487-2742

and by CRC Press
4 Park Square, Milton Park, Abingdon, Oxon, OX14 4RN

CRC Press is an imprint of Taylor & Francis Group, LLC

Library of Congress Cataloging-in-Publication Data
Names: Kumar, Ajay, Dr., editor.
Title: Handbook of smart manufacturing : forecasting the future of industry 4.0 / edited by Ajay, Hari Singh, Parveen, Bandar AlMangour.
Description: First edition. | Boca Raton : CRC Press, 2023. | Includes bibliographical references and index.
Identifiers: LCCN 2022060155 (print) | LCCN 2022060156 (ebook) | ISBN 9781032363431 (hbk) | ISBN 9781032367866 (pbk) | ISBN 9781003333760 (ebk)
Subjects: LCSH: Industry 4.0. | Manufacturing processes. | Automation.
Classification: LCC T59.6 .H36 2023 (print) | LCC T59.6 (ebook) | DDC 629.8/90285--dc23/eng/20230210
LC record available at https://lccn.loc.gov/2022060155
LC ebook record available at https://lccn.loc.gov/2022060156

ISBN: 978-1-032-36343-1 (hbk)
ISBN: 978-1-032-36786-6 (pbk)
ISBN: 978-1-003-33376-0 (ebk)

DOI: 10.1201/9781003333760

Typeset in Times
by MPS Limited, Dehradun

Contents

Preface

Smart manufacturing is a broad category of manufacturing that employs computer-integrated manufacturing, high levels of adaptability and rapid design changes, digital information technology, and more flexible technical workforce training. Other goals sometimes include fast changes in production levels based on demand, optimization of the production system, efficient production, and recyclability. This book is expected to provide the compiled smart manufacturing knowledge to the involved stakeholders from users in the industry to the academics and researchers. The detailed description of various smart manufacturing techniques like 3D printing, 4D printing, and approaches that makes manufacturing digital such as Industry 4.0 techniques are included in this book along with various smart materials like polymers, metals, ceramics, composites, biomaterials, biodegradable materials, responsive materials, functionally graded materials, and various applications of smart industries. In addition to above various case studies of smart manufacturing specially related to production, mechanical, industrial, computer science, electronics engineering, and biomedical are covered to help undergraduates, postgraduates, and research scholars. This compilation will bring to its readers the latest developments/advancements in smart manufacturing, which is currently not available at one place. This book will bridge the gap between R&D in smart manufacturing and professionals.

This book is an outcome of the extensive research accomplished by various researchers, academicians, scientists, and industrialists in smart manufacturing. The area is under-explored and the outcomes are worth the research effort. Experimentation, modeling, characterization, and simulation techniques are the powerful tools for developing new concepts, approaches, and solutions to devise valuable information on the process which led to need of compiling this work. Since, the information related to smart manufacturing is scattered into patents and research publications and not at one place in a systematic form, editors recognize their ethical responsibility to compile, share, and spread the knowledge accumulated and technology developed, with the students, researchers, and industry people, to draw the benefits of this work in the form of a book and gain technical competence in the frontal area.

The book consists of 18 chapters that describe perspectives of smart manufacturing with Industry 4.0 aspects. "Smart Manufacturing and Industry 4.0: State-of-the-Art Review" determines the dimensions that help in implementing smart manufacturing in the I4.0 environment in SMEs. "Study and Analysis of IOT (Industry 4.0): A Review" provides an overview of analysis of IOT adaptation effect in past and present industries. "Recent advances in Cybersecurity in Smart Manufacturing Systems in the Industry" discusses advances in the security of smart manufacturing systems, strengths and weaknesses of the manufacturing systems, existing threats, and preparedness for future cyber-attacks. "Integration of Circular Supply Chain and Industry 4.0 to Enhance Smart Manufacturing Adoption" focuses to develop a framework based on the Fuzzy Analytical Hierarchy Process (FAHP), one of the multi-criteria decision-making methods (MCDM), which computes and assigns weights to each identified circular supply chain practice within an industry. "Artificial Intelligence with Additive Manufacturing" summarizes the role of artificial intelligence in additive manufacturing in a manner to show how innovative additive manufacturing plays a role in today's industry. "Robotic Additive Manufacturing Vision towards Smart Manufacturing and Envisage the Trend with Patent Landscape" foresees robotic additive manufacturing in various aspects covering basic system step to futuristic vision. "Smart Materials for Smart Manufacturing" demonstrates the various kinds of smart nanomaterials and their advantageous properties in both historical and future applications. "Smart Biomaterials in Industry and Healthcare" explores smart materials for bringing advancement in biomedical engineering and other industrial applications. "Ferroelectric Polymer Composites and Evaluation of Their Properties" presents the results of a study of ferroelectric polymer composites with a high permittivity. "4D Print Today and Envisaging the Trend with Patent Landscape for Versatile Applications" analyzes the future trend with the aid of patent landscape analysis. "Investigating the Work Generation Potential of SMA Wire Actuator" aims to develop an elementary mathematical model of an SMA wire actuator in biasing conditions on the SIMULINK platform while including the traveled path history of SMA material and thereby, carries out a parametric study that determines the most influencing parameter that affects the work generation capability of wire actuator. "Troubleshooting on the Sample Preparation during Fused Deposition Modeling" deals in-depth with the challenges and issue during the printing and suitable solution for better print quality. "Hybrid Additive Manufacturing Technologies" focuses to produce and compare research studies on the different hybrid additive manufacturing processes. "Smart Manufacturing Using 4D Printing" deals with the concepts and regulations governing 4D printing, along with the materials utilized, applications, and obstacles that still need to be removed. "Developments in 4D Printing and Associated Smart Materials" provides a brief overview of 4D printing processes using various composites and intelligent materials including smart materials, and shape memory alloys. "Role of Smart Manufacturing Systems in Improving Electric Vehicle Production" explores the current market need for manufacturing that closely relates to an electric vehicle and develops a structural framework that utilizes the interpretative structural modeling approach and MICMAC approach to achieve the objective. "Safety Management with Application of Internet of Things, Artificial Intelligence, and Machine Learning for Industry 4.0 Environment" evaluates the major issues regarding health and safety integration with the Industry 4.0 revolution. "CPM/PERT-Based Smart Project Management: A Case Study" optimize the design and installation time of the 8-megawatt power plant to be built for a cement factory in Barbados.

This book is intended for both the academia and the industry. The postgraduate students, PhD students, and researchers in universities and institutions, who are involved in the areas of smart manufacturing with Industry 4.0 perspectives, will find this compilation useful.

The editors acknowledge the professional support received from CRC Press and express their gratitude for this opportunity.

Reader's observations, suggestions, and queries are welcome,

Dr. Ajay
Dr. Hari Singh
Mr. Parveen
Dr. Bandar AlMangour

Acknowledgments

The editors are grateful to the CRC Press for showing their interest to publish this book in the area of smart manufacturing and Industry 4.0. The editors express their personal adulation and gratitude to Ms. Cindy Renee Carelli (Executive Editor) CRC Press, for giving consent to publish our work. She undoubtedly imparted the great and adept experience in terms of systematic and methodical staff who have helped the editors to compile and finalize the manuscript. The editors also extend their gratitude to Ms. Christina Graben, CRC Press, for support during her tenure.

The editors wish to thank all the chapter authors for contributing their valuable research and experience to compile this volume. The chapter authors, corresponding authors in particular, deserve special acknowledgments for bearing with the editors, who persistently kept bothering them for deadlines and with their remarks.

The editors, Dr. Ajay, Mr. Parveen and Dr. Bandar AlMangour, wish to thank Prof. Hari Singh for his unreserved guidance, valuable suggestions, and encouragement in nurturing this work. Prof. Hari Singh is a wonderful person and the epitome of simplicity, forthrightness, and strength and is a role model for editors.

Dr. Ajay also wishes to express his gratitude to his parents, Sh. Jagdish and Smt. Kamla, and his loving brother Sh. Parveen for their true and endless support. They have made him able to walk tall before the world, regardless of sacrificing their happiness and living in a small village. He cannot close these prefatory remarks without expressing his deep sense of gratitude and reverence to his life partner, Mrs. Sarita Rathee, for her understanding, care, support, and encouragement to keep his morale high all the time. No magnitude of words can ever quantify the love and gratitude the he feels in thanking his daughters, Sejal Rathee and Mahi Rathee, and son Kushal Rathee who are the world's best children.

Finally, the editors obligate this work to the divine creator and express their indebtedness to the "ALMIGHTY" for gifting them power to yield their ideas and concepts into substantial manifestation. The editors believe that this book would enlighten the readers about each feature and characteristics of smart manufacturing and Industry 4.0.

Dr. Ajay
Dr. Hari Singh
Mr. Parveen
Dr. Bandar AlMangour

Editor's Biography

Dr. Ajay is currently serving as an associate professor in the Mechanical Engineering Department, School of Engineering and Technology, JECRC University, Jaipur, Rajasthan, India. He received his PhD in the field of advanced manufacturing from Guru Jambheshwar University of Science & Technology, Hisar, India after B.Tech. (Hons.) in mechanical engineering and M.Tech. (Distinction) in manufacturing and automation. His areas of research include artificial intelligence, materials, incremental sheet forming, additive manufacturing, advanced manufacturing, Industry 4.0, waste management, and optimization techniques. He has over 60 publications in international journals of repute including SCOPUS, Web of Science, and SCI indexed database and refereed international conferences. He has also co-authored and co-edited many books and proceedings including: *Incremental Sheet Forming Technologies: Principles, Merits, Limitations, and Applications*, CRC Press, ISBN: 9780367276744; *Advancements in Additive Manufacturing: Artificial Intelligence, Nature Inspired and Bio-manufacturing*, ISBN: 9780323918343, ELSEVIER; *Handbook of Sustainable Materials: Modelling, Characterization, and Optimization*, CRC Press, ISBN: 9781032286327; and *Waste Recovery and Management: An Approach Towards Sustainable Development Goals*, CRC Press, ISBN: 9781032281933.

He has organized various national and international events, including an international conference on Mechatronics and Artificial Intelligence (ICMAI-2021) as a conference chair. He has more than 20 national and international patents to his credit. He has supervised more than 8 MTech, PhD scholars, and numerous undergraduate projects/ thesis. He has a total of 15 years of experience in teaching and research. He is a guest editor and review editor of the reputed journals including Frontiers in Sustainability. He has contributed to many international conferences/symposiums as a session chair, expert speaker, and member of the editorial board. He has won several proficiency awards during the course of his career, including merit awards, best teacher awards, and so on.

He is adviser of the QCFI, Delhi Chapter student cell at JECRC University and has also authored many in-house course notes, lab manuals, monographs, and chapters in books. He has organized a series of Faculty Development Programs, International Conferences, workshops, and seminars for researchers, PhD, UG, and PG students. He is associated with many research, academic, and professional societies in various capacities.

Prof. Hari Singh is currently a professor in the Mechanical Engineering Department at NIT Kurukshetra, India. He completed his bachelor's degree in mechanical engineering with Honours in 1987, master in mechanical engineering in 1994, and PhD in 2001 from Regional Engineering College, Kurukshetra (now, NIT Kurukshetra). He has 34 years of teaching and research experience in the present institute. He has published 150 research papers in various journals of repute and national/international conference proceedings. He has attended a number of conferences abroad in Brazil, Australia, Hong Kong, Singapore, France, Italy, and the United Kingdom. He has supervised 15 PhDs and 8 are in progress. He has also supervised 42 MTech students for their dissertations. He has three patents to his credit. His areas of interest include conventional and unconventional manufacturing processes, product and process improvement, single and multi-response optimization, design of experiments, Taguchi methods, response surface methodology, etc. He has memberships in various professional societies like the Institution of Engineers (India), Indian Society of Theoretical and Applied Mechanics (ISTAM), Indian Society of Technical Education (ISTE), and International Association of Engineers (IAENG).

Mr. Parveen is currently serving as an assistant professor and head in the Department of Mechanical Engineering at Rawal Institute of Engineering and Technology, Faridabad, Haryana, India. Currently, he is pursuing a PhD from the National Institute of Technology, Kurukshetra, Haryana, India. He completed his BTech (Hons.) in mechanical engineering and MTech (Distinction) in manufacturing and automation. His areas of research include materials and their synthesis and characterization techniques, die-less forming, bio-additive manufacturing, CAD/CAM and artificial Intelligence, machine learning and Internet of Things in manufacturing, and multi-objective optimization techniques. He has over 20 publications in international journals of repute including SCOPUS, Web of Science, and SCI indexed database and refereed international conferences. He has more than eight national and international patents in his credit. He has supervised more than 2 MTech scholars and numerous undergraduate projects/theses. He has a total of 12 years of experience in teaching and research. He has co-authored/co-edited the following books:

- Waste Recovery and Management: An Approach Towards Sustainable Development Goals, CRC Press, Taylor and Francis, ISBN: 9781032281933
- Handbook of Sustainable Materials: Introduction, Modelling, Characterization and Optimization, CRC Press, Taylor and Francis, ISBN: 9781032295874

He has organized a series of faculty development Programs, workshops, and seminars for researchers, UG-level students. He is associated with many research, academic, and professional societies in various capacities.

Bandar AlMangour is currently an Assistant Professor at King Fahd University of Petroleum and Minerals (KFUPM) in the Department of Mechanical Engineering. Prior to joining KFUPM, AlMangour has served in multiple operational and research roles within SABIC, Saudi Arabia, and has pursued his Postdoctoral Fellowship at Harvard University. Dr. AlMangour received his Ph.D./MSc, with distinction, in Materials Science and Engineering from UCLA in 2017 and 2014 respectively, an MEng in Materials Engineering from McGill University in 2012, and a BSc in Mechanical Engineering from KFUPM in 2005. He sits on the Editorial (Advisory) Board of several international journals and he is a member of several professional organizations. His research is generally concerned with the investigation of fundamental industrial problems related to the processing, microstructure and behavior of materials, ranging from metals, to (nano-) composites, for advanced structural applications. His research focuses on advanced materials processing (additive manufacturing, friction joining, laser processing, powder metallurgy, bulk nanostructured alloys and composites), mechanical behavior at multiple length scales, and surface engineering. AlMangour has authored more than 40 peer-reviewed papers in internationally recognized journals, and his research works have been presented at well over two dozen international conferences, including invited talks in UAE, Turkey, and Greece.

Contributors

B. Arulmurugan
Department of Mechanical Engineering
KPR Institute of Engineering and Technology
Coimbatore, Tamil Nadu, India

Dhanasekaran Arumugam
Center for NC Technologies
Department of Mechanical Engineering
Chennai Institute of Technology Madras
Chennai, Tamil Nadu, India

Dinesh Kumar Atal
Department of Biomedical Engineering
Deenbandhu Chhotu Ram University of Science and Technology
Murthal, Sonipat, Haryana, India

Devarajan Balaji
Department of Mechanical Engineering
KPR Institute of Engineering and Technology
Coimbatore, Tamil Nadu, India

Nisha Bhatt
Thapar Institute of Engineering and Technology
Patiala, Punjab, India

Bhavna
Advanced Sensors Lab Department of Physics
Deen-Bandhu Chhotu Ram University of Science and Technology
Murthal, Sonipat, Haryana, India

V. Bhuvaneswari
Department of Mechanical Engineering
KPR Institute of Engineering and Technology
Coimbatore, Tamil Nadu, India

Don Biswas
Departmnet of Instrumentation Engineering
Hemvati Nandan Bahuguna Garhwal University
Srinagar Garhwal, Uttarakhand, India

Aryan Boora
Advanced Sensors Lab
Department of Physics
Deen-Bandhu Chhotu Ram University of Science and Technology
Murthal, Sonipat, Haryana, India

Ganesh P. Borikar
School of Mechanical Engineering
MIT World Peace University
Pune, Maharashtra, India

Nidhi Chaubey
Department of Biotechnology
Deenbandhu Chhotu Ram University of Science and Technology
Murthal, Sonipat, Haryana, India

Sandeep Chhillar
Department of Mechanical Engineering
JECRC University
Jaipur, Rajasthan, India

S. Dharanikumar
Department of Mechanical Engineering
KPR Institute of Engineering and Technology
Coimbatore, Tamil Nadu, India

Surender Duhan
Advanced Sensors Lab
Department of Physics
Deen-Bandhu Chhotu Ram University of Science and Technology
Murthal, Sonipat, Haryana, India

Olga S. Gefle
National Research Tomsk Polytechnic University
Tomsk, Russia

Manoj Kumar Gupta
Department of Mechanical Engineering
Hemvati Nandan Bahuguna Garhwal University
Srinagar Garhwal, Uttarakhand, India

Tarun Gupta
Department of Mechanical Engineering
GL Bajaj Institute of Technology & Management
Greater Noida, UP, India

Snehal B. Kolekar
School of Mechanical Engineering
MIT World Peace University
Pune, Maharashtra, India

Tanush Kumaar
Department of Mechanical Engineering
Chennai Institute of Technology Madras
Chennai, Tamil Nadu, India

Ajay
Department of Mechanical Engineering
School of Engineering and Technology
JECRC University
Jaipur, Rajasthan, India

Dharmender Kumar
Department of Biotechnology
Deenbandhu Chhotu Ram University of Science and Technology
Murthal, Sonipat, Haryana, India

Love Kumar
National Institute of Technology Hamirpur
H.P., India

Parveen
Department of Mechanical Engineering
Rawal Institute of Engineering and Technology
Faridabad, Haryana, India

M. Kumaran
Department of Production Engineering
National Institute of Technology
Tiruchirappalli, Tamil Nadu, India

Sergey M. Lebedev
National Research Tomsk Polytechnic University
Tomsk, Russia

Dharamvir Mangal
Department of Mechanical Engineering
Gautam Buddha University
Greater Noida, UP, India

Ravindra Mohan
Department of Mechanical Engineering
Samrat Ashok Technological Institute (Engineering College)
Vidisha, MP, India

Ashutosh Patil
School of Mechanical Engineering
MIT World Peace University
Pune, Maharashtra, India

Ajay John Paul
School of Mechanical Engineering
Kyungpook National University
Daegu, South Korea

Priya
Advanced Sensors Lab
Department of Physics
Deen-Bandhu Chhotu Ram University of Science and Technology
Murthal, Sonipat, Haryana, India

M. Priyadharshini
School of Computer Science and Engineering
Vellore Institute of Technology
AP University
India

S. Rajkumar
School of Mechanical and Electrochemical Engineering
Institute of Technology
Hawassa University
Hawassa, Ethiopia

Akash Rai
Swarnim Startup and Innovation University
Gandhinagar, Gujarat, India

Supriya Sehrawat
Advanced Sensors Lab
Department of Physics
Deen-Bandhu Chhotu Ram University of Science and Technology
Murthal, Sonipat, Haryana, India

J.P. Shakya
Department of Mechanical Engineering
Samrat Ashok Technological Institute (Engineering College)
Vidisha, MP, India

Pankaj Sharma
Department of Mechanical Engineering
JECRC University
Jaipur, Rajasthan, India

Rajiv Kumar Sharma
National Institute of Technology
Hamirpur, H.P., India

Pradeep Singh
Department of Mechanical Engineering
Samrat Ashok Technological Institute (Engineering College)
Vidisha, MP, India

Ranbir Singh
Department of Mechanical Engineering
BML Munjal University
Haryana, India

Ashish Singla
Thapar Institute of Engineering and Technology
Patiala, Punjab, India

Sanjeev Soni
Central Scientific Instruments Organization (CSIR-CSIO)
Chandigarh, India

Christopher Stephen
Department of Mechanical Engineering
Vel Tech Rangarajan Dr. Sagunthala R&D Institute of Science and Technology
Chennai, Tamil Nadu, India

Prashant Thapliyal
Departmnet of Instrumentation Engineering
Hemvati Nandan Bahuguna Garhwal University
Srinagar Garhwal, Uttarakhand, India

Vishal Tiwari
Department of Biomedical Engineering
Deenbandhu Chhotu Ram University of Science and Technology
Murthal, Sonipat, Haryana, India

Arunpillai Viswanathan
Department of Mechanical Engineering
Chennai Institute of Technology Madras
Chennai, Tamil Nadu, India

Monika Vyas
L. D. College of Engineering
Ahmedabad, Gujarat, India

Gunjan Yadav
Swarnim Startup and Innovation University
Gandhinagar, Gujarat, India

1 Smart Manufacturing and Industry 4.0

State-of-the-Art Review

Love Kumar
Department of Mechanical Engineering, NIT Hamirpur, Himachal Pradesh, India

Ajay
Department of Mechanical Engineering, School of Engineering and Technology, JECRC University, Jaipur, Rajasthan, India

Rajiv Kumar Sharma
Department of Mechanical Engineering, NIT Hamirpur, Himachal Pradesh, India

Parveen
Department of Mechanical Engineering, Rawal Institute of Engineering and Technology, Faridabad, Haryana, India

CONTENTS

DOI: 10.1201/9781003333760-1

1.1 INTRODUCTION

The term "smart manufacturing," or SM, was used to describe a new approach to production that is in step with the trends of "industry 4.0". The potential of I4.0 in data networking and information technology have considerable effect on manufacturing operations [1]. With big data processing, artificial intelligence, and intelligent robots, "smart manufacturing" improves factory output while cutting costs in energy and labour. This allows for the machines and tools that are used in smart manufacturing to be interconnected with one another [2]. Smart manufacturing is a subset of I4.0 that seeks to improve production by applying AI techniques including machine learning, big data analytics, and computer simulations. [3]. It also refers to the adoption of cutting-edge cyber technologies by corporate executives, including enhanced sensing, control, modelling, and platform technologies in I4.0 environment [4].

The manufacturing plans of most influential countries in the global market are focused on the industries of the future. This is a major development in the implementation of intelligent production methods. Germany was the first country to launch the I4.0 initiative for smart manufacturing, which is now recognized across the globe as the generational leap in technology that will revolutionize the manufacturing industry [2]. I4.0 was recognised for the first time in German industries, where it has led to increased productivity with the adoption of SM practices and more effective utilisation of the resources that are available [5,6]. It encouraged the businesses to adopt smart manufacturing practices, IIoT, cloud analytics and big data etc. Notably, it has resulted in high performance and, in comparison to the past, significantly greater positive effects. Role of work, management, and social ecosystem elements in SM need to be considered. [7]. The job profiles are undergoing transformations as a direct result of the opportunities made available by I4.0. Leaders in the corporate and public sectors, as well as academics and practitioners, need to collaborate to learn more about smart manufacturing's potential and its necessary infrastructure. [8,9].

At present, investigations into smart manufacturing are underway, but lacks Industry 4.0 adoption. Also, the present literature is devoid of assessment of the different facets of Industry 4.0 [10]. Several supplementary investigations have delved into the current literature and identified hindrances to the adoption of intelligent manufacturing [10–13]. Zhou et al. [14] examine the main topics of smart factories and SM within the context of Industry 4.0, while also presenting obstacles, prospects, strategic planning, and crucial technologies.

The adoption of I4.0 concept among practitioners is popular as a means of putting smart manufacturing practises into effect. Industry 4.0 is not yet ready to be implemented because there is a lack of technical know-how and terminology [15]. Currently, there is limited knowledge about the possible advantages and disadvantages associated with the introduction of intelligent manufacturing [16,15]. Companies are unable to implement I4.0 because they are unable to comprehend the particulars of the implementation that are relevant to their own organisation [17]. Organizations have not conducted any specific research in their business operations, as such research has not been found [17,15]. As a direct consequence of this, the vast majority of businesses are unable to devise an appropriate strategy for implementing industry 4.0 [18].

For in-depth understanding of the dimensions that influence the success of I4.0 implementation, a Petersen et al. [19] guidelines-based SLR was carried out. The participants in this study will be provided with a tool that will assist them in determining which aspects of their organization are the most important and in formulating a strategy to improve those aspects using the findings of this study.

The entire research paper is broken up into six sections. After the introduction has been finished for section 1, the related work, which is contained in section 2, is presented. The discussion of the methodology can be found in Section 3. Section 4 contains a presentation of the results. In section 5, the validity threats are detailed, and in section 6, the conclusions are discussed.

1.2 RELATED WORK

Numerous research has been performed to identify and evaluate dimensions. All the studies are distinct from one another in terms of the approaches and procedures that were utilized to give the facets of putting the I4.0 concept into practice in manufacturing firms. The contributions that were made by the related work can be found in this section.

In order to give the reader a full picture of what SMSs are and how they work, Qu et al. [20] summarises the background, definition, goals, functional requirements, business needs, and technological requirements of SM. At the same time, it describes where things stand in terms of progress. Based on this, we present a model of autonomous smart manufacturing that is driven by fluctuating demand and key performance measures. The technological differences between traditional production and smart manufacturing were explored by Phuyal et al. [2]. In this study, as well as its current implementation status. Recent developments and their consequences were reviewed for this field.

To ascertain the dimensions, Elibal and Ozceylan [21] conducted a relevant literature study. The authors employed various database search to examine 90 distinct studies. There is a discussion on the categorization of the studies, which includes things like meta-models, combinations of models, comparison models, and so on. To conduct this research, qualitative analysis is performed. In the study, the descriptive evaluation model for the review study was presented, and the limitations of the existing model were discussed. The study has some problems with its validity. Ghobakhloo et al. [7] in their work determined drivers of I4.0 concept. They also described a descriptive evaluation of the 745 studies that were eligible for consideration. The study found that organisational factors, technology factors, and environmental factors are the primary determinants of I4.0. The author discussed both the study's theoretical underpinnings and its potential real-world applications. However, the study does not address whether the model being used is accurate. Hizam-Hanafiah et al. [22] and Kamble et al. [23] analysed various studies and proposed numerous dimensions. The authors identified technology as the most crucial dimension, with 44 percent of the 158 dimensions pertaining to technology. Consequently, technology is deemed the most significant factor. They refer to three critical dimensions, including technology, based on aspects, and process integration. The author put forth strategic guidelines in their proposal. The study does not include the validation component. In their presentation on I4.0 dimensions, Liao et al. [24] used SLR. The author went over the repercussions of putting industry 4.0 into practice as well as the research agenda for the foreseeable future. However, many of the studies suffer from the same types of shortcomings.

Despite the significance and potential opportunities in SM fields with I4.0 adoption, there has been limited research conducted on identifying the dimensions of SM. In this study, the dimensions are identified from selected studies and ranked according to their importance [25]. However, previous studies lack validation aspects as they mostly rely on qualitative analysis without strong evidence to support the SLR. Therefore, the dimensions are identified quantitatively from the selected studies to provide stronger evidence.

1.3 METHODOLOGY

This section presents details about the research techniques that are employed. Following Petersen et al. [19] guidelines for conducting SLR and Aria and Cuccurullo [26] for conducting bibliometric mapping, methodology based on five phases is adopted. In the first phase, the authors define the parameters of the work and formulated research questions. Three research questions are formulated. RQ1 and RQ2 were answered using SLR and RQ3 was answered following the bibliometric analysis. Search methodology is the focus of the next phase. In this step, the author searched for relevant articles in a scientific database by entering a string of search terms that makes use of "AND" and "OR" Operators. In accordance with the approach proposed by Kitchenham and Charters [27], a search string was constructed by combining relevant keywords

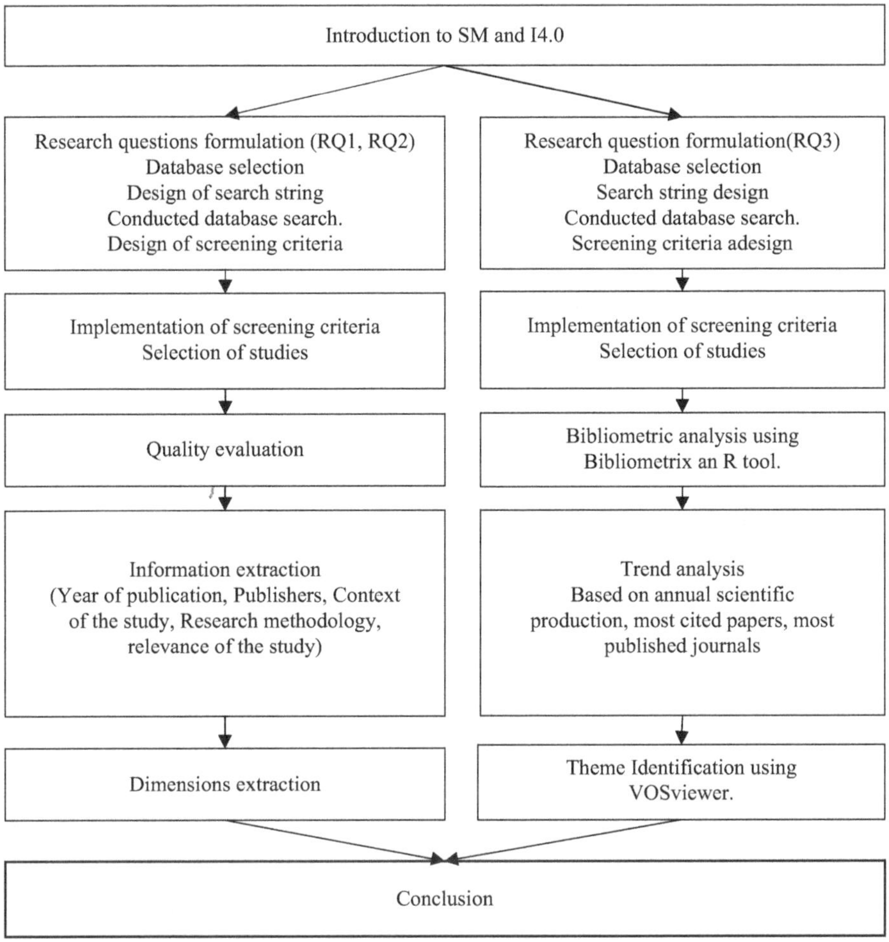

FIGURE 1.1 Methodology flowchart.

using PICO approach. The third phase presents screening of the studies. In the fourth phase, the work is reviewed on the basis of their titles and keywords. The final phase of the study involved mapping research questions. To minimize bias in the study, a quality evaluation was conducted. The results were analyzed and presented in Figure 1.1, which illustrates the research process.

1.3.1 Research Questions

The following research questions are framed to meet the research objectives:

RQ1: What are I4.0 dimensions which help in smart manufacturing?
RQ2: What are those dimensions which are critical for SM systems?
RQ3: What are the future research directions for SM systems?

1.3.2 Search strategy

The work utilized the PICO (Population, intervention, comparison, and outcome) approach recommended by Kitchenham and Charters [27] to guide the search process.

1.3.2.1 Population

It refers to the targeted group. In this study, the population was Smart Manufacturing or Industry 4.0 elements.

1.3.2.2 Intervention

Various maturity models and industry reports served as interventions.

1.3.2.3 Comparison

Empirical studies were conducted to compare the different smart manufacturing techniques and methods.

1.3.2.4 Outcomes

The study's outcomes involved the extensive literature review and empirical analysis of dimensions.

To conduct the study, the author searched for studies in popular databases utilizing search strings tailored to each database. The author utilized the "AND" operator to form the search string. Search string for the databases is as follows:

("smart manufacturing" AND "industry 4.0")

1.3.3 Selection of studies

Our study used screening criteria to select relevant articles. The inclusion year (2007–2022) for articles in English language. We identified seven criteria for selecting articles, labelled as C1 to C7, which are listed in Table 1.1, we provide the criteria that will be used to determine what will and will not be considered for screening.

1.3.4 Data extraction

The study has gathered information regarding smart manufacturing. Additionally, a data attributes sheet has been generated that outlines the research questions (See Table 1.2).

1.3.5 Quality evaluation

According to the standards in Table 1.3, a quality evaluation has been performed. The purpose of this step was to ensure that only relevant studies were considered and to strengthen the validity of our findings. For a comprehensive breakdown of the quality evaluation scores, please refer to Table A1 in appendix section.

TABLE 1.1
Criteria for Screening

Criteria Id	Criteria Type	Explanation
C1	Inclusion	Publication year: 2007–2022
C2	Inclusion	English language articles
C3	Exclusion	Books, book chapters, dissertations, and theses
C4	Exclusion	Articles containing less than six pages
C5	Inclusion	Articles including the terms "SM," "I4.0," and "digital transformation".
C6	Inclusion	Articles that address SM and I4.0.

TABLE 1.2
The Data Attributes in the Present Study

Class	Elements	Address Research Question
Intuitive information	Publication year, publisher, author, title and abstract	1
Methodology	Study design, case study, survey, evaluation study	2
Setting	Work motivation, validity threats, subjective research	2
Results and conclusion	Work limitations, address challenges, future work discussion	1, 2

TABLE 1.3
Quality Evaluation Parameters (QEP)

QEP ID	Description
QEP1	Does the article state its research aims clearly?
QEP2	Does the study explore dimensions related to its research aims?
QEP3	Does the article provide a literature review and outline the main contributions?
QEP4	Does the work describe the design of research methodology?
QEP5	Does the study present research findings?
QEP6	Are the research objectives and conclusions of the work explained in a clear manner?
QEP7	Does the article discuss future research or areas for further investigation?

1.4 RESULTS AND DISCUSSION

1.4.1 Discussion on RQ1 and RQ2

In accordance with Petersen et al. [19], the results of SLR are discussed. A pilot study was conducted using PICO criteria [28] to develop the search string. Applying the search criteria to the databases identified 3001 studies, which were screened based on selection criteria and publication date from 2007 to 2022. This resulted in 115 studies, with 40 duplicates removed, leaving 75 studies for quality assessment. Based on quality scores less than 3, approximately 13.37% of studies were removed, and 65 final articles were selected for the study. The search results are shown in Table 1.4. The number of academic studies showed a constant trend from 2007 to 2015, with only 1 or 2 studies, but the trend increased continuously from 2016–2022. The distribution of primary research according to year is presented in Figure 1.2.

The author aimed to respond to RQ1 by identifying dimensions. Additionally, the author calculated the frequency/percentage of each dimension. The authors identified 8 dimensions. To address RQ2, the author selected five dimensions that had a percentage equal to or greater than 25%. Table 1.5 provides the order and variables of the identified dimensions.

1.4.1.1 Technology (E_8)

The study found that technology is the most significant dimension in SMEs' digital readiness, with 92.31% of the selected studies focusing on various technological areas. Cyber-physical systems, as noted by Rahamaddula et al. [29] and Swarnima et al. [30], are particularly critical for integrating digital and physical entities. Saad et al. [31] outlines the importance of technology in smart manufacturing in industry 4.0 environment has further highlighted the growing significance of AI, ML, and cloud computing [32–34].

1.4.1.2 Organizational strategy (E_1)

The study identifies the second most significant dimension as strategy with a frequency of 30. 46.15% studies considered organization strategy as a dimension for smart manufacturing. Strategy involves planning for products and services that align with smart manufacturing requirements. Additionally, the strategy includes the funding strategy, which refers to the financial investment made for implementing smart manufacturing practices.

1.4.1.3 People/culture/employees (E_3)

The current research reveals that people/culture is identified as crucial dimensions. It is ranked at 5 with frequency 17 and 26.15%. This dimension relates to the employee attitude towards use of digital means for work [35,36,29]. As a final recommendation, it is essential to foster a learning culture to achieve smart manufacturing culture [37]. People/culture/employees are crucial to the success of SM. A skilled workforce and a culture of collaboration and continuous improvement are essential for implementing and optimizing advanced technologies.

TABLE 1.4
Search Results

Name of Database Library	Search Count	Articles Left after Applying Screening Criteria						Articles Left after Removing Duplicates	Articles Remaining after Quality Evaluation
		C1	C2	C3	C4	C5	C6		
ScienceDirect	2430	2428	2421	1315	785	286	50	75	65
ACM	115	108	133	109	93	50	22		
Wiley	209	209	200	143	123	56	25		
JSTOR	22	22	20	16	10	8	3		
DOAJ	225	225	172	150	96	76	15		
Total	3001	2992	2946	1733	1107	351	115	75	65

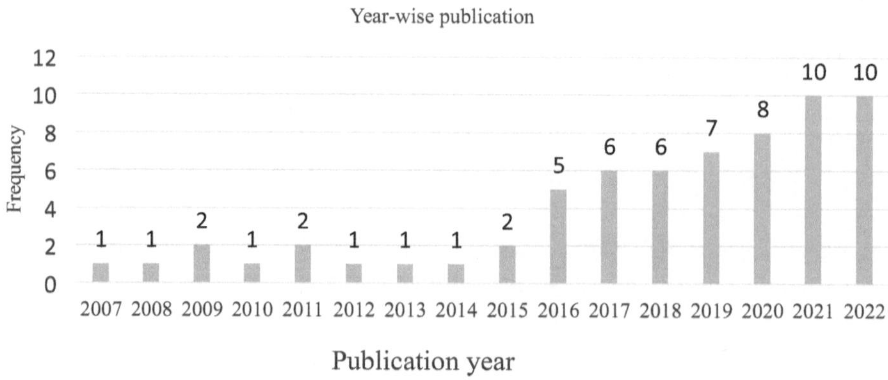

FIGURE 1.2 Year-wise studies.

1.4.1.4 Processes (E4)

smart manufacturing processes are crucial to the success of modern manufacturing operations. By optimizing processes, manufacturers can increase efficiency, reduce waste, improve product quality, and be more responsive to customer needs, ultimately leading to greater competitiveness in the marketplace. Findings reveals that, processes with frequency 27 and 41.54 % were ranked third. This dimension encompasses various types of processes such as business, communication, and management processes.

1.4.1.5 Products (E_6)

Product as a dimension (Frequency 21, percentage 32.31) ranked at four. One of the main goal of SM is to produce high-quality products that meet or exceed customer expectations while minimizing waste and reducing costs. It requires in-depth knowledge of the product and its manufacturing process [34]. Digital features in products enhance customer interest and promote association with the industry 4.0 ecosystem [38–41].

1.4.1.6 Customers (E_5)

According to the study, the sixth-ranked dimension is customers (Frequency 12 and percentage 18.46). Customers are important in smart manufacturing because they have high expectations for the products they purchase. They want products that are high-quality, reliable, and meet their specific needs. By understanding customer preferences and behaviours, manufacturers can tailor their products to meet those needs more effectively. This can lead to increased customer satisfaction and loyalty, which is essential for the long-term success of any business [5,29].

1.4.1.7 Innovation (E_2)

According to the information provided in Table 1.5, innovation (Frequency 10, percentage 16.67) is ranked seventh among the critical dimensions for industry 4.0.

TABLE 1.5
Identified Variables

S. no.	Dimensions	Definition	References	Frequency	Percentage	Ranking
E_1	Organisational Strategy	Organisational Strategy refers to the approach and plan of action adopted by an organization to align its resources, goals, and capabilities with the opportunities and challenges posed by the new technologies and digitalization. It involves the development of a clear roadmap for Industry 4.0 adoption, infrastructure, organizational structure, workforce planning, and continuous improvement strategies. Organisational strategy is critical for SMEs in their implementation of smart manufacturing practices ensuring its long-term sustainability and competitiveness.	[3,17,42–51,30,52–61,35–41]	30	46.15	2
E_2	Innovation	Innovation in this context includes the exploration and implementation of new business models, supply chain strategies, and value creation opportunities to stay competitive in the rapidly evolving digital landscape.	[33,39,40,43,44,47,51,59,62, 63]	10	16.67	7
E_3	People/culture/ Employees	People/culture/employees refer to the human aspect of the organization, including its employees' skills, knowledge, behaviour, and attitudes towards digital transformation. It involves the organization's culture, values, leadership, and change management strategies necessary for successful adoption of smart manufacturing practices. It also includes the development of a skilled workforce capable of operating and maintaining advanced digital technologies.	[17,29,58,60,62,64–67,33, 36–39,44–46]	17	26.15	5

(Continued)

TABLE 1.5 (Continued)
Identified Variables

S. no.	Dimensions	Definition	References	Frequency	Percentage	Ranking
E_4	Processes	Processes refer to the methods and procedures used in making products. It includes adoption of technology with process automation, industrial robots. Processes in I4.0 involve the integration of various systems across the entire value chain, including design, production, supply chain, logistics, and customer service. Smart processes are designed to be flexible, adaptable, and responsive to changes in customer demand, market trends, and technological advancements.	[3,29,50,55,60–63,65, 66,68,69,33,70–75,35–38,43, 38,43,45,46]	27	41.54	3
E_5	Customers	This dimension includes understanding and meeting customer expectations, providing customer-tailored products, and using data analytics to improve customer experience. Customer retention is a key focus, and companies strive to create a customer-centric environment to improve their competitiveness in the market.	[29,36,64,76,38,39,42,44, 55,57,60,62]	12	18.46	6
E_6	Products	Products refers to the physical goods or digital services produced by a company or organization using advanced technologies and processes. These products may be customized or personalized for individual customers and may incorporate digital features and connectivity to enhance their functionality and value.	[3,35,62,64,65,69–73,75,77, 36,78,38,42,43,46,57,59,60]	21	32.31	4

E_7	Services	Services refers to the digital features and capabilities that are integrated with the products offered by an organization. This includes data-driven services, remote monitoring, product configuration, product simulation, and condition monitoring, among others. By leveraging digital technologies and integrating them with their products, organizations can improve their competitiveness and customer satisfaction.	[35,36,44,50,59,60,64,71,79]	9	13.85	8
E_8	Technology	It involves the integration of physical and cyber systems, resulting in smart and efficient manufacturing processes. Technology is a critical dimension in Industry 4.0 as it enables companies to improve their operations, increase productivity, reduce costs, and develop new products and services.	[3,7,38–40,42–46,50,51,17, 52–55,57–60,62,63,29, 64–73,30,74–83,31,84–90, 33,35–37,91,92]	60	92.31	1

Innovation allows manufacturers to improve their processes and operations, resulting in reduced costs and improved quality. For example, the implementation of predictive maintenance systems can reduce downtime and prevent costly equipment failures. Similarly, the use of robotics and automation can increase production efficiency and reduce labour costs.

Innovation can also enable manufacturers to create new products and services, opening new markets and revenue streams. For example, smart products equipped with sensors can provide valuable data used to improve customer satisfaction.

1.4.1.8 Services (E_7)

According to the findings presented in Table 1.5, the dimension of services ranks at the eighth position (frequency 9, percentage 13.85. In I4.0, services refer to the digital features offered by an organization that attract end-users to adapt to the industry 4.0 environment. Such services may include product configuration, development, and simulation, as well as remote and condition monitoring, and other data-driven services that are integrated with digital technologies.

Sections 4.1.1 through 4.1.8 presents discussion on variables. Based on the analysis, it is apparent that Technology (E8), Organizational strategy (E1), Processes (E4), Products (E6), and People/culture/employees (E3) are the most crucial dimensions for SMEs. Dimensions reporting a frequency of more than 25% considered critical. The selection of these critical dimensions is subjective, and further input from experts in the field may be necessary to support the findings.

1.4.2 Discussion on RQ3

For the identification of future research theme, the authors have conducted bibliometric analysis. The work is done on Bibliometrix and Vosviewer. The authors selected the WoS database. Authors designed the search string such that it covers the objective of the chapter. The search string is as follows: ("smart manufacturing" AND "industry 4.0"). the authors conducted the search from the web of science database using the string followed by screening of the articles (Table 1.6).

186 articles are selected for further analysis. The authors have gone through the most relevant sources. The analysis shows that the IJPR journal receives a total of 598 citations on the topic followed by Journal of manufacturing systems with 386 citations. Figure 1.3 shows the most cited sources form the smart manufacturing and industry 4.0.

The authors now conducted the keyword cooccurrence analysis using VOSviewer. Which result in making the five clusters. Figure 1.4 shows the keyword cooccurrence analysis and clusters. These clusters are identified by the colors. Each cluster contain several elements. The details of the clusters are given in Table 1.7.

TABLE 1.6
Screening Results

Description	Type	Article Count
Search using keywords ("smart manufacturing" AND "industry 4.0")	Inclusion	609
Review article	Exclusion	504
Editorial material	Exclusion	495
Proceeding paper	Exclusion	492
Web of science categories such as engineering industrial, operation management, Engineering manufacturing, computer science and AI, automation control systems, engineering mechanical	Inclusion	233
Citation topics meso such as design and manufacturing, supply chain and logistics, management, AI and ML, manufacturing, human computer interaction, knowledge engineering and representation	Inclusion	186
Analysis of title, abstract, keywords	Inclusion	184

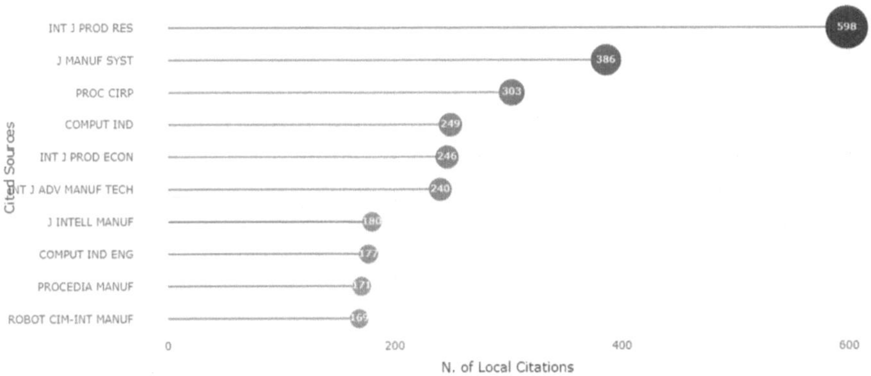

FIGURE 1.3 Most cited sources.

1.4.2.1 Theme based on blue cluster.

"Enhancing Smart Manufacturing through CPS and Industrial Internet: Investigating Decision-Making in Smart Factories."

The theme directs researchers and academicians to exploring the challenges and barriers to the implementation of SM systems in I4.0 environment and developing strategies to overcome them. The research will investigate how CPS

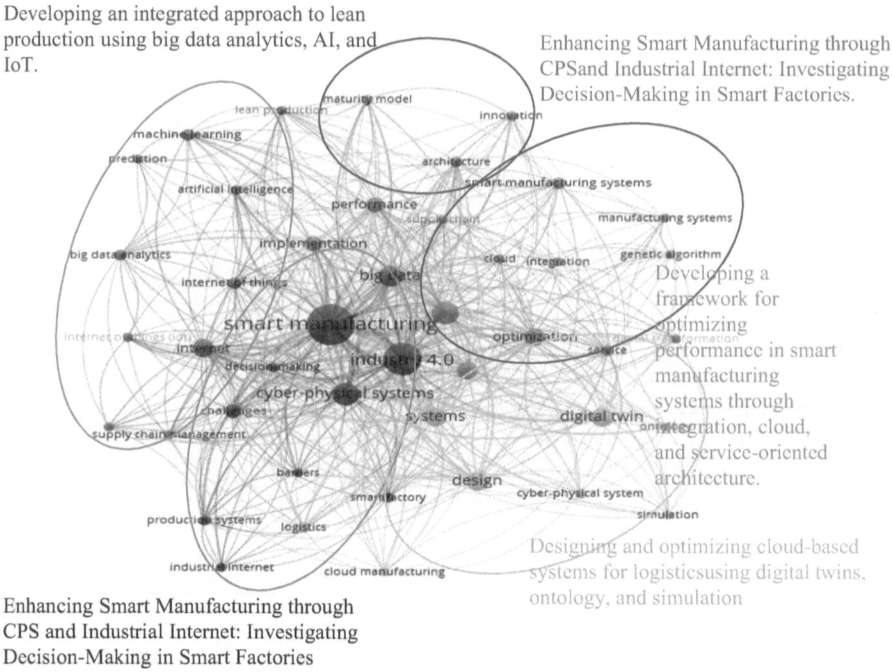

FIGURE 1.4 keyword cooccurrence network analysis with research themes.

and IIoT can be integrated to create a smart factory environment. The decision-making processes involved in the implementation of such systems will also be studied to ensure effective implementation. Finally, the proposed strategies will be evaluated in terms of their effectiveness in addressing the identified challenges and barriers and their ability to facilitate the implementation of SM systems in I4.0 environment.

1.4.2.2 Theme based on yellow cluster.

> ***"Designing and optimizing cloud-based systems for logistics using digital twins, ontology, and simulation."***

This covers areas on utilizing cloud manufacturing, digital twins, and simulation technologies to design and optimize manufacturing and logistics systems. The design aspect of this theme emphasizes the importance of considering both product and process design in the development of such systems. Ontology can be used to standardize data and communication protocols, while logistics can help ensure efficient material flow and inventory management. Finally, the use of digital twins allows for virtual testing and optimization of the proposed systems before implementation, reducing risks and costs.

TABLE 1.7
Cluster Items with Proposed Themes

Cluster colour	Items (Occurrence)	Sub-Items (Occurrence)	Proposed Theme	References
Blue	Smart manufacturing (116) Industry 4.0 (78)	Barriers (5), challenges (15), cyber-physical systems (36), decision-making (7), industrial internet (5), industry 4.0 (78), production systems (7), smart factory (8), smart manufacturing (116)	Enhancing Smart Manufacturing through CPS and Industrial Internet: Investigating Decision-Making in Smart Factories.	[93–99]
Yellow	Systems (26) Digital twin (25) Design (25)	cloud manufacturing (5), cloud manufacturing (7), design (25), digital twin (25), logistics (8), ontology (7), simulation (5), systems (26)	Designing and optimizing cloud-based systems for logistics using digital twins, ontology, and simulation	[93,94]
Red	Framework (37), Optimization (20)	Architecture (8), cloud (5), digital transformation (5), framework (37), genetic algorithm (5), integration (5), manufacturing systems (6), optimization (20), performance (17), service (8), smart manufacturing systems (9), supply chain (6)	Developing a framework for optimizing performance in smart manufacturing systems through integration, cloud, and service-oriented architecture.	[95,98,99,100]
Green	Big data analytics (40) Internet (22)	AI (7), big data analytics (40), cloud computing (5), internet (22), IoT (14), lean production (5), machine learning (10), prediction (5), supply chain management (6)	Developing an integrated approach to lean production using big data analytics, AI, and IoT.	[93,97,98,101]
Purple	Management (23) Implementation (17)	Implementation (17), innovation (7), management (23), maturity model (7)	Developing a maturity model for effective innovation management implementation	[93,95,96]

1.4.2.3 Theme based on red cluster.

Developing a framework for optimizing performance in smart manufacturing systems through integration, cloud, and service-oriented architecture.

This theme focuses on developing framework that incorporates elements of cloud computing, service-oriented architecture, and integration to optimize performance in smart manufacturing systems. The framework will be developed through digital transformation of existing manufacturing systems and supply chain networks. Optimization techniques such as genetic algorithms will be used to fine-tune the system performance. The use of SM systems can help ensure effective data collection, decision-making. Finally, the framework will be evaluated to improve the performance of SM systems in real-world settings.

1.4.2.4 Theme based on green cluster.

"Developing an integrated approach to lean production using big data analytics, AI, and IoT."

The theme focuses on utilizing big data analytics to improve SCM by developing a framework that incorporates IoT, ML, and AI. The framework will be deployed on cloud computing platforms and will leverage lean production principles to reduce waste and improve efficiency. The internet will play a key role in facilitating data exchange and communication between different supply chain partners. The use of ML will enable prediction of supply chain events and help in making data-driven decisions.

1.4.2.5 Theme based on purple cluster.

"Developing a maturity model for effective innovation management implementation".

It focuses on developing a maturity model for effective innovation management implementation. The research points towards identification of factors that help in innovation management implementation. The maturity model can help organizations assess their current innovation management capabilities and identify areas for improvement. The research can use case studies to evaluate the effectiveness of model in different organizational contexts. The findings of the research can help organizations improve their innovation management capabilities and drive growth and competitiveness.

1.4.2.6 Co-citation network analysis

The authors have conducted the co-citation network analysis for deeper understanding of the emerging themes. Analysis was done using the tool VOSviewer. The result of co-citation network reveals that there are four clusters with minimum cluster size 17 and minimum link strength 2. Figure 1.5 shows the co-citation

FIGURE 1.5 Co-citation network analysis of articles (VOS viewer).

network analysis of the selected articles. Table 1.8 presents the co-citation network details with top terms and themes.

The themes discussed can be used for research in smart manufacturing by providing a framework for investigating and improving various aspects of manufacturing operations, including maintenance, supply chain management, logistics, quality control, and performance optimization. For example, researchers can explore how data-driven approaches, such as condition-based monitoring and predictive maintenance, can be leveraged to improve equipment reliability and reduce downtime. Collaborative networks and cyber-physical systems can be studied to enable effective communication and coordination between different parts of the manufacturing process, from suppliers to customers. Augmented reality and machine vision can be investigated to enhance human-centered manufacturing, while blockchain and security monitoring can be explored to ensure data integrity and cyber resilience. Optimization techniques, such as simulation modelling and parameter identification, can be applied to improve manufacturing efficiency and quality.

Overall, the discussed themes provide a comprehensive framework for investigating various research questions related to smart manufacturing and offer opportunities for improving manufacturing processes through the integration of advanced technologies and data-driven approaches.

1.5 VALIDITY THREATS

Despite the use of SLR guidelines there are still potential validity issues with the findings. To further minimize bias, we followed the selection criteria guidelines and invited a guest researcher to validate the study selections, resulting in a 70% agreement

TABLE 1.8
Co-Citation Network Details with Themes

Cluster (Colour)	Name of Themes	Items Count	Year (Mean)	Top Terms
1 (Red)	Optimizing Industrial Maintenance and Supply Chain through Collaborative Networks and Big Data Ecosystems	42	2021	Collaborative network, success factors, industrial maintenance, operational performance, bigdata ecosystems, predictive maintenance, smart manufacturing, supply chain, inventory management, security, collaborative robots, toe framework, cps, lean supply chain, big data analytics, cyber-security challenges, sustainable I4.0, manufacturing scm, hybrid manufacturing
2 (Green)	Enabling Agile and Customizable Manufacturing through Data-Driven Approaches and Augmented Reality	31	2019	Data-driven scheduling, changeability, reconfigurability, augmented reality, cloud manufacturing, product-service system, feasibility evaluation, optimization, self-tuning model, operationalization framework, strategic response, cyber-physical production network, industrial management, management approach, human-cantered additive manufacturing, enterprise information systems, SM systems requirements, augmented reality-based SM, machine vision, mass personalization.
3 (Blue)	Enhancing Manufacturing Performance and Security through Digital Twins and Autonomous Collaboration	29	2021	Optimization, onto-based modelling, digital twin, CPS, autonomous collaboration, reinforcement learning, human-in-the-loop SM, web-based digital twin, security monitoring, quality, mes integration, cognitive twin, blockchain, data formats, human decision models, simulation modelling, product management, iot enabled, simulation-based approach, edge computing.
4 (Yellow)	Maximizing Efficiency and Quality in Industry 4.0 Ecosystems through Condition-Based Monitoring and Data-Driven Approaches	21	2020	Condition-based monitoring, synchroperation, bottleneck detection model, data-driven models, data analytics, SM systems, cyber-physical factories, logistics management, quality improvement, I4.0 ecosystem, parameter identification, machine learning, training, anomaly detection model, dynamic environment.

on the selections. However, there is still the possibility of publication bias, where researchers may only present positive results from their research. To address this, we utilized multiple search libraries. Finally, there is a possibility of data extraction bias, but we minimized it by using a data extraction template to record unbiased data.

Also, there may be the possibility of biasness while conducting bibliometric analysis such as biasness in database selection, incompleteness of data, citation practices, self-citations, inaccurate citation data, changing publication practices etc. However, to avoid such threats, the authors have conducted the word co-occurrence network analysis and co-citation network analysis for exploring future research directions.

1.6 CONCLUSION

Smart manufacturing has emerged as a transformative approach to manufacturing that integrates cutting-edge technologies and innovative processes to enable more efficient, flexible, and sustainable production. The success of smart manufacturing depends on several dimensions that are critical for achieving operational excellence and competitive advantage. To this end, the author conducted review of studies published from 2007 to 2022, eight dimensions that aid manufacturing firms in opting smart manufacturing practices. RQ1 was addressed by extracting data from 65 studies to identify 8 key dimensions, namely technology, organizational strategy, people/culture/employees, processes, products, customers, innovation, and services. To address RQ2, dimensions (Count percentage > 25%), which included technology, organizational strategy, products, processes, and people/culture/employees. The RQ3 was addressed by conducting a bibliometric analysis resulting in five potential research directions (keyword co-occurrence network analysis) and another four themes (using co-citation network analysis). This finding underscores the high demand for research on smart manufacturing. These findings can inform future smart manufacturing research. Further work on these themes can add value to research domain and ultimately lead to achieve socio-economic goals.

REFERENCES

[1] J. Davis *et al.*, "Smart Manufacturing," *Annu. Rev. Chem. Biomol. Eng.*, vol. 6, no. April, pp. 141–160, 2015, doi: 10.1146/annurev-chembioeng-061114-123255.

[2] S. Phuyal, D. Bista, and R. Bista, "Challenges, Opportunities and Future Directions of Smart Manufacturing: A State of Art Review," *Sustain. Futur.*, vol. 2, no. March, p. 100023, 2020, doi: 10.1016/j.sftr.2020.100023.

[3] F. Bachinger, G. Kronberger, and M. Affenzeller, "Continuous improvement and adaptation of predictive models in smart manufacturing and model management," *IET Collab. Intell. Manuf.*, vol. 3, no. 1, pp. 48–63, 2021, doi: 10.1049/cim2.12009.

[4] M. Shahin, F. F. Chen, H. Bouzary, and K. Krishnaiyer, "Integration of Lean practices and Industry 4.0 technologies: smart manufacturing for next-generation enterprises," *Int. J. Adv. Manuf. Technol.*, vol. 107, no. 5–6, pp. 2927–2936, 2020, doi: 10.1007/s00170-020-05124-0.

[5] J. M. Müller, O. Buliga, and K. I. Voigt, "The role of absorptive capacity and innovation strategy in the design of industry 4.0 business Models - A comparison between SMEs and large enterprises," *Eur. Manag. J.*, vol. 39, no. 3, pp. 333–343, 2021, doi: 10.1016/j.emj.2020.01.002.

[6] A. Felsberger, F. Hasan Qaiser, A. Chaudhary, and G. Reiner, "Felsberger et al. The impact of Industry 4.0 on the reconciliation of dynamic capabilities evidence from the European manufacturing indu-annotated.pdf," *Prod. Plan. Control*, vol. 33, no. 2–3, pp. 277–300, 2022, doi: 10.1080/09537287.2020.1810765.
[7] M. Ghobakhloo and M. Iranmanesh, "Digital transformation success under Industry 4.0: a strategic guideline for manufacturing SMEs," *J. Manuf. Technol. Manag.*, vol. 32, no. 8, pp. 1533–1556, 2021, doi: 10.1108/JMTM-11-2020-0455.
[8] M. Ghobakhloo, "Industry 4.0, digitization, and opportunities for sustainability," *J. Clean. Prod.*, vol. 252, p. 119869, 2020, doi: 10.1016/j.jclepro.2019.119869.
[9] S. V. Shet and V. Pereira, "Proposed managerial competencies for Industry 4.0 – Implications for social sustainability," *Technol. Forecast. Soc. Change*, vol. 173, no. March, p. 121080, 2021, doi: 10.1016/j.techfore.2021.121080.
[10] B. Bajic, A. Rikalovic, N. Suzic, and V. Piuri, "Industry 4.0 Implementation Challenges and Opportunities: A Managerial Perspective," *IEEE Syst. J.*, vol. 15, no. 1, pp. 546–559, 2021, doi: 10.1109/JSYST.2020.3023041.
[11] A. Raj, G. Dwivedi, A. Sharma, A. B. Lopes de Sousa Jabbour, and S. Rajak, "Barriers to the adoption of industry 4.0 technologies in the manufacturing sector: An inter-country comparative perspective," *Int. J. Prod. Econ.*, vol. 224, no. August 2019, p. 107546, 2020, doi: 10.1016/j.ijpe.2019.107546.
[12] G. Orzes, E. Rauch, S. Bednar, and R. Poklemba, "Industry 4.0 Implementation Barriers in Small and Medium Sized Enterprises: A Focus Group Study," *IEEE Int. Conf. Ind. Eng. Eng. Manag.*, vol. 2019-Decem, pp. 1348–1352, 2019, doi: 10.1109/IEEM.2018.8607477.
[13] A. G. Frank, L. S. Dalenogare, and N. F. Ayala, "Industry 4.0 technologies: Implementation patterns in manufacturing companies," *Int. J. Prod. Econ.*, vol. 210, no. September 2018, pp. 15–26, 2019, doi: 10.1016/j.ijpe.2019.01.004.
[14] K. Zhou, T. Liu, and L. Zhou, "Industry 4.0: Towards future industrial opportunities and challenges," *2015 12th Int. Conf. Fuzzy Syst. Knowl. Discov. FSKD 2015*, pp. 2147–2152, 2016, doi: 10.1109/FSKD.2015.7382284.
[15] A. Issa, B. Hatiboglu, A. Bildstein, and T. Bauernhansl, "Industrie 4.0 roadmap: Framework for digital transformation based on the concepts of capability maturity and alignment," *Procedia CIRP*, vol. 72, pp. 973–978, 2018, doi: 10.1016/j.procir.2018.03.151.
[16] K. Y. Akdil, A. Ustundag, and E. Cevikcan, *Maturity and Readiness Model for Industry 4.0 Strategy*. 2018.
[17] L. Bibby and B. Dehe, "Defining and assessing industry 4.0 maturity levels–case of the defence sector," *Prod. Plan. Control*, vol. 29, no. 12, pp. 1030–1043, 2018, doi: 10.1080/09537287.2018.1503355.
[18] K. Lichtblau *et al.*, "IMPULS-Industrie 4.0-Readiness," 2015.
[19] K. Petersen, S. Vakkalanka, and L. Kuzniarz, "Guidelines for conducting systematic mapping studies in software engineering: An update," *Inf. Softw. Technol.*, vol. 64, pp. 1–18, 2015, doi: 10.1016/j.infsof.2015.03.007.
[20] Y. J. Qu, X. G. Ming, Z. W. Liu, X. Y. Zhang, and Z. T. Hou, "Smart manufacturing systems: state of the art and future trends," *Int. J. Adv. Manuf. Technol.*, vol. 103, no. 9–12, pp. 3751–3768, 2019, doi: 10.1007/s00170-019-03754-7.
[21] K. Elibal and E. Özceylan, "A systematic literature review for industry 4.0 maturity modeling: state-of-the-art and future challenges," *Kybernetes*, vol. 50, no. 11, pp. 2957–2994, 2021, doi: 10.1108/K-07-2020-0472.
[22] M. Hizam-hanafiah, M. A. Soomro, and N. L. Abdullah, "Industry 4.0 Readiness Models.Pdf," 2020.

[23] S. S. Kamble, A. Gunasekaran, and S. A. Gawankar, "Sustainable Industry 4.0 framework: A systematic literature review identifying the current trends and future perspectives," *Process Saf. Environ. Prot.*, vol. 117, pp. 408–425, 2018, doi: 10.1016/j.psep.2018.05.009.
[24] Y. Liao, F. Deschamps, E. de F. R. Loures, and L. F. P. Ramos, "Past, present and future of Industry 4.0 - a systematic literature review and research agenda proposal," *Int. J. Prod. Res.*, vol. 55, no. 12, pp. 3609–3629, 2017, doi: 10.1080/00207543.2017.1308576.
[25] A. A. Khan, J. W. Keung, Fazal-E-Amin, and M. Abdullah-Al-Wadud, "Spiimm: Toward a model for software process improvement implementation and management in global software development," *IEEE Access*, vol. 5, no. c, pp. 13720–13741, 2017, doi: 10.1109/ACCESS.2017.2728603.
[26] Aria, M. and Cuccurullo, C. "Bibliometrix: An R-tool for comprehensive science mapping analysis," *Journal of Informetrics*, vol. 11, no. 4, pp. 959–975, 2017.
[27] B. Kitchenham and S. M. Charters, "Guidelines for performing Systematic Literature Reviews in Software Engineering Guidelines for performing Systematic Literature Reviews in Software Engineering EBSE Technical Report EBSE-2007-01 Software Engineering Group School of Computer Science and Ma," no. January, 2007.
[28] M. B. Eriksen and T. F. Frandsen, "The impact of patient, intervention, comparison, outcome (PICO) as a search strategy tool on literature search quality: a systematic review," vol. 106, no. October, pp. 420–431, 2018.
[29] S. R. Bin Rahamaddulla, Z. Leman, B. T. H. T. Bin Baharudin, and S. A. Ahmad, "Conceptualizing smart manufacturing readiness-maturity model for small and medium enterprise (Sme) in malaysia," *Sustain.*, vol. 13, no. 17, pp. 1–18, 2021, doi: 10.3390/su13179793.
[30] C. Swarnima, P. Mehra, and A. Dasot, "India's Readiness for Industry 4.0 – A Focus on Automotive Sector," *Grant Thornt. Indian*, p. 46, 2017, [Online]. Available: http://www.grantthornton.in/insights/articles/indias-readiness-for-industry-4.0--a-focus-on-automotive-sector/.
[31] S. M. Saad, R. Bahadori, and H. Jafarnejad, "The smart SME technology readiness assessment methodology in the context of industry 4.0," *J. Manuf. Technol. Manag.*, vol. 32, no. 5, pp. 1037–1065, 2021, doi: 10.1108/JMTM-07-2020-0267.
[32] M. M. Queiroz, S. C. F. Pereira, R. Telles, and M. C. Machado, "Industry 4.0 and digital supply chain capabilities: A framework for understanding digitalisation challenges and opportunities," *Benchmarking*, 2019, doi: 10.1108/BIJ-12-2018-0435.
[33] E. Rauch, P. Dallasega, and M. Unterhofer, "Requirements and Barriers for Introducing Smart Manufacturing in Small and Medium-Sized Enterprises," *IEEE Eng. Manag. Rev.*, vol. 47, no. 3, pp. 87–94, 2019, doi: 10.1109/EMR.2019.2931564.
[34] Z. Suleiman, S. Shaikholla, D. Dikhanbayeva, E. Shehab, and A. Turkyilmaz, "Industry 4.0- Clustering of concepts and characteristics," *cogent Eng.*, vol. 9, no. 1, p. 2034264, 2022.
[35] F. Tao, Q. Qi, L. Wang, and A. Y. C. Nee, "Digital Twins and Cyber–Physical Systems toward Smart Manufacturing and Industry 4.0: Correlation and Comparison," *Engineering*, vol. 5, no. 4, pp. 653–661, 2019, doi: 10.1016/j.eng.2019.01.014.
[36] K. Traganos, P. Grefen, I. Vanderfeesten, J. Erasmus, G. Boultadakis, and P. Bouklis, "The HORSE framework: A reference architecture for cyber-physical systems in hybrid smart manufacturing," *J. Manuf. Syst.*, vol. 61, no. October, pp. 461–494, 2021, doi: 10.1016/j.jmsy.2021.09.003.
[37] E. Pessl, S. R. Sorko, and B. Mayer, "Roadmap industry 4.0 - Implementation guideline for enterprises," *Int. J. Sci. Technol. Soc.*, vol. 5, no. 6, pp. 193–202, 2017, doi: 10.11648/j.ijsts.20170506.14.

[38] D. Guo *et al.*, "Synchroperation in industry 4.0 manufacturing," *Int. J. Prod. Econ.*, vol. 238, no. March, p. 108171, 2021, doi: 10.1016/j.ijpe.2021.108171.

[39] S. Krishnan, S. Gupta, M. Kaliyan, V. Kumar, and J. A. Garza-Reyes, "Assessing the key enablers for Industry 4.0 adoption using MICMAC analysis: a case study," *Int. J. Product. Perform. Manag.*, vol. 70, no. 5, pp. 1049–1071, 2021, doi: 10.1108/IJPPM-02-2020-0053.

[40] D. Horvat, T. Stahlecker, A. Zenker, C. Lerch, and M. Mladineo, "A conceptual approach to analysing manufacturing companies' profiles concerning Industry 4.0 in emerging economies," *Procedia Manuf.*, vol. 17, pp. 419–426, 2018, doi: 10.1016/j.promfg.2018.10.065.

[41] J. C. Serrano-Ruiz, J. Mula, and R. Poler, "Development of a multidimensional conceptual model for job shop smart manufacturing scheduling from the Industry 4.0 perspective," *J. Manuf. Syst.*, vol. 63, no. January, pp. 185–202, 2022, doi: 10.1016/j.jmsy.2022.03.011.

[42] G. Liu *et al.*, "Two-stage Competitive Particle Swarm Optimization Based Timing-driven X-routing for IC Design under Smart Manufacturing," *ACM Trans. Manag. Inf. Syst.*, 2022, doi: 10.1145/3531328.

[43] D. Li, "Perspective for smart factory in petrochemical industry," *Comput. Chem. Eng.*, vol. 91, pp. 136–148, 2016, doi: 10.1016/j.compchemeng.2016.03.006.

[44] A. Iyer, "Moving from Industry 2.0 to Industry 4.0: A case study from India on leapfrogging in smart manufacturing," *Procedia Manuf.*, vol. 21, pp. 663–670, 2018, doi: 10.1016/j.promfg.2018.02.169.

[45] F. Pirola, C. Cimini, and R. Pinto, "Digital readiness assessment of Italian SMEs: a case-study research," *J. Manuf. Technol. Manag.*, vol. 31, no. 5, pp. 1045–1083, 2019, doi: 10.1108/JMTM-09-2018-0305.

[46] S. Parhi, K. Joshi, T. Wuest, and M. Akarte, "Factors affecting Industry 4.0 adoption – A hybrid SEM-ANN approach," *Comput. Ind. Eng.*, vol. 168, no. March 2021, p. 108062, 2022, doi: 10.1016/j.cie.2022.108062.

[47] F. Nwaiwu, M. Duduci, F. Chromjakova, and C. A. F. Otekhile, "Industry 4.0 concepts within the czech sme manufacturing sector: An empirical assessment of critical success factors," *Bus. Theory Pract.*, vol. 21, no. 1, pp. 58–70, 2020, doi: 10.3846/btp.2020.10712.

[48] H. Wang and B. Lin, "Pipelined van Emde Boas tree: Algorithms, analysis, and applications," *Proc. – IEEE INFOCOM*, pp. 2471–2475, 2007, doi: 10.1109/INFCOM.2007.303.

[49] M. Lassnig, J. M. Müller, K. Klieber, A. Zeisler, and M. Schirl, "A digital readiness check for the evaluation of supply chain aspects and company size for Industry 4.0," *J. Manuf. Technol. Manag.*, vol. 33, no. 9, pp. 1–18, 2018, doi: 10.1108/JMTM-10-2020-0382.

[50] M. K. Kazi, F. Eljack, and E. Mahdi, "Data-driven modeling to predict the load vs. displacement curves of targeted composite materials for industry 4.0 and smart manufacturing," *Compos. Struct.*, vol. 258, no. October 2020, p. 113207, 2021, doi: 10.1016/j.compstruct.2020.113207.

[51] S. Kumar, M. Suhaib, and M. Asjad, "Narrowing the barriers to Industry 4.0 practices through PCA-Fuzzy AHP-K means," *J. Adv. Manag. Res.*, 2020, doi: 10.1108/JAMR-06-2020-0098.

[52] L. Li, F. Su, W. Zhang, and J. Y. Mao, "Digital transformation by SME entrepreneurs: A capability perspective," *Inf. Syst. J.*, vol. 28, no. 6, pp. 1129–1157, 2018, doi: 10.1111/isj.12153.

[53] A. G. Khanzode, P. R. S. Sarma, S. K. Mangla, and H. Yuan, "Modeling the Industry 4.0 adoption for sustainable production in Micro, Small & Medium

Enterprises," *J. Clean. Prod.*, vol. 279, p. 123489, 2021, doi: 10.1016/j.jclepro.2020.123489.

[54] S. M. Kannan *et al.*, "Towards industry 4.0: Gap analysis between current automotive MES and industry standards using model-based requirement engineering," *Proc. - 2017 IEEE Int. Conf. Softw. Archit. Work. ICSAW 2017 Side Track Proc.*, vol. 0, pp. 29–35, 2017, doi: 10.1109/ICSAW.2017.53.

[55] N. Grufman, S. Lyons, and E. Sneiders, "Exploring Readiness of SMEs for Industry 4.0," *Complex Syst. Informatics Model. Q.*, no. 25, pp. 54–86, 2020, doi: 10.7250/csimq.2020-25.04.

[56] M. Estensoro, M. Larrea, J. M. Müller, and E. Sisti, "A resource-based view on SMEs regarding the transition to more sophisticated stages of industry 4.0," *Eur. Manag. J.*, no. xxxx, 2021, doi: 10.1016/j.emj.2021.10.001.

[57] S. R. Hamidi, A. A. Aziz, S. M. Shuhidan, A. A. Aziz, and M. Mokhsin, "SMEs maturity model assessment of IR4.0 digital transformation," *Adv. Intell. Syst. Comput.*, vol. 739, pp. 721–732, 2018, doi: 10.1007/978-981-10-8612-0_75.

[58] D. Zuhlke, *SmartFactory - A vision becomes reality*, vol. 13, no. PART 1. IFAC, 2009.

[59] J. El Baz, S. Tiwari, T. Akenroye, A. Cherrafi, and R. Derrouiche, "A framework of sustainability drivers and externalities for Industry 4.0 technologies using the Best-Worst Method," *J. Clean. Prod.*, vol. 344, no. October 2021, p. 130909, 2022, doi: 10.1016/j.jclepro.2022.130909.

[60] L. T. Letchumanan *et al.*, "Analyzing the Factors Enabling Green Lean Six Sigma Implementation in the Industry 4.0 Era," *Sustain.*, vol. 14, no. 6, pp. 1–15, 2022, doi: 10.3390/su14063450.

[61] Z. Liu *et al.*, "The architectural design and implementation of a digital platform for Industry 4.0 SME collaboration," *Comput. Ind.*, vol. 138, p. 103623, 2022, doi: 10.1016/j.compind.2022.103623.

[62] F. Longo, L. Nicoletti, and A. Padovano, "Smart operators in industry 4.0: A human-centered approach to enhance operators' capabilities and competencies within the new smart factory context," *Comput. Ind. Eng.*, vol. 113, pp. 144–159, 2017, doi: 10.1016/j.cie.2017.09.016.

[63] F. Dillinger, O. Bernhard, M. Kagerer, and G. Reinhart, "Industry 4.0 implementation sequence for manufacturing companies," *Prod. Eng.*, no. 0123456789, 2022, doi: 10.1007/s11740-022-01110-5.

[64] V. Jain and P. Ajmera, "Modelling the enablers of industry 4.0 in the Indian manufacturing industry," *Int. J. Product. Perform. Manag.*, 2020, doi: 10.1108/IJPPM-07-2019-0317.

[65] A. Amaral and P. Peças, "A framework for assessing manufacturing smes industry 4.0 maturity," *Appl. Sci.*, vol. 11, no. 13, 2021, doi: 10.3390/app11136127.

[66] L. Bosman, N. Hartman, and J. Sutherland, "How manufacturing firm characteristics can influence decision making for investing in Industry 4.0 technologies," *J. Manuf. Technol. Manag.*, vol. 31, no. 5, pp. 1117–1141, 2019, doi: 10.1108/JMTM-09-2018-0283.

[67] D. Lucke, C. Constantinescu, and E. Westkämper, "Smart Factory - A Step towards the Next Generation of Manufacturing," *Manuf. Syst. Technol. New Front.*, no. Sfb 627, pp. 115–118, 2008, doi: 10.1007/978-1-84800-267-8_23.

[68] S. C. Lee, T. G. Jeon, H. S. Hwang, and C. S. Kim, "Design and implementation of wireless sensor based-monitoring system for smart factory," *Lect. Notes Comput. Sci. (including Subser. Lect. Notes Artif. Intell. Lect. Notes Bioinformatics)*, vol. 4706 LNCS, no. PART 2, pp. 584–592, 2007, doi: 10.1007/978-3-540-74477-1_54.

[69] J. C. Serrano-Ruiz, J. Mula, and R. Poler, "Toward smart manufacturing scheduling from an ontological approach of job-shop uncertainty sources," *IFAC-PapersOnLine*, vol. 55, no. 2, pp. 150–155, 2022, doi: 10.1016/j.ifacol.2022.04.185.
[70] S. Paasche and S. Groppe, *Enhancing data quality and process optimization for smart manufacturing lines in industry 4.0 scenarios*, vol. 1, no. 1. Association for Computing Machinery, 2022.
[71] Y. Lu and F. Ju, "Smart Manufacturing Systems based on Cyber-physical Manufacturing Services (CPMS)," *IFAC-PapersOnLine*, vol. 50, no. 1, pp. 15883–15889, 2017, doi: 10.1016/j.ifacol.2017.08.2349.
[72] G. Shao, S. Jain, C. Laroque, L. H. Lee, P. Lendermann, and O. Rose, "Digital Twin for Smart Manufacturing: The Simulation Aspect," *Proc. - Winter Simul. Conf.*, vol. 2019-Decem, no. Bolton 2016, pp. 2085–2098, 2019, doi: 10.1109/WSC40007.2019.9004659.
[73] M. Trstenjak, D. Lisjak, T. Opetuk, and D. Pavković, "Application of multi criteria decision making methods for readiness factor calculation," *EUROCON 2019 - 18th Int. Conf. Smart Technol.*, pp. 1–6, 2019, doi: 10.1109/EUROCON.2019.8861520.
[74] C. Faller and D. Feldmúller, "Industry 4.0 learning factory for regional SMEs," *Procedia CIRP*, vol. 32, no. Clf, pp. 88–91, 2015, doi: 10.1016/j.procir.2015.02.117.
[75] S. I. Shafiq, G. Velez, C. Toro, C. Sanin, and E. Szczerbicki, "Designing intelligent factory: Conceptual framework and empirical validation," *Procedia Comput. Sci.*, vol. 96, pp. 1801–1808, 2016, doi: 10.1016/j.procs.2016.09.351.
[76] B. Hameed *et al.*, "Functional Thinking for Value Creation," *Funct. Think. Value Creat.*, pp. 326–331, 2011, doi: 10.1007/978-3-642-19689-8.
[77] M. G. Seok, W. J. Tan, B. Su, and W. Cai, "Hyperparameter Tunning in Simulation-based Optimization for Adaptive Digital-Twin Abstraction Control of Smart Manufacturing System," in *SIGSIM Con-ference on Principles ofAdvanced Discrete Simulation*, 2022, pp. 61–68, doi: 10.1145/3518997.3531024.
[78] G. Saravanan, S. S. Parkhe, C. M. Thakar, V. V. Kulkarni, H. G. Mishra, and G. Gulothungan, "Implementation of IoT in production and manufacturing: An Industry 4.0 approach," *Mater. Today Proc.*, vol. 51, pp. 2427–2430, 2022, doi: 10.1016/j.matpr.2021.11.604.
[79] J. Lee, M. Azamfar, J. Singh, and S. Siahpour, "Integration of digital twin and deep learning in cyber-physical systems: Towards smart manufacturing," *IET Collab. Intell. Manuf.*, vol. 2, no. 1, pp. 34–36, 2020, doi: 10.1049/iet-cim.2020.0009.
[80] F. Simetinger and Z. Zhang, "Deriving secondary traits of industry 4.0: A comparative analysis of significant maturity models," *Syst. Res. Behav. Sci.*, vol. 37, no. 4, pp. 663–678, 2020, doi: 10.1002/sres.2708.
[81] S. Wang, J. Wan, D. Zhang, D. Li, and C. Zhang, "Towards smart factory for industry 4.0: A self-organized multi-agent system with big data based feedback and coordination," *Comput. Networks*, vol. 101, pp. 158–168, 2016, doi: 10.1016/j.comnet.2015.12.017.
[82] James, T. "Smart factories," *Engineering & technology*, vol. 7, no. 6, pp. 64–67, 2012.
[83] F. E. Gruber, "Industry 4.0: A best practice project of the automotive industry," *IFIP Adv. Inf. Commun. Technol.*, vol. 411, pp. 36–40, 2013, doi: 10.1007/978-3-642-41329-2_5.
[84] H. Lasi, P. Fettke, H. G. Kemper, T. Feld, and M. Hoffmann, "Industry 4.0," *Bus. Inf. Syst. Eng.*, vol. 6, no. 4, pp. 239–242, 2014, doi: 10.1007/s12599-014-0334-4.
[85] T. P. Cunha, M. P. Méxas, A. Cantareli da Silva, and O. L. Gonçalves Quelhas, "Proposal guidelines to implement the concepts of industry 4.0 into information technology companies," *TQM J.*, vol. 32, no. 4, pp. 741–759, 2020, doi: 10.1108/TQM-10-2019-0249.

[86] N. Chonsawat, "Defining SMEs' 4.0 Readiness Indicators Nilubon," *Appl. Sci.*, pp. 124–134, 2018.
[87] P. Dhamija and S. Bag, "Role of artificial intelligence in operations environment: a review and bibliometric analysis," *TQM J.*, vol. 32, no. 4, pp. 869–896, 2020, doi: 10.1108/TQM-10-2019-0243.
[88] D. Chen, T. Kjellberg, and A. Von Euler, "Software tools for the digital factory - An evaluation and discussion," *Adv. Intell. Soft Comput.*, vol. 66 AISC, pp. 803–812, 2010, doi: 10.1007/978-3-642-10430-5_62.
[89] M. Asif Rashid, H. Qureshi, M.-D. Shami, and N. Khan, "ERP Lifecycle Management for Aerospace Smart Factory: A Multidisciplinary Approach," *Int. J. Comput. Appl.*, vol. 26, no. 11, pp. 55–62, 2011, doi: 10.5120/3171-4372.
[90] B. J. Van Putten, M. Kuestner, and M. Rosjat, "The future factory initiative at SAP research," *ETFA 2009 - 2009 IEEE Conf. Emerg. Technol. Fact. Autom.*, pp. 5–8, 2009, doi: 10.1109/ETFA.2009.5347257.
[91] Weyer, S., Schmitt, M., Ohmer, M., and Gorecky, D. "Towards Industry 4.0-Standardization as the crucial challenge for highly modular, multi-vendor production systems," *Ifac-Papersonline*, vol. 48, no. 3, pp. 579–584, 2015.
[92] S. Chou, "The fourth industrial revolution: digital fusion with internet of things," *J. Int. Aff.*, vol. 72, no. 1, pp. 107–120, 2016.
[93] Frank, A. G., Dalenogare, L. S., and Ayala, N. F. "Industry 4.0 technologies: Implementation patterns in manufacturing companies," *International Journal of Production Economics*, vol. 210, pp. 15–26, 2019.
[94] F. Tao, H. Zhang, A. Liu and A. Y. C. Nee, "Digital Twin in Industry: State-of-the-Art," in *IEEE Transactions on Industrial Informatics*, vol. 15, no. 4, pp. 2405–2415, April 2019, doi: 10.1109/TII.2018.2873186.
[95] Lopes de Sousa Jabbour, A.B., Jabbour, C.J.C., Godinho Filho, M. et al., "Industry 4.0 and the circular economy: a proposed research agenda and original roadmap for sustainable operations," *Ann Oper Res*, vol. 270, pp. 273–286, 2018. 10.1007/s10479-018-2772-8
[96] Moeuf, A., Pellerin, R., Lamouri, S., Tamayo-Giraldo, S., and Barbaray, R. "The industrial management of SMEs in the era of Industry 4.0," *International journal of production research*, vol. 56, no. 3, pp. 1118–1136, 2018.
[97] Buer, S. V., Strandhagen, J. O., and Chan, F. T. "The link between Industry 4.0 and lean manufacturing: mapping current research and establishing a research agenda," *International journal of production research*, vol. 56, no. 8, pp. 2924–2940, 2018.
[98] Mittal, S., Khan, M. A., Romero, D., and Wuest, T. "Smart manufacturing: Characteristics, technologies and enabling factors." *Proceedings of the Institution of Mechanical Engineers, Part B: Journal of Engineering Manufacture*, vol. 233, no. 5, 1342–1361, 2019.
[99] Fatorachian, H., and Kazemi, H. "A critical investigation of Industry 4.0 in manufacturing: theoretical operationalisation framework." *Production Planning & Control*, vol. 29, no. 8, 633–644, 2018.
[100] Qin, J., Liu, Y., and Grosvenor, R. "A categorical framework of manufacturing for industry 4.0 and beyond," *Procedia cirp*, vol. 52, 173–178, 2016.
[101] Jung, K., Kulvatunyou, B., Choi, S., and Brundage, M. P. "An overview of a smart manufacturing system readiness assessment. In Advances in Production Management Systems," Initiatives for a Sustainable World: IFIP WG 5.7 International Conference, APMS 2016, Iguassu Falls, Brazil, September 3–7, 2016, Revised Selected Papers (pp. 705–712), 2016. Springer International Publishing.

APPENDIX

TABLE A1
Quality Evaluation Score (QES) of Some Studies

Author/Year	QEP1	QEP2	QEP3	QEP4	QEP5	QEP6	QEP7	QES Score
Lee et al., 2007	0	1	0	0	1	1	1	4
Lucke et al., 2008	1	1	0	1	1	0	1	5
Zühlke, 2009	1	1	0	1	0	0	0	3
James, 2012	1	1	1	0	0	1	1	5
Gruber, 2013	0	1	1	1	1	1	0	5
Lasi et al., 2014	0	1	1	1	0	1	1	5
Faller and Feldmüller, 2015	0	1	1	1	0	1	0	4
Weyer et al., 2015	0	1	1	1	1	1	0	5
Li, 2016	0	1	1	1	1	1	1	6
Lu and Ju, 2017	0	1	1	1	1	1	0	5
Iyer, 2018	1	1	0	1	1	1	0	5
Bibby and Dehe, 2018	1	1	0	1	1	1	1	6
Hamidi et al., 2018	1	1	1	1	1	1	1	7
Horvat et al., 2018	1	1	1	1	1	0	0	5
Chou, 2018	1	1	1	1	1	1	0	6
Simetinger and Zhang, 2020	1	1	1	1	1	0	0	5
Bachinger and Kronberger, 2021	1	1	0	1	1	0	1	5
Paasche and Groppe, 2022	1	1	1	1	1	1	0	6
Liu et al., 2022	0	1	1	1	1	0	1	5

2 Study and Analysis of IoT (Industry 4.0)

A Review

Manoj Kumar Gupta
Department of Mechanical Engineering, Hemvati Nandan Bahuguna Garhwal University, Srinagar Garhwal, Uttarakhand, India

Tarun Gupta
Department of Mechanical Engineering, GL Bajaj Institute of Technology & Management, Greater Noida, India

Dharamvir Mangal
Department of Mechanical Engineering, Gautam Buddha University, Greater Noida, India

Prashant Thapliyal and Don Biswas
Departmnet of Instrumentation Engineering, Hemvati Nandan Bahuguna Garhwal University, Srinagar Garhwal, Uttarakhand, India

CONTENTS

2.1 INTRODUCTION

Automation has emerged in recent years following the fourth industrial revolution, or the concept of Industry 4.0. The definition of "Industry 4.0" was developed in 2011 when the idea was promoted by an institution composed of industry, politics, and academia called "Industry 4.0" new concepts of the manufacturing industry [1]. Industry 4.0 represents the fourth industrial revolution. Industry 4.0 is the most efficient and accurate real-time data system in the manufacturing industry today. The Industrial Revolution is depicted in Figure 2.1. Industry 4.0 is a new production model that combines industrial automation and integrates new production technologies to

DOI: 10.1201/9781003333760-2

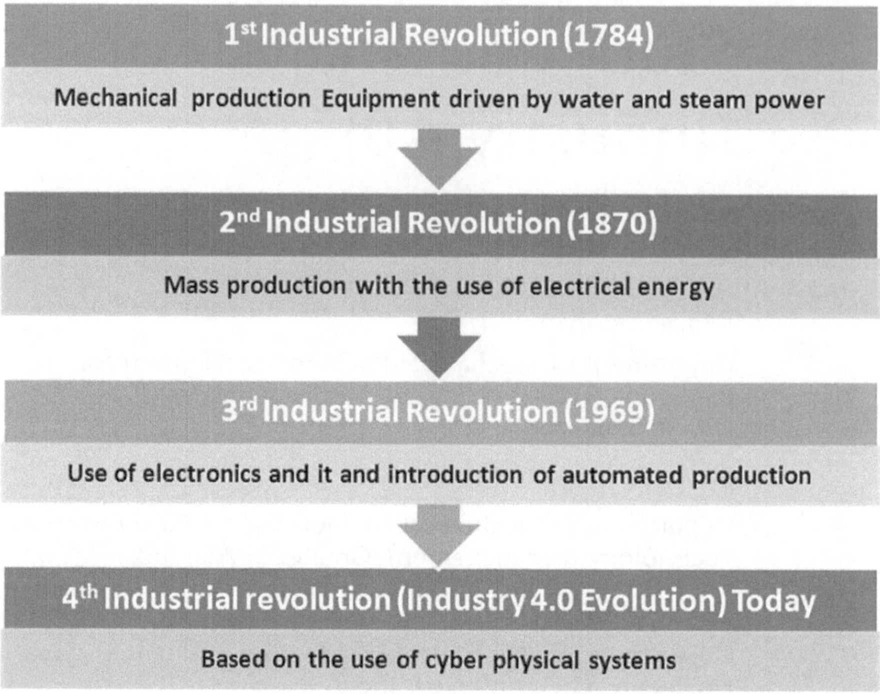

FIGURE 2.1 Industrial revolutions [2].

renovate working conditions and increase productivity and quality [2]. In the first industry, it was introduced (1784) as a mechanical production device powered by water and steam power. After that, the industry grew and then the second revolution (1870) of industry came into the picture as the concept of mass production by using electrical energy. This is the real beginning of industrial production. The industry evolved and the demand for production increased as then it required efficient mass production, which fulfilled the requirement of increased demands and then the third industrial revolution (1969) introduced the concept of automated production by the use of electronics and IT. In today's period, the industry needs the most efficient production with high productivity, high quality, and fulfillment of today's required demands and then the fourth revolution came. The fourth revolution of industry (Industry 4.0) is the most demanding topic in both professional industries and academics [2,3]. Industry 4.0 interfaces humans and machines, machine to machine (M2M) in an extravagant network using computerization technologies such as cyber-physical systems, the Internet of Things (IoT), and cloud computing [3]. Cyber-physical systems are the integration of computational and physical processes. The computers and networks monitor physical processes [4,5]. This revolution forced the industry to allocate manufacturing processes to more places. It creates powerful virtualization of reality and its active performance monitoring and remote management make the industry smart and efficient. This chapter discusses the automation technologies used in Industry 4.0 and explores the advantages, disadvantages, and future scope of Industry 4.0.

2.2 AUTOMATION TECHNIQUES USED IN INDUSTRY 4.0

Various techniques and concepts are used in the industry to digitize manufacturing activities and related services. Kusiak [6] discussed the concept of Industry 4.0 and its main element. The smart industrialized system coordinates production through cloud computing, sensors, communication technologies, control system, simulation, modeling, and analytical engineering. Frank et al. [7] stated that Industry 4.0 was considered a new industrial phase in which the rise of automation provides digital solutions. Nardo et al. [1] described the evolution of interactions between workers and machines, or man-machine interaction ("MMI" or human-machine interaction "HMI"). It is a dynamic manufacturing system including automation, verdict support equipment, and software. This new concept of manufacturing control systems at different levels is shown in Figure 2.2. Automation helps people work in the phase of production, logistics, and data collection for ERP and management. A systematic study of the cyber-physical system (CPS) and the discovery of a new framework were proposed by the company of the interaction of CPS and cyber-human system (CHS), as shown in Figure 2.3. It shows that human skills are essential in the decision-making stage and their skills are an accelerator in Industry 4.0.

Kumar et al. [9], Ferdows et al. [10], Almada et al. [11], and Rosa et al. [12] reported that industrial robots are acquiring the main manpower element in industries instead of human interference due to their accuracy, higher productivity rate than

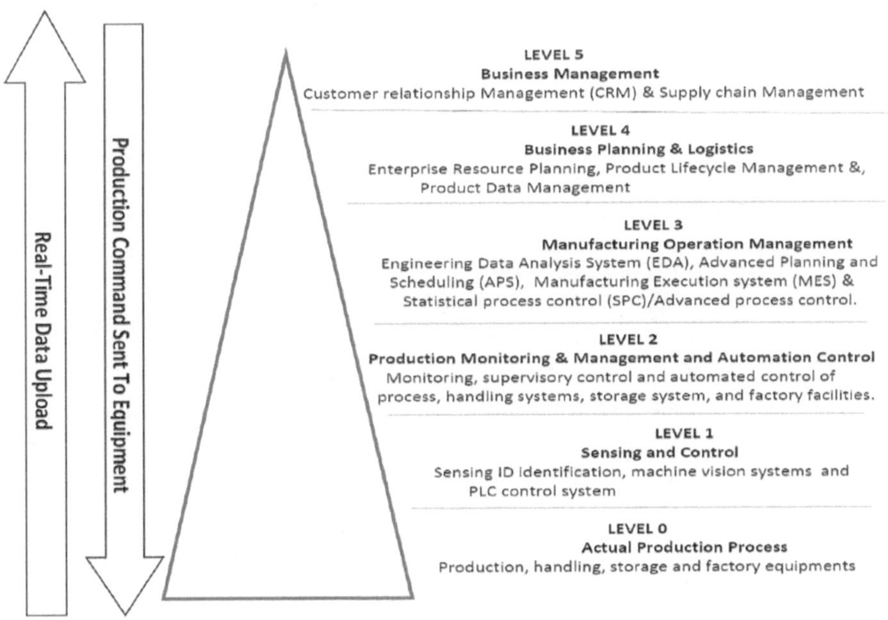

FIGURE 2.2 Control system levels in manufacturing [1].

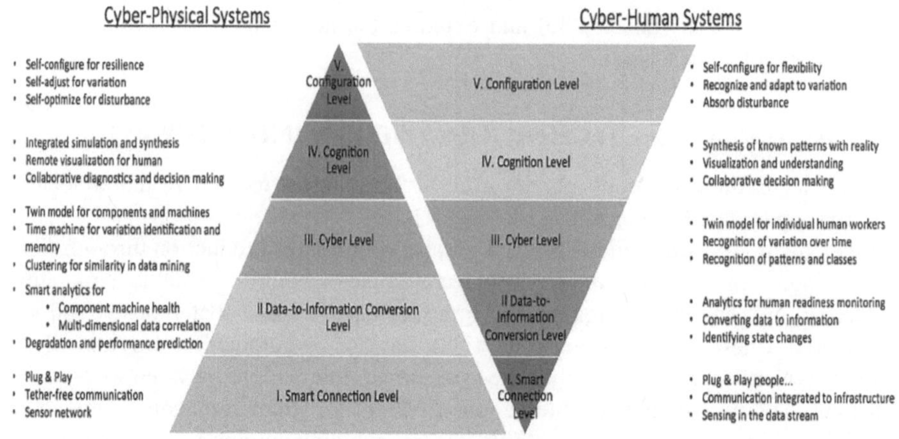

FIGURE 2.3 Cyber-physical and human systems [8].

humans, better quality, no emotional interference, and less physical efforts. This is offering drastic changes in today's industry. We require only a few skilled human workers instead of various unskilled human workers. Due to this, today's industry grows very fast, which helps the development of a country. It gets easier for humans because robots perform most routine work and maintenance work. It will replace human interference day by day in future industries. For example, as shown in Figure 2.4, if we require heavy lifting work in industries, then robots can perform easily on the command of a single human operator.

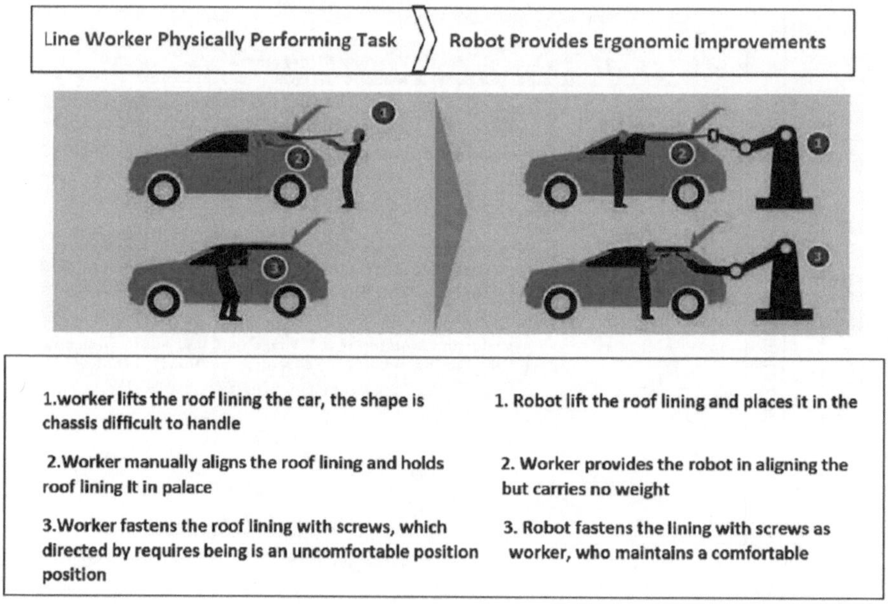

FIGURE 2.4 Robotic production system [1].

Ferdows et al. [10] and Nardo et al. [1] automated guided vehicle systems (AVGSs) that built an essential part of today's industry. An AGVS is a driverless vehicle that is connected wirelessly with controllers and sensors on the concept of IIOT (Industrial Internet of Things) and cloud computing. It performs the entire task with high efficiency in today's industries. Its main benefit is that it will perform the task by a single skilled worker. It reduced manpower in industries and we cannot compare its accuracy and efficiency with humans, which helps to produce the best-quality products or we can say it also helped in total quality management (TQM), which helps to get a place in the list of top-quality companies in local and international markets, which are major supports in the development of a company's economy as well as the country. Today's industrial revolution in technologies such as cyber-physical systems (CPS), the Internet of Things (IoT), and data mining changes the way of controlling manufacturing activities. Big data helps TQM by following a rule of "PLAN–DO–CHECK–ACT," which was developed with today's technologies, as shown in Figure 2.5.

Machado et al. [13] reported the sustainability of today's manufacturing industries on various aspects of Industry 4.0, like technology pillars, the scope of Industry 4.0, and models developed in the industry. They observed the fourth industrial revolution and checked the sustainability of today's manufacturing industries for future aspects of improvements in industries. They link Industry 4.0 and sustainability manufacturing. Rosin et al. [14] and Sony [15] reported the changes in the lean manufacturing process due to the fourth revolution of industries. They describe the linkage between the lean principle, Industry 4.0, and capability levels, as shown in Figure 2.6.

Balland et al. [16] and Rosa et al. [12] describe the use of industrial 4.0 technologies in today's manufacturing systems based on circular economic aspects due to the implementation of Industry 4.0. They use 14T (14 technologies of Industry 4.0) for various reasons for observations of increasing implementations of Industry 4.0, as shown in Figure 2.7 and hybrid circular and digital circular economy in Figure 2.8.

Moeuf et al. [17] reported that the opportunities for small and medium industries (SMIs) like IoT, CPS, and cloud computing [15] as well as risk also involved like

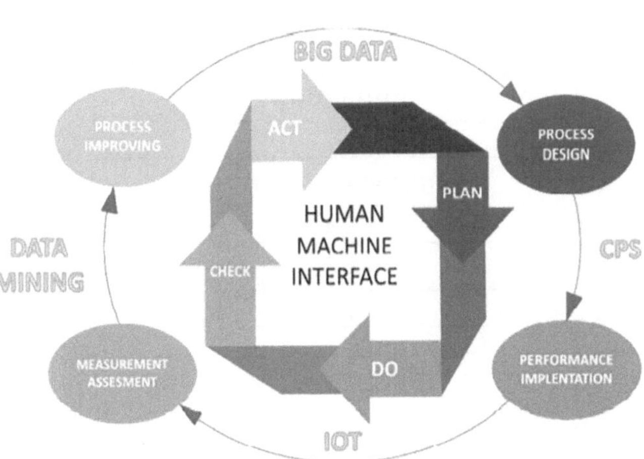

FIGURE 2.5 Man-machine collaboration in Industry 4.0 [1].

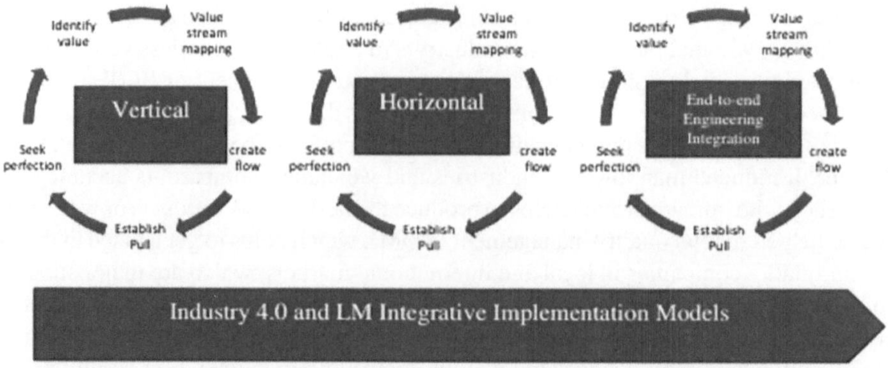

FIGURE 2.6 Integration between Industry 4.0 and lean management [15].

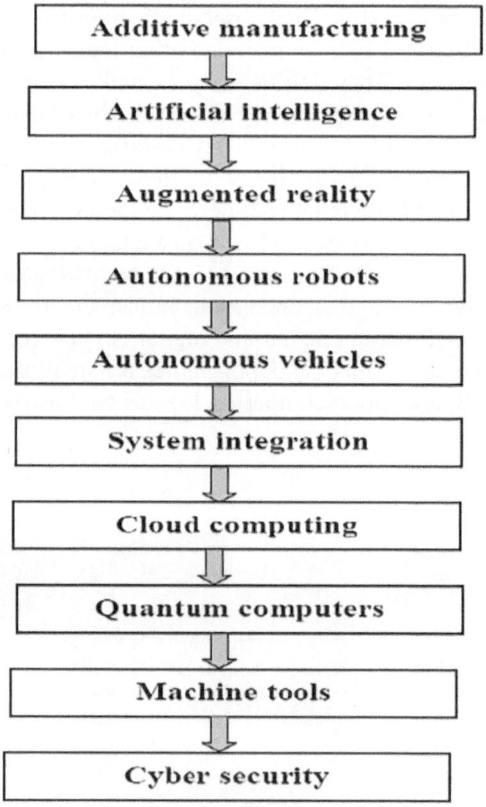

FIGURE 2.7 Technologies used in Industry 4.0 [16].

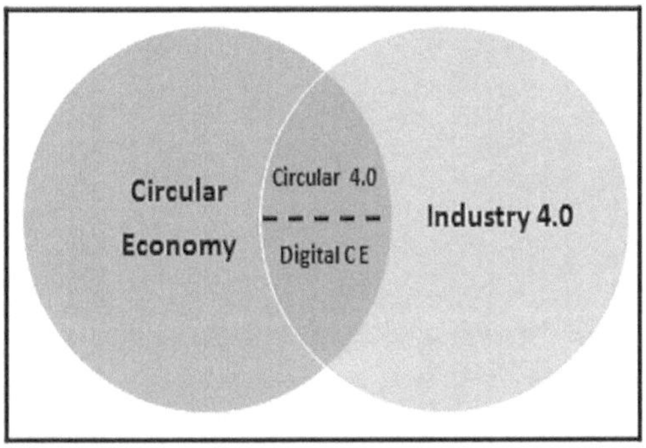

FIGURE 2.8 Hybrid categories of circular and digital circular economy [12].

lack of skilled workers in SMIs. Tan et al. [18] and Wang et al. [19] describe big data analysis and cloud computing implementation in Industry 4.0 for efficient use of resources, which improves the efficiency of today's industries. Also described are innovative devices based on these technologies like drones and unmanned aerial vehicles (UAVs), which have a very high potential for performing efficient work in today's industries. Valeske [20] described nondestructive evaluation (NDE4.0) sensors, which are the elements of IIOT used with 5C architecture and IIOT for multiple optimizations by feedback loop command under various steps during the manufacturing process. Sony [3] describes the adaptation of Industry 4.0 and finds seven pros and cons of Industry 4.0, as depicted in Figure 2.9.

The data collected from the different research papers and internet sources were summarized and reported in table 3 to get the output visible. Table 2.1 reveals the advantages and drawbacks of various technologies used in Industry 4.0.

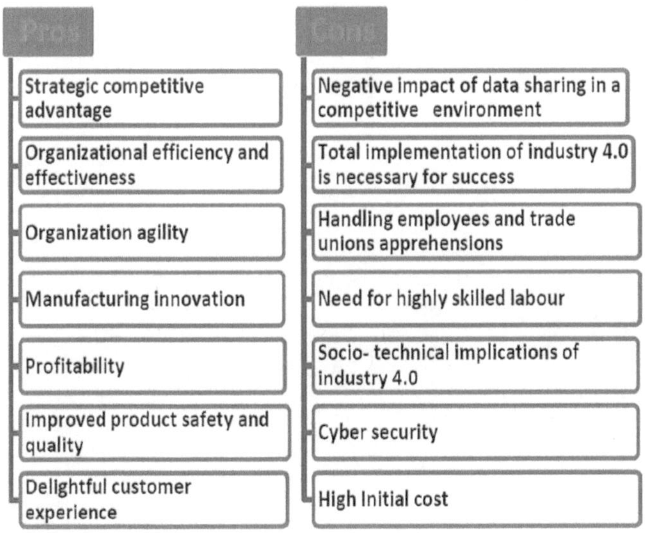

FIGURE 2.9 Pros and cons of Industry 4.0.

TABLE 2.1
Effects of Technologies on Industry 4.0

S. No.	Technologies used in Industry 4.0	Advantage	Disadvantage	References
1	Big Data and Analytics	It helps to understand and target customers and optimize business processes. One platform carries an unlimited amount of information.	Higher data storage cost. The results of big data analysis are sometimes misleading. Rapid updates of big data may not correspond to actual numbers.	[1,2,4,6,7]
2	Autonomous Robots	Cost efficiency, improved quality, certainty, and increased productivity.	Potential job losses, high initial investment costs, and hiring qualified personnel.	[5,8,9]
3	Simulation	It can avoid danger and loss of life. Conditions may vary and results may be reviewed. The investigation of critical situations can be done without any risk. It is cost-effective.	Errors may be in the simulation or model. The simulation model cost is high. The cost of running several different simulations can be high. It may take time for the results to be understood.	[10,11]
4	Horizontal and Vertical System Integration	Greater market share. Higher revenues.	Stopping the economic growth of a new business. Reduced flexibility, destroying value rather than creating it. It forces the business to operate within economies of scale. It reduces flexibility. When entering a new market, there may be unforeseen obstacles.	[13,14]
5	The Industrial Internet of Things	Decreased operating costs, improve productivity, improve customer satisfaction.	Issues related to security, privacy, and connectivity.	[16]

6	Cyber Security	Protection against data theft, hackers, and viruses. It provides users with privacy.	Firewalls can be challenging to configure properly. Improper configured firewall configuration may block users from performing a certain action. Makes the system slower than before. New software needs to be constantly updated to keep security up to date.	[12]
7	Cloud	Backup for data storage, Better cooperation, first-rate accessibility, lower maintenance cost, and big storage capacity.	Could be costly for an average user. The problems are Internet connectivity, vendor lock-in, limited control, and security.	[17]
8	Additive Manufacturing	Accelerated prototyping, customization, environment benefits, inventory stock reduction.	Cost of entry, production costs, additional materials, it's slow post-processing	[15,18]
9	Augmented Reality	Communication and interaction, business improvements.	Privacy issues, hazards of reality, requirements for changes and implementation.	[3,19,20]

2.3 CONCLUSION

This paper discussed the evolution of the industrial revolution. The detail about Industry 4.0 technology with the analysis or various methods was discussed. The use of the new technologies and their impacts on the implementation of the Industry 4.0 are explained in detail. The adaption, pros, and cons of the fourth revolution of industry were elaborated on, such as how it can make ordinary things advance subsidiary industrial items like home automation with IIOT, cloud computing, and cyber security. The 14 technologies (14T) of Industry 4.0 used for increasing the implementations of Industry 4.0 were also discussed. Future research can focus on the implementation of various opportunities to link both approaches, allowing industries to achieve the full objectives of Industry 4.0. Studies show that the most frequently proposed technologies to improve lean principles are IoT and simulation [–].

REFERENCES

[1] Nardo, M., Forino, D., & Murino, T. 2020. The evolution of man-machine interaction: The role of human in Industry 4.0 paradigm. *Production & Manufacturing Research* 8(1):20–34.

[2] Drath, R., & Horch, A. 2014. Industrie 4.0: Hit or hype? [industry forum]. *IEEE Industrial Electronics Magazine* 8(2):56–58.

[3] Sony, M. 2020. Pros and cons of implementing Industry 4.0 for the organizations: A review and synthesis of evidence. *Production & Manufacturing Research* 8(1):244–272.

[4] Lee, E. A. 2008. Cyber physical systems: Design challenges. In 2008 11th IEEE International Symposium on Object and Component-Oriented Real-Time Distributed Computing (ISORC):363–369). IEEE.

[5] Gubbi, J., Buyya, R., Marusic, S., & Palaniswami, M. 2013. Internet of things (IoT): A vision, architectural elements, and future directions. *Future Generation Computer Systems* 29(7):1645–1660.

[6] Kusiak, A. 2018. Smart manufacturing. *International Journal of Production Research* 56(1–2):508–517.

[7] Frank, A. G., Dalenogare, L. S., & Ayala, N. F. 2019. Industry 4.0 technologies: Implementation patterns in manufacturing companies. *International Journal of Production Economics* 210:15–26.

[8] Krugh, M., & Mears, L. 2018. A complementary cyber-human systems framework for Industry 4.0 cyber-physical systems. *Manufacturing Letters* 15(partB):89–92.

[9] Kumar, R., Haleem, A., Garg, S. K., & Singh, R. K. 2015. Automated guided vehicle configurations in flexible manufacturing systems: A comparative study. *International Journal of Industrial and Systems Engineering* 21(2):207.

[10] Ferdows, K., & De Meyer, A. 1990. Lasting improvements in manufacturing performance: In search of a new theory. *Journal of Operations Management* 9(2):168–184.

[11] Almada-Lobo, F. 2015. The Industry 4.0 revolution and the future of Manufacturing Execution Systems (MES). *Journal of Innovation Management* 3(4):16–21.

[12] Rosa, P., Sassanelli, C., Urbinati, A., Chiaroni, D., & Terzi, S. 2020. Assessing relations between Circular Economy and Industry 4.0: A systematic literature review. *International Journal of Production Research* 58(6):1662–1687.

[13] Machado, C. G., Winroth, M. P., & Ribeiro da Silva, E. H. D. 2020. Sustainable manufacturing in Industry 4.0: An emerging research agenda. *International Journal of Production Research* 58(5):1462–1484.

[14] Rosin, F., Forget, P., Lamouri, S., & Pellerin, R. 2020.Impacts of Industry 4.0 technologies on Lean principles. *International Journal of Production Research* 58(6):1644–1661.

[15] Sony, M. (2018). Industry 4.0 and lean management: A proposed integration model and research propositions. *Production & Manufacturing Research* 6(1):416–432.

[16] Balland, P. A., & Boschma, R. 2021. Mapping the potentials of regions in Europe to contribute to new knowledge production in Industry 4.0 technologies. *Regional Studies* 55(10–11):1652–1666.

[17] Moeuf, A., Lamouri, S., Pellerin, R., Tamayo-Giraldo, S., Tobon-Valencia, E., & Eburdy, R. 2020. Identification of critical success factors, risks and opportunities of Industry 4.0 in SMEs. *International Journal of Production Research* 58(5):1384–1400.

[18] Li, G., Tan, J., & Chaudhry, S. S. 2019. Industry 4.0 and big data innovations. *Enterprise Information Systems* 13(2):145–147.

[19] Li, G., Wang, H., & Hardjawana, W. 2020. New advancement in information technologies for Industry 4.0. *Enterprise Information Systems* 14(4):402–405.

[20] Valeske, B., Osman, A., Römer, F., & Tschuncky, R. 2020. Next generation NDE sensor systems as IIoT elements of Industry 4.0. *Research in Nondestructive Evaluation* 31(5–6):340–369.

3 Recent Advances in Cybersecurity in Smart Manufacturing Systems in the Industry

Dinesh Kumar Atal and Vishal Tiwari
Department of Biomedical Engineering, Deenbandhu Chhotu Ram University of Science and Technology, Murthal, Sonipat, Haryana, India

Dharmender Kumar
Department of Biotechnology, Deenbandhu Chhotu Ram University of Science and Technology, Murthal, Sonipat, Haryana, India

CONTENTS

DOI: 10.1201/9781003333760-3

3.1 INTRODUCTION TO SMART MANUFACTURING SYSTEMS

The need for greater customization, higher efficacy, and effectiveness is increasing [1,2] among manufacturers. Every product manufactured today in industries has already undergone planned smart manufacturing processes. When the processes defined by corresponding process parameters are executed, their outcomes inevitably deviate from the desired outputs, giving undesired results such as surface roughness, processing time, and other vital characteristics [3].

Industrialization and informatization are advancing using industrial, manufacturing, information, and communication technologies. Because of technical advantages, different countries resume their manufacturing competitiveness in terms of innovativeness, cost, and reliable products to meet customer demands [4,5]. To be successful, other manufacturers and companies must choose and incorporate technologies that will help in rapid changes while maintaining higher effectiveness and efficacy, greater customization, and reducing energy usage. Combining such technologies forms the core of an information-centric system known as smart manufacturing systems (SMS) [1]. While manufacturing involves multiple phases to create a product out of raw materials, smart manufacturing is a large-scale complex system aiming to enable flexibility in processes using information and manufacturing technologies to create an even better product. It is so because smart manufacturing relates to ICT, management technology, and industrial and management technology [4].

Thus, smart manufacturing results as a key component of the fourth industrial revolution, proving a broader thrust. SMS mainly relies on creating a bridge between the physical and digital environments with the Internet of Things (IoT) technologies coupling to the improvements in the digital world through machine learning, data analytics, cloud systems, etc. [6].

The manufacturing sectors have a significant effect on the growth of a country, and their performance and effectiveness have a direct impact on the country's economy. Therefore, various technologies are developing to get the desired sound quality, product yield, and performance [5,7].

In the smart manufacturing industries, deploying more intelligent technologies affects investment. In terms of percentage, smart manufacturing systems have achieved a quality improvement of 15%–20%, with productivity improvement of 17%–20% through improved optimization of energy and utilization of machines [5,6].

The manufacturing industry underwent many industrial transformations called the industrial revolution. Industry 1.0 (the 1760s), known as the first industrial revolution, includes steam-powered machines, textile industry, iron production, mining, and metallurgy. Industry 2.0 (the 1870s), known as the second industrial revolution, involves electrification, mass production, engines/turbines, railroads, and electrical telegraph. Industry 3.0 (the 1970s), known as the third industrial revolution, involves PLC/robotics, computer/Internet, digital manufacturing, automation, electronic digital networks, IT, and OT digitization [7]. The manufacturing industry is currently going through the fourth industrial revolution by merging the cyber, physical, and virtual worlds at the core of its transformation. It includes IoT, AI, ML, CPS, and real-time data processing. Smart factory Industry 4.0 aims to provide even better customer satisfaction, meet customer needs, improve quality, lower costs, and provide a more sustainable and productive system. Smart manufacturing indicates "a data-intensive application enabling a more effective, efficient, accurate system thus, performing intelligent and responsive operations" while focusing on the data and technology. SMS emphasizes the knowledge of manufacturing and the importance of human ingenuity such that desired outputs can be achieved and may result in attaining a good percentage of quality and productivity gains [3,8].

Smart manufacturing systems can more accurately predict product requirements, thus leading to quick identification of errors and re-creating an essential function in enhancing sustainability and quality while reducing costs and providing better stability [7]. SMS offers flexibility in information, i.e., information is available when needed and, in the form, found to be more beneficial throughout the engineering life cycle, production, design, and planning [9]. The development of ICT, cyber-physical systems (CPSs), and related technologies have given rise to the following:

- IoT
- Cloud computing
- Digital twin
- Next-generation ML and AI

Various advanced manufacturing paradigms are proposed to develop processes and systems with a higher intelligence level, i.e., smart intelligence systems. Different definitions of smart manufacturing systems proposed in the past years are as follows:

- From an engineering point of view, SMS is defined as the technology that uses advanced smart/intelligent technologies, enabling stable and rapid manufacturing along with real-time optimization and dynamic response to personalized product demands. Smart manufacturing platforms can integrate designs, products, and operations that span any industry, multi-national company, and enterprise.
- From the networking point of view, smart manufacturing involves using cyber-physical systems, industrial IoT and IoT powered by various sensors, and ICT that gather information at all manufacturing levels. These systems get smarter as time goes on and as there is an increase in productivity following reduced production waste and errors.
- From a decision-making point of view, smart manufacturing systems use domain data to help industries maintain production processes and procedures, thus improving productivity. These systems optimize various control processes of manufacturing operations based on big data analytics [10].
- It is also defined as the technology that utilizes tools and involves machines for providing improved performance and optimizing the workforce required AI, ML, advanced robotics, and big data processing [5].
- It enables us to obtain real-time data with improved accuracy, efficacy, and effectiveness, increasing overall productivity [11].

Further, characteristics defining intelligent/smart manufacturing systems involve data analytics, smart products, the Internet of Things, cyber-physical systems, cybersecurity, cloud manufacturing, etc. The design factors of SMS are modularity, heterogeneity, context awareness, etc. Also, there are some critical enabling factors for such systems. Examples include innovative training, laws and regulations, and data-sharing standards [6]. Smart manufacturing systems show some key capabilities vital for adding value to these systems. Key capabilities for intelligent manufacturing systems are as follows:

- Agility – It is the capability to survive and prosper in an environment, that is, changing dynamically and reacting effectively. Agility's success impacts enabling technologies, including distributed intelligence, supply chain integration, simulation, and modeling.
- Quality – It indicates measures of product customization and innovation. It shows how well the final product conforms to the design specifications.
- Productivity – The ratio of output to input within production by using vital parameters such as cost, labor, materials, energy efficiency, and manufacturing time. Responsiveness to customer demands.
- Sustainability – It is defined as the effect of manufacturing on society, the environment and the well-being of employees, and economic viability. More critical to smart manufacturing systems in comparison to the productivity measure.

Smart manufacturing systems are often when every control process of the manufacturing system is digitized, having features such as real-time controlling and monitoring, with more advanced technologies having quicker response time to the market, more efficient manufacturing systems along with enhanced productivity, and achieving a higher level of sustainability. Smart manufacturing systems are connected with manufacturing processes, product delivery systems, real-time machining units, design, and customization, and then to the end customers by means of cloud computing, enabling more efficient manufacturing processes, customization, and meeting customer demands [5]. The smart manufacturing systems have two working modes:

- Semi-autonomous (In this, the manager will define all the goals to be achieved and will set the parameters)
- Fully autonomous (In this, the SMS will itself define the goals and parameters for its effective working that will be implemented automatically in all the units of production)

As in such competitiveness between the different industries and countries, the manufacturer mainly checks for quality, sustainability, time, cost of SMS, and its ability to withstand quality assurance [5]. Smart manufacturing, thus, has more advantages over traditional manufacturing systems as it offers a higher degree of automation and has a cluster of control processes while, despite all these benefits, every time system is more vulnerable to cyber-attacks, thus making it more challenging [12].

3.2 CYBERSECURITY AND ITS NEED IN SMART MANUFACTURING SYSTEMS

With the increase in automation in the industry, there arises a risk of malicious cyber-attacks. Nowadays, cybersecurity risks (cyber-attacks or cyber-threats) are considered to be the foremost and the most concerning challenge that manufacturers face. Due to the increase in cyber-attacks since the start of Industry 4.0, i.e., the 2010s, sometimes also referred to as the fourth industrial revolution, targeting smart manufacturing systems, industries are required to treat this as a matter of major concern as these attacks are severe business risks. Thus, as far as the security of smart manufacturing systems is concerned, cybersecurity measures need to be taken care of. Therefore, in the context of smart planning, the identification of cyber-attacks and analyzing their impact is an important step. Standards of cybersecurity are crucial for systems and industries that take digital transformation, providing a baseline for understanding cybersecurity in cyber-physical systems. Over the last decade, developing and maintaining cybersecurity standards have been its best practice that tackles and prevent any cyber-attack, which mainly focuses on control systems and industrial applications [13].

In past years, the manufacturing systems were either designed ignoring the important aspect of security or with the idea of security, i.e., making it isolated from external access such that it is not subject to any vulnerable attacks. Therefore, isolation based on control of physical access was the only method of security for traditional manufacturing systems. Nowadays, keeping in view the important parameters such as convenience and cost, IP stack and Ethernet work as a core plant of the manufacturing system as connecting to such networks is much easier. Similarly, extending infrastructure in terms of the network to every remote area, mobility in handling, and lowering costs, leads to an increase in the deployment of wireless networks. Cybersecurity is the process of securing any cyber-physical system or manufacturing system from any malicious cyber-attacks to which it is more vulnerable. This process of cybersecurity is often known to start at or before the design phase and continues further [6].

Securing software development focuses on offering prevention against software vulnerabilities. As many attempts to ensure cybersecurity for manufacturing systems, in view of the cyber-attacks, resulted in poor output and reinforced the fact that if new security systems are developed, then it is hard to ensure whether they have been tested as the original design and results as overly conservative while introducing new bugs in the systems, thereby failing to secure it. Thus, it is important to understand that securing a smart manufacturing system is of higher importance [6]. Every development comes with a risk, so there is always a dire need to fight against those risks, and having a security system proves to be a boon for it.

Cyber/data security is defined as a dynamic process that fights against any vulnerable attack. Irrespective of having an extent of preparation, risks will always be there, and attackers take advantage of the weak parts of the system and will attack the system. So, to cope with this situation, it is, therefore, crucial to observe the dynamic behavior, thereby determining any errors in the system [6].

As documented, many incidents have existed over the past years, which indicates that evaluation of the impact of cyber-attacks on smart manufacturing systems is on the rise. A short literature survey depicting threats to SMS, as well as some proposed methods by the researchers to deal with these threats in past years, is as follows:

- In 2014, Wells et al. depicted specific threats to the systems through the potential approaches for analyzing and dealing with such concerns [14,15].
- In 2015, Vincent et al. proposed a real-time detection approach for increasing the quality and ensuring quality assurance affected by Trojan attacks in manufacturing systems [14,16].
- In 2016, Zeltmann et al. stated the importance of modification in designing the products in manufacturing control processes [14,17].
- In 2017, Sturm et al. focused on the threats of additive manufacturing (AM) technologies, especially while using STL files [14,18].
- In 2017, DeSmit et al. used the decision tree method of analysis and approach for assessing CPS in smart manufacturing companies [14,19].

In summary of the above-proposed techniques, we can say that researchers have been focused on the attack and not on the quantification of the effects of such attacks in smart manufacturing systems. Also, a quantitative model to check the effects of cyber-attacks has not been found in the literature survey [14].

Traditionally, manufacturing systems use some cyber/data security architectures that integrate different security architectures. These architectures provide services that include maintenance of confidentiality, integrity, non-repudiation, access control, and authenticity, which offers a system cyber/data security by preventing computer and network intrusions and attacks [20].

Smart manufacturing systems, i.e., are often known as the next-generation manufacturing paradigm with numerous technologies providing a more comprehensive range of automation. These systems offer enhanced optimization, artificial intelligence, and sensing capabilities. SMS uses ICT technologies to provide an even better flow and use of digital information from one system to another. With real-time analysis, a system could produce better responses in terms of quality and economy. However, the increased use of intelligent technologies such as smart sensors, big data, and cloud computing imposes a severe risk of cyber-threats. According to IBM's report, the manufacturing industry was in second place as the most frequently attacked industry in 2015. Therefore, as a boosting industry, the manufacturing industry must undertake cyber/data security measures per their ICT budgets, respectively [21].

Thus, cybersecurity is defined as keeping it under the control of data security. Also, security that cannot be provided to systems and devices working on the Internet results in data loss with compromised confidentiality. The main aim of providing cybersecurity is to ensure the security of the company prevention from any loss to the physical world, maintain confidentiality, and ensure that physical communication is carried out in the most appropriate environment [22].

3.3 CYBER-THREATS TO SMART MANUFACTURING

Over a wide range of industrial networks, there are various sources for the generation of cyber-attacks to which the system is more vulnerable. Also, there are a number of common attacks to which smart manufacturing systems are more vulnerable. These attacks adversely affect the manufacturing systems in the way of compromising productivity, sustainability, profit margin, and safety and finally affecting the reputation of the industry [13].

In a smart manufacturing system, there is the use of integrated network systems for the exchange of information between one unit to another and to customers as well. Thus arises a need for network connectivity arranged through the Internet, and the exchange of information through it requires information security, i.e., end-to-end encryption such that it could be protected from any malicious activity. Therefore, ensuring the security of the system is a matter of main concern in the case of every smart manufacturing industry, i.e., countermeasures to mitigate risks for the system by identifying all the possible cyber-attacks at an early stage [5,13]. However, it is a key enabler for determining all the possible cyber-threats, but it still needs to evolve so as to cope with deploying digital twins and tools for deploying. To a large extent, they are limited to simulation [13].

Some common attacks to which smart manufacturing systems are more vulnerable are as follows:

a. Denial of Service (DoS) attack – It involves the denial of the availability of assets [6,13,23]. An asset can be a system device, any memory, process, file system, or network. Distributed DoS attacks employ many compromised systems infected with malicious malware that further attacks a target. In past years, the largest distributed DoS attack was caused by Mirai botnets (high profile botnet) proved to be compromising a great number of IoT devices. Brickerbot and Raeper botnets, like Mirai botnets, demonstrate the danger of setup and forget the approach to IoT devices [6,24]. Some research studies have shown that among various security solutions, honeypot-based botnet detection is much more effective [25].
b. Deception attack – Such an attack is termed a false data injection attack. In such attacks, the adversary sends malicious data to a network as in the field bus [6,24].
c. Time-delay attack – In this type of attack, the attacker will adversely affect the stability of the system by injecting delays in time into the control values and measurements of the system. This attack can cause the equipment to crash [6].
d. Man-in-the-middle attack – In such an attack, the adversary relays the communication between different communicating devices in industrial systems and thus sends malicious traffic. For example, it could sabotage the key exchange protocol between communication devices such as an actuator device and a control system [6,13].
e. Spoofing attack – In these types of attacks, an attacker's node is impersonated by a system entity leading to a lack of appropriate authentication between control mechanisms, thereby leading to illegitimate access [6].
f. Eavesdropping attack – An adversary can gain any sensitive information by monitoring it so as to perpetrate attacks. If any attacker performs the network traffic analysis, even if they have packets with encryption

and reveal confidential information and the security of the network gets compromised [6].

g. Replay attacks – In replay attacks, an adversary retransmits legitimate packets. It can be defined as replaying false information from legitimate traffic. For example, intercepting an authentication message [6,13].
h. Data tampering attack – It aims at the unauthorized alteration of data in storage or in transit. This attack can cause severe damage to the working of the industry [6].
i. Covert-channel attacks – This aims at leaking sensitive information from a sensitive environment. It uses legitimate communication channels and some compromised devices. Such an attack involves bypassing all the security measures of a protected environment [6].
j. Ransomware attack – It involves the spread of malicious code [13].
k. Side channel attack – This attack analyzes the leakage of information from hardware to software. It uses various techniques, for example, an optical signal, analysis of power consumption, traffic flow, time for an operation to be conducted, and light, thermal, acoustic, and electromagnetic emission [6].
l. Zero-day attacks – These attacks have an average duration of 312 days. However, in some cases, the discovery, fixation, and distribution of a software patch can take time of somewhere up to 30 months [6].
m. Attacks against data analytics and machine learning – As machine learning is used in cybersecurity architecture, security through ML ranges from biometrics to network security monitoring [6].
n. Physical attack – In this, if an attacker gets physical access to smart manufacturing devices or systems, then they can also have the capability to manipulate those devices. Examples include de-calibrating sensors and changing the location of a sensor [6].

Also, attacks stem due to a large number of process deviations, such as lack of training, malfunctioning machinery, environmental factors, statistical variances, operation errors, and poor conditioning of the machinery [3].

Smart manufacturing industries also experience some hard times due to budget constraints [7]. Also, sometimes the process of risk analysis becomes far more challenging as the visibility of malicious cyber threats has decreased. In every smart manufacturing industry, risk analysis or the process of estimating the behavior of attackers to find out the possible threat is more difficult [8,26].

3.4 STRENGTHS AND WEAKNESSES OF SMART MANUFACTURING SYSTEMS

3.4.1 Strengths of Smart Manufacturing Systems

As the fourth industrial revolution allows us to achieve a higher state of operational abilities and rapid growth in agility, productivity, and sustainability parameters, it encompasses several new technologies, i.e., IoT, and IIoT, to name a few, along with the new 21st-century associated paradigms. The human-to-machine collaboration leads to new industrial processes and advanced manufacturing [20]. The sustainable competitiveness of manufacturers depends on their capabilities on vital parameters such as security, cost, flexibility, delivery, and reliability. The main objectives of a smart manufacturing system involve [1]:

- Increased decision-making quality
- Avoidance of media discontinuities, waste losses, and rejects
- Increased resource efficiency and delivery performance
- Reduced inventory, quality, and maintenance costs
- Increased flexibility and new customer benefits, such as mass customization
- Increased sustainability and transparency over the value-creation process
- Reduced use of energy, labor, and materials used for high-quality customized production, i.e., Quality products for timely delivery
    - Develop innovative products and technologies
    - Support intelligent marketing for better production planning
    - Rapid response to market and supply chain demand

3.4.2 Weaknesses of Smart Manufacturing Systems

Technology selection can be one of the main challenges in transforming existing systems into smart production systems [5], which can ultimately cause huge losses in production. Some industrial sectors face higher commercial and insurance risks [1]. However, with the increasing tendency to use networked devices, production systems become more vulnerable to cyber-attacks, and as malicious cyber-attacks become less visible, they become more difficult to detect [8]. The increasing adoption of intelligent machine maintenance in manufacturing faces challenges involving IIoT concerning robustness, scalability, and security [27]. Some major drawbacks of smart manufacturing systems are listed below:

- Uncertainties regarding data protection
- Changed requirements profile for employees
- Human conflicts in the transformation process (as with every transformation >> change management)
- Concrete approach and complex decision making regarding strategy
- Cost-intensive and lengthy transformation process

3.5 PREPAREDNESS AGAINST THE THREATS USING THE PROPOSED TECHNIQUES

3.5.1 Threat Modeling in ICPS

Threat modeling is a proposed technique that analyzes all the possible cyber-attacks that industries face. It addresses all the expected cyber-attacks that can affect a manufacturing system under consideration. In this technique, firstly, all the attack vectors will be analyzed along with a complete detailed understanding of the system's architecture, thereby providing an even better way to offer security from these threats via these security measures. To perform threat modeling in ICPS, demonstration of its models, methodologies, taxonomies, and characterization followed by the identification of appropriate security control measures is the method. In threat characterization, the determination of profiles of adversaries and their behavior also takes place. This characterization consists of motivation, capabilities, and types of adversaries to perform an attack on the system or network [13].

Adversaries can be classified as either insider or outsider. Insider adversaries are adversaries who perform some unauthorized actions through a level of authorized access to ICPS. These unauthorized actions may be due to some ego, revenge, etc. Outsider adversaries are adversaries who are highly skilled and aim to attack with political or economic objectives. For example, criminals and terrorists are often known as outsider adversaries. Detailed information and surveys regarding TTBs (tactics, techniques, and behaviors) of an adversary are called threat taxonomies. Examples of such threat taxonomies include [13]:

- The Reference Incident classification taxonomy – This taxonomy aims to focus on incident response so that handlers can deal with cyber incidents in a much more organized way.
- The AVOIDIT Cyberattack taxonomy – It involves how to classify all the vulnerabilities that can lead to the cyber-attack.
- The common attack pattern enumeration and classification – It aims to provide taxonomy patterns of classification to enhance cybersecurity against a cyber-attack.

There are different threat methodologies that identify processes to describe the practices, tools, and principles. Some of the threat methodologies are explained in Figure 3.1. These threat methodologies help in analyzing threats to enhance cyber defense.

The cyber kill chain methodology is developed by Lockheed Martin and aims at determining the steps that an attacker uses to initiate malicious attacks. At the same time, ICS cyber kill chain methodology is the customization of cyber kill chain

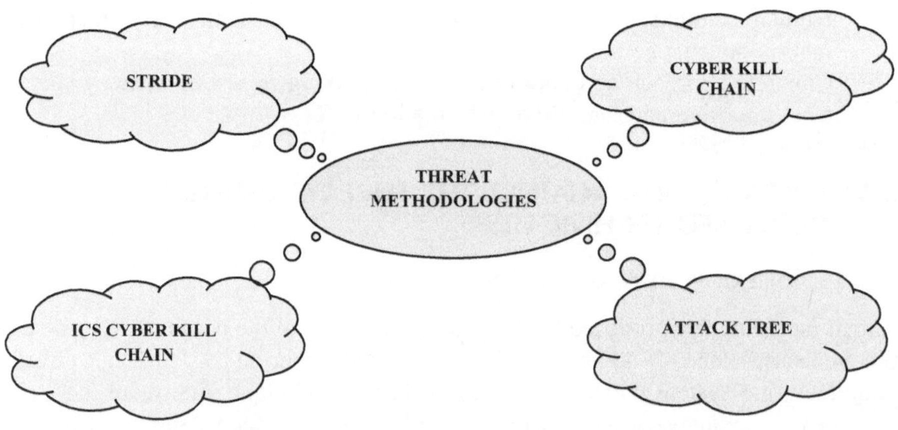

FIGURE 3.1 Different proposed threat methodologies.

methodology to include industrial control systems. ICS cyber kill chain technology uses two stages, i.e., ICS attack development and execution and cyber intrusion preparation and execution. The attack tree methodology is a methodology that represents a series of attacks on a particular system. The STRIDE methodology defines six types: elevation of privilege, information disclosure, tampering, repudiation, and DoS attack [13].

Threat modeling involves the different models used in identifying and analyzing threats with threat capabilities and the proposed security technique. These threat models can be classified into [13]

- Asset-centric – device of systems that needs security
- System-centric – software being targeted in an attack
- Attacker-centric – aims at the attacker's motivations and goals

3.5.2 Digital Twins

One of the proposed techniques, i.e., digital twin, aims to create ICPS asset representation so that it can reflect its physical environment with the help of tools and methods. The representation of ICPS assets is often known as cyber models. The required onboard simulations and data analysis are done using the cyber model obtained using the method of digital twins, allowing real-time data analysis on the system. In smart manufacturing systems, the digital twin works as a helping hand in the optimization and forecasting of production and the ICPS system [13].

3.5.3 Regulations and Standards

There are some standards set for maintaining cybersecurity in smart manufacturing systems, which are very helpful for guiding organizations. Over the past decade, several bodies have put great efforts into developing and enhancing standards and the best practices to be undertaken to help in cyber defense against any vulnerability to the system [13]. Industrial standards govern the performance and quality of the product [28]. Regulators may dictate compliance with a particular standard to make the standard a part of the regulation. It is a matter of debate, i.e., of the advantages of regulation and how effective they are in the way it offers cybersecurity to the system. When security is regulated, the industries will forward for an escape. For example, the NAERC (North American Electric Reliability Corporation) data stated that industries removed black-start capabilities to avoid paying for compliance [6,29,30]. Also, if heavy regulations are to be imposed, then it may lead to an increased probability of the generation of attacks [6].

There are some standards that are developed mainly for the security of smart manufacturing industries. Some of them are:

- ISO/IEC 29180:2012 – It provides a security framework for information exchange [6,31] for sensor networks.
- ISO/IEC 27019-2013 – It provides guidance in the field of process control and automation [6,32].

- ISA/IEC 62443 – According to ISA99 [6,33], the current state of work is a set of sub-standards grouped into policies, procedures, systems, and components [6,34]. The domain of this standard is industrial-automation-and-control systems (IACS).
- ISO/IEC 27033-1:2015 – It provides network security guidance [6,35] for IT network security.
- IEC 61508 – It provides functional safety standards applicable in every industry [6,36].
- IEC 61784 – It provides functional safety and security [6,37,38], especially in industrial communication networks.
- ISO/IEC 27000-series – It provides maintenance and improvement of information security [6,39] in all domains of information security management.

Contradicting the ability of standards and of the aim for which they are developed, the evaluation process is often harsh as there are no success metrics to check how a standard is working well and to analyze its role in the security of the systems. Also, these standards can prove to be a hypothetical sense of security [6].

3.5.4 Cryptographic Techniques

Considering the smart manufacturing systems, a wide variety of devices are presently used that use networking and software protocols for smooth working. Another technique, i.e., cryptographic countermeasures, is used in corporate networks. Cryptography techniques aim at integrity and confidentiality. To ensure authorized entities only, the public key infrastructure, the key agreement, digital signatures, cryptographic hash functions, distribution protocols, hybrid encryption schemes, and symmetric encryption algorithms are crucial [6,40]. Many studies [6,41–44] do not meet the performance requirements of the industry and the diverse deployment challenges. Therefore, a key management system must have the capability to work with limited computational power [6].

3.5.5 Intrusion Detection Systems

These systems are used in the smart manufacturing industry to contribute to the system's security. Every time, even if good quality preparations are made, there is still a chance of developing new attacks, i.e., every system has a weakness, and attackers take advantage of the system. To cope with this situation, the proposed intrusion detection system (IDS) is developed especially to take this endeavor of ICS [6].

These systems are classified on two bases, namely,

- Audit source (source of fata) – network or host based and
- Detection technique (data required for analysis) – behavior and knowledge based

In the audit source classification of intrusion detection systems, the host-based IDS system uses the data gathered by the host to determine suspicious or unauthorized behavior. This system works on the principle of gathering data/information by a single host. While another type of system, i.e., network-based IDS, collects the evidence from the whole network or its segment. In SMS, the attacker can attack any IoT node; therefore, IDS systems will have to run on each resource-poor host. Also, it becomes a challenge as we are in the early stage of developing IDS, and they are relatively computing-intensive [6]. For example, see [45,46].

Based on the detection technique classification of the intrusion detection system, the system is classified into behavior-based and network-based IDS. Knowledge-based IDS are often known by the name pattern or signature-based detection systems. It works based on the detailed study of the previous intrusions and attacks, while behavior-based IDS is based on the variations, anomalies in the system's behavior, and deviations from the normal behavior; thus, called anomaly-based IDS. Here, the system's normal behavior means the system's behavior under no attack situation [6].

A receiver operating characteristics (ROC) curve is used to illustrate the probability of detection to the probability of the pseudo-alarms. This is, therefore, used to rate the performance characteristics of the IDS systems.

3.5.6 Human Factors and Training of Security Skills

As we all know, the M3, i.e., men, money, and materials, are all crucial for establishing and developing smart manufacturing systems. But, in establishing secure systems and offering the system a good security defense, the main challenge is the shortage of manpower, i.e., the skilled manpower to empower the security. So, manpower planning is eventually a need for smart industries.

The lack of appropriate manpower applies to the complete industry, from the corporate level to the lower levels, such as the factory level, to engineers that are involved in the building and managing of these services. Moving towards IoT systems makes it more complex. If developing security policies will prove to be a burden to its users, history states that users will try not to use or comply with those policies [47,48] and may misuse the system deliberately. Nowadays, most incidents are due to the lack of skilled manpower. Due to this, there is a dire need for personnel training [49]. Some regulations and standards are given in Table 5.1, which gives some directions on personnel training and awareness. It is also indicated that security is a technical and human concern.

3.5.7 Blockchain—The Emerging Technology

Blockchain is the next technology and technique that enable us to perform more efficient transactions with more transparency and security [27,50]. Key cryptography and digital signatures are needed for authorizing transactions and recognizing accounts [27,50]. The integrity of the transaction is guaranteed using cryptography techniques [50,51].

3.5.7.1 Key Highlights of Blockchain

- It results in new thinking, a new paradigm for the industry, and proves to be beneficial in the way that the information of these systems can be stored with integrity and security among different companies and stakeholders that unnecessarily trust each other and facilitates transactions between them [50,52].

- A significant advantage of blockchain is that it helps reduce the "trust tax," benefiting the enterprises at all levels that tolerate a heavier trust tax than the established manufacturers [50].
- It is an emerging distributed ledger technology with greater attention in both fields, i.e., industry and academia [50,52].
- Blockchain can enhance the trust between the collaborative participants and not rely on any third party in IoT-based intelligent manufacturing systems [50,53].
- Also, the records of each transaction are managed in a peer-to-peer network, so they are immutable, traceable, and verifiable [50,54]. This technology is vital in the context of data security, transmission, transparency, and fault tolerance [1].
- Blockchain, in other words, can be defined as a structure focusing on the security of the systems with transparency, excellent potential, and decentralization [55].
- Blockchain technology is considered a tremendous boost to company digitalization [56].

3.5.7.2 Applications of Blockchain

Using the blockchain mechanism, the information generated in SMS can be leveraged to reduce the higher trust tax during transactions among different manufacturers, customers, distributors, and suppliers. Blockchain is the method to tackle the security issue and bring excellent efficiency to smart manufacturing systems, which are bringing a revolution in the digitized world [57,58]. Blockchain offers a decentralized and transparent working mechanism of transactions in both areas, i.e., the industry and the business [57,59,60].

There are specific properties of blockchain which ensures cybersecurity and especially shows promising enhancements for smart manufacturing systems and other automated industries. Blockchain pursuing robust cybersecurity features avoids intervention from third-party parties, thereby enabling a lower transaction cost. Industrial applications of blockchain include but are not limited to manufacturing management, resource planning, and operation scheduling using blockchain [57].

In [27,61], a private hyperledger fabric platform is used to implement the blockchain mechanism to secure SMS. Blockchain-based solutions are widely used for decentralizing medical device manufacturing and supply chain management [62]. Many cryptocurrencies like Bitcoin use blockchain technology to increase the integrity of manufacturing information and procedures of data science analysis [63]. The blockchain can also effectively and potentially be used as a disruptive method to significantly enhance drug life cycle management [64,65].

3.5.8 Artificial Neural Network

Architectures such as artificial neural networks (ANN) contribute to the security tasks in manufacturing systems and other cyber-physical systems. It helps in the identification of the advantages of CPS systems. ANN architectures' application involves solving the attack vulnerability to the system [66].

There are many different types of ANN architectures. Some of the architectures represented in a literature survey are explained here.

Much research has been directed to recurrent neural networks (RNN) [67,68]. Benefits include network "memory" mechanisms, solving complex problems, and high accuracy in their results. However, it has its drawbacks. For example, the network has a longer learning curve and requires a larger data set to train the network.

Recently, spiking neural networks have gained popularity [69,70]. There are many studies on neural networks using fuzzy logic [71–73]. Researchers point out the pros and cons. Pros are using the fuzzy logic mechanism and the results of your work will have a higher degree of accuracy. Cons include it requires a non-standard data generation mechanism.

In T. Bakir et al. [74], neural networks using wavelet transform are used. This architecture possesses a simple architecture that allows it to perform classification tasks with lesser running time. In addition, much research interest has been focused on RNN with short-term long-term memory (LSTM RNN) [75–77]. Benefits include containing a "memory" network mechanism, solving complex problems, and the result having a high level of accuracy.

There are also many studies on generative adversarial neural networks [78,79] that can work with data of a complex nature (sounds, images), which can be used to solve security problems by providing authentication.

V. Sze et al. [80] examine deep neural networks (DNNs) and point out the advantages, such as having a deep learning structure and moderate accuracy of task results. The main drawback is the dependency of accuracy on the complexity of the task.

Also, the authors in [81,82] investigated meta-neural networks and heterogeneous ensembles of neural networks. Benefits include having average values for all parameters of artificial neural networks that are part of the ensemble.

ANN architectures for perceptron, RNNs, DNNs, wavelet ANNs, LSTM RNNs, and heterogeneous ensembles of meta-neural networks/neural networks have been simulated to determine the optimal ANN architecture suitable for solving the task of detecting attacks in the network [66].

3.5.9 Management Post-Incident

Successful attacks will always happen even after such proposed techniques and security measures. In this case, once an attack has taken place, there is a dire need to make the necessary preparations to cope with the attack, i.e., the need to respond safely to the incident, letting the conditions of the system such that it returns to its normal state. After the system has returned to the initial state, to cope with the loss caused due to the attack, the system services need to be resumed as soon as possible [6].

The attack may have big severe consequences, and it may extend to the control processes of the industry. There can be a loss to the physical world as well. After a system vulnerable attack, the system needs to revive to its natural conditions, and its recovery should mitigate all the production losses and reputation losses. These require a committee or a group of people so that they can act immediately on the losses that occurred and may lead to an effective working team. Delegation of powers related to such attack conditions is a vital step in every smart manufacturing system and automation industry, to name a few. There are some regulations following which the administration of the industries needs to report the incident, and having a quicker response plan ready is a boon to the system [6]. Examples of such regulations are as follows:

- North-American-Electrical-Reliability Corporation Critical Infrastructure Protection (NERC CIP) [83]
- Chemical-Facility-Anti-Terrorism-Standards (CFATS) [84]

In practice, it is far more challenging to determine how much an industry system and its concerned administration are prepared for such vulnerable attacks and whether they have a quicker response plan or not. Several studies show that common management measures such as [6,85] training, awareness, and documentation are limited in an industrial system. There are also fundamental differences among the operators of an incident, irrespective of the regulations and standards. ICS provides a large amount of data used to identify not just the cause of a particular attack and the attacker-centric information but also in providing the knowledge of the pinhole, i.e., why it happened. Current data collection practices could be modified to provide forensic analysis and incident management information [6,86].

3.6 SUMMARY

Smart manufacturing systems were developed as a new paradigm possessing higher degrees of automation with the start of Industry 4.0. Smart manufacturing systems ensure production at low costs, reliable products, more productivity, sustainability, and meeting the end-customer demands. SMS offers a wide range of advantages, such as higher efficiency, efficacy, and control processes with higher accuracy. While providing such an increased range of abilities, the new evolving threats to smart manufacturing systems to which these systems are more vulnerable make information security a crucial matter of concern. In other words, a major and foremost challenge to smart manufacturing systems is the need for cybersecurity. While adopting new technologies, manufacturing industries are leading toward a threat landscape. Hence, to offer cybersecurity to the SMS systems, there are some proposed techniques, namely threat modeling in ICPS systems, some methodological solutions for threat modeling in CPS systems, digital twins, regulations and standards, cryptographic countermeasures, intrusion detection systems, human factors and training of security skills, blockchain technology, artificial neural networks (ANNs), and the management post-incident techniques. Apart from all these techniques, researchers are doing further studies and researching new architectures and techniques for threat modeling and offering cybersecurity to manufacturing systems. Irrespective of how much cybersecurity is provided to the system, some malicious activity will take place at least occasionally. Any risks encountered, whenever noticed, can be managed at that instant with the help of whatever means is available.

3.7 CONCLUSION AND FUTURE PERSPECTIVES

Nowadays, manufacturing systems are directly interconnected to information and communication technologies. Also, production systems have become stronger through ICT technologies. Manufacturing involves heavy investment in producing large quantities of products in anticipation of a steady trade flow. Changes in these conditions not only disrupt operations and slow down the production of goods but also make it difficult to manage investments. It is a challenge that manufacturers face every day. A big part of this challenge is understanding how to respond to changes in production facilities and plan for new production facilities discussed in this chapter.

It is discussed that by analyzing existing data in manufacturing systems, manufacturers can make more informed decisions about how to respond to change. Advances in the technological infrastructure in manufacturing systems enable a more reliable and timely flow of information at each manufacturing level. The effective use of operational data will help a more automated and faster response to changing conditions, i.e., smart manufacturing. It has been seen in studies that intelligent systems include the use of automatic intelligent systems and advanced technologies, which utilizing end-user data and information.

Further, in future scopes, interdisciplinary, social integration, and cross-domain research areas can include privacy and security, ethics and standards, data storage, and multi-physics modeling. Also, human-machine symbiosis and the development of key technology increase the smartness of the smart manufacturing system.

REFERENCES

[1] Kosmowski, K. T., Śliwiński, M., & Piesik, J. (2019). Integrated functional safety and cybersecurity. Analysis method for smart manufacturing systems. *Task Quarterly*, 23(2), 177–207. 10.17466/tq2019/23.2/c

[2] ISO/IEC TR 27019:2013: Information Technology – Security Techniques – Information Security Management Guidelines Based on ISO/IEC 27002 for Process Control Systems Specific to the Energy Utility Industry. 2013 https://www.iso.org/standard/43759.html

[3] Lenz, J., MacDonald, E., Harik, R., & Wuest, T. (2020). Optimizing smart manufacturing systems by extending the smart products paradigm to the beginning of life. *Journal of Manufacturing Systems*, 57, 274–286. 10.1016/j.jmsy.2020.10.001

[4] Li, Q., Tang, Q., Chan, I., Wei, H., Pu, Y., Jiang, H., Li, J., & Zhou, J. (2018). Smart manufacturing standardization: Architectures, reference models and standards framework. *Computers in Industry*, 101, 91–106. 10.1016/j.compind.2018.06.005

[5] Phuyal, S., Bista, D., & Bista, R. (2020). Challenges, Opportunities and Future Directions of Smart Manufacturing: A State of Art Review.*Sustainable Futures,* Vol. 2, p.100023. 10.1016/j.sftr.2020.100023. Elsevier Ltd.

[6] Tuptuk, N., & Hailes, S. (2018). Security of smart manufacturing systems. *Journal of Manufacturing Systems*, 47, 93–106. 10.1016/j.jmsy.2018.04.007

[7] Sahoo, S., & Lo, C.-Y. (2022). Smart manufacturing powered by recent technological advancements: A review. *Journal of Manufacturing Systems*, 64, 236–250. 10.1016/j.jmsy.2022.06.008

[8] Zarreh, A., Saygin, C., Wan, H., Lee, Y., Bracho, A., & Nie, L. (2018). Cybersecurity analysis of smart manufacturing system using game theory approach and quantal response equilibrium. *Procedia Manufacturing*, 17, 1001–1008. 10.1016/j.promfg.2018.10.087

[9] Shukla, M., & Shankar, R. (2022). An extended technology-organization-environment framework to investigate smart manufacturing system implementation in small and medium enterprises. *Computers and Industrial Engineering*, 163. 10.1016/j.cie.2021.107865

[10] Wang, B., Tao, F., Fang, X., Liu, C., Liu, Y., & Freiheit, T. (2021). Smart Manufacturing and Intelligent Manufacturing: A Comparative Review. In *Engineering* (Vol. 7, Issue 6, pp. 738–757). Elsevier. 10.1016/j.eng.2020.07.017

[11] Junior, A. A. de S., Pio, J. L. de S., Fonseca, J. C., de Oliveira, M. A., Valadares, O. C. de P., & da Silva, P. H. S. (2021). The state of cybersecurity in smart manufacturing systems: A systematic review. *European Journal of Business and Management Research*, 6(6), 188–194. 10.24018/ejbmr.2021.6.6.1173

[12] Choi, S. S., Jung, K., Kulvatunyou, B., & Morris, K. C. (2016). An analysis of technologies and standards for designing smart manufacturing systems. *Journal of Research of the National Institute of Standards and Technology*, 121, 422–433. 10.6028/jres.121.021

[13] Jbair, M., Ahmad, B., Maple, C., & Harrison, R. (2022). Threat modelling for industrial cyber physical systems in the era of smart manufacturing. *Computers in Industry*, 137. 10.1016/j.compind.2022.103611

[14] Bracho, A., Saygin, C., Wan, H., Lee, Y., & Zarreh, A. (2018). A simulation-based platform for assessing the impact of cyber-threats on smart manufacturing systems. Procedia Manufacturing. 26, 1116–1127. 10.1016/j.promfg.2018.07.148

[15] L. J. Wells, J. A. Camelio, C. B. Williams, & J. White (2014). Cyber-physical security challenges in manufacturing systems. *Manufacturing Letters*, 2(2), 74–77.

[16] Vincent, H., Wells, L., Tarazaga, P., & Camelio, J. (2015). Trojan detection and side-channel analyses for cyber-security in cyber-physical manufacturing systems. *Procedia Manuf.*, 1, 77–85.

[17] Zeltmann, S. E., Gupta, N., Tsoutsos, N. G., Maniatakos, M., Rajendran, J., & Karri, R. (2016). Manufacturing and security challenges in 3D printing. *Jom*, 68(7), 1872–1881.

[18] Sturm, L. D., Williams, C. B., Camelio, J. A., White, J., & Parker, R. (2017). Cyber-physical vulnerabilities in additive manufacturing systems: A case study attack on the. STL file with human subjects. *J. Manuf. Syst.*, 44, 154–164.

[19] DeSmit, Z., Elhabashy, A. E., Wells, L. J., & Camelio, J. A. (2017). An approach to cyber-physical vulnerability assessment for intelligent manufacturing systems. *Journal of Manuf. Syst.*, 43, 339–351.

[20] Schaefer, D. (n.d.). Lane Thames Cybersecurity for Industry 4.0 Springer Series in Advanced Manufacturing Analysis for Design and Manufacturing. http://www.springer.com/series/7113.

[21] Ren, A., Wu, D., Zhang, W., Terpenny, J., & Liu, P. (2017). Cyber security in smart manufacturing: Survey and challenges. In H. B. Nembhard, K. Coperich, & E. Cudney (Eds.), *67th Annual Conference and Expo of the Institute of Industrial Engineers 2017*. Institute of Industrial Engineers. pp. 716–721.

[22] Gerekli, İ., Çelik, T. Z., & Bozkurt, İ. (2021). Industry 4.0 and smart production. *TEM Journal*, 10(2), 799–805. 10.18421/TEM102-37

[23] Almaraz-Rivera, J. G., Perez-Diaz, J. A., & Cantoral-Ceballos, J. A. (2022 Apr 28). Transport and application layer DDoS attacks detection to iot devices by using machine learning and deep learning models. *Sensors (Basel)*, 22(9), 3367. doi: 10.3390/s22093367. PMID: 35591056; PMCID: PMC9103313.

[24] Zhang, D., Wang, Q. G., Feng, G., Shi, Y., & Vasilakos, A. V. (2021 Oct). A survey on attack detection, estimation and control of industrial cyber-physical systems. *ISA Trans*, 116, 1–16. doi: 10.1016/j.isatra.2021.01.036. Epub 2021 Jan 28. PMID: 33581894.

[25] Lee, S., Abdullah, A., Jhanjhi, N., & Kok, S. (2021 Jan 25). Classification of botnet attacks in IoT smart factory using honeypot combined with machine learning. *PeerJ Comput Sci.*, 7, e350. doi: 10.7717/peerj-cs.350. PMID: 33817000; PMCID: PMC7924422.

[26] Zarreh, A., Wan, H. da, Lee, Y., Saygin, C., & Janahi, R. al. (2019). Cybersecurity concerns for total productive maintenance in smart manufacturing systems. *Procedia Manufacturing*, 38, 532–539. 10.1016/j.promfg.2020.01.067

[27] Moens, P., Bracke, V., Soete, C., Vanden Hautte, S., Nieves Avendano, D., Ooijevaar, T., Devos, S., Volckaert, B., & Van Hoecke, S. (2020 Aug 2). Scalable fleet monitoring and visualization for smart machine maintenance and industrial IoT applications. *Sensors (Basel)*, 20(15), 4308. doi: 10.3390/s20154308. PMID: 32748809; PMCID: PMC7435597.

[28] Tambare, P., Meshram, C., Lee, C. C., Ramteke, R. J., & Imoize, A. L. (2021 December 29). Performance measurement system and quality management in data-driven Industry 4.0: A Review. *Sensors (Basel)*, 22(1), 224. doi: 10.3390/s2201 0224. PMID: 35009767; PMCID: PMC8749653.

[29] Anderson R., & Fuloria S. (2010). Security Economics and Critical National Infrastructure. In: Moore T., Pym D., Ioannidis C., (Eds.), *Economics of Information Security and Privacy*. Springer US. pp. 55–66.

[30] Is NERC CIP Compliance a Game? 2008 http://www:controlglobal:com/blogs/unfettered/electric-power-2008-is-nerc-cip-compliance-a-game

[31] ISO/IEC 29180: 2012 Information Technology – Telecommunications and Information Exchange between Systems – Security Framework for Ubiquitous Sensor Networks. 2012 https://www.iso.org/standard/45259.html.

[32] ISO/IEC 27019 Information Technology – Security Techniques – Information Security Controls for the Energy Utility Industry. 2017 https://www.iso.org/standard/68091.html

[33] ISA99: Developing the ISA/IEC 62443 Series of Standards on Industrial Automation and Control Systems (IACS). 2017 http://isa99.isa.org/ISA9920Wiki/Home.aspx/

[34] ISA99: Developing the Vital ISA/IEC 62443 Series of Standards on Industrial Automation and Control Systems (IACS) Security. 2001 http://isa99.isa.org/.

[35] ISO/IEC 27033-1:2015 Preview Information technology-Security techniques – Network Security – Part 1: Overview and Concepts. 2015 https://www.iso.org/standard/63461.html.

[36] Functional Safety and IEC 61508. 2010 http://www.iec.ch/functionalsafety/.

[37] IEC 61784-1:2014: Industrial Communication Networks – Profiles – Part 1: Fieldbus Profiles. 2014 https://webstore.iec.ch/publication/5878

[38] Qu, Y. J., Ming, X. G., Liu, Z. W., Zhang, X. Y., & Hou, Z. T. (2019). Smart manufacturing systems: State of the art and future trends. *International Journal of Advanced Manufacturing Technology*, 103(9–12), 3751–3768. 10.1007/s00170-019-03754-7

[39] ISO/IEC 27000 Family – Information Security Management Systems. 2014 https://webstore.iec.ch/publication/5878

[40] Piètre-Cambacédès L., & Sitbon P. (2008). Cryptographic key management for SCADA systems – Issues and perspectives. Proceedings of the 2008 International Conference on Information Security Assurance (ISA 2008), ISA '08. Washington, DC, USA: IEEE Computer Society. pp. 156–161. 10.1109/IS.A.2008.77. ISBN 978-0-7695-3126-7.

[41] Choi, D., Lee, S., Won, D., & Kim, S. (2010). Efficient secure group communications for SCADA. *IEEE Trans. Power Deliv.*, 25(2), 714–722. 10.1109/TPWRD.2009.2036181. ISSN 0885-8977.

[42] Choi, D., Kim, H., Won, D., & Kim, S. (2009). Advanced key-management architecture for secure SCADA communications. *IEEE Trans. Power Deliv.*, 24(3), 1154–1163. 10.1109/TPWRD.2008.2005683. ISSN 0885-8977.

[43] Pal, O., Saiwan, S., Jain, P., Saquib, Z., & Patel, D. (2009). Cryptographic key management for SCADA system: An architectural framework. 2009 International Conference on Advances in Computing, Control, and Telecommunication Technologies. pp. 169–174. 10.1109/ACT.2009.51

[44] Jiang, R., Lu, R., Lai, C., Luo, J., & Shen, X. (2013). Robust group key management with revocation and collusion resistance for SCADA in smart grid. 2013 IEEE Global Communications Conference (GLOBECOM). pp. 802–807. 10.1109/GLOCOM.2013.6831171. ISSN 1930-529X.

[45] Mrugala, K., Tuptuk, N., & Hailes, S. (2017). Evolving attackers against wireless sensor networks using genetic programming. *IET Wirel. Sens. Syst.*, 7(4), 113–122. 10.1049/iet-wss.2016.0090

[46] Mrugala, K., Tuptuk, N., & Hailes, S. (2016). Evolving attackers against wireless sensor networks. Genetic and Evolutionary Computation Conference, GECCO 2016, Denver, CO, USA. pp. 107–108. 10.1145/2908961.2908974

[47] Beautement, A., Sasse, M. A., & Wonham, M. (2008). The compliance budget: Managing security behaviour in organisations. Proceedings of the 2008 Workshop on New Security Paradigms, NSPW '08. New York, USA. ACM. pp. 47–58. 10.1145/1595676.1595684. ISBN 978-1-60558-341-9.

[48] Herath, T., & Rao, H. R. (2009). Protection motivation and deterrence: A framework for security policy compliance in organizations. *Eur J Inf. Syst.*, 18(2), 106–125. 10.1057/ejis.2009.6. ISSN 1476-9344.

[49] White Paper: Protecting your organisation from itself, Tech. Rep. 2016.

[50] Zhang, Y., Xu, X., Liu, A., Lu, Q., Xu, L., & Tao, F. (2019). Blockchain-based trust mechanism for iot-based smart manufacturing system. *IEEE Transactions on Computational Social Systems*, 6(6), 1386–1394. 10.1109/TCSS.2019.2918467

[51] Xu, J. H., Tian, Y., Ma, T. H., & Al-Nabhan, N. (2020 Aug 24). Intelligent manufacturing security model based on improved blockchain. *Math Biosci. Eng.*, 17(5), 5633–5650. doi: 10.3934/mbe.2020303. PMID: 33120570.

[52] Pop, C. *et al.* (Jan. 2018). Blockchain based decentralized management of demand response programs in smart energy grids. *Sensors*, 18(1), 162.

[53] Hawlitschek, F., Notheisen, B., & Teubner, T. (May/Jun. 2018). The limits of trust-free systems: A literature review on blockchain technology and trust in the sharing economy. *Electron. Commerce Res. Appl.*, 29, 50–63.

[54] Fernández-Caramés, T. M. & Fraga-Lamas, P. (2018). A review on the use of blockchain for the Internet of Things. *IEEE Access*, 6, 32979–33001.

[55] Shahbazi, Z., & Byun, Y. C. (2021). Integration of blockchain, IoT, and machine learning for multistage quality control and enhancing security in smart manufacturing. *Sensors*, 21(4), 1–21. doi: 10.3390/s21041467

[56] Bellavista, P., Esposito, C., Foschini, L., Giannelli, C., Mazzocca, N., & Montanari, R. (2021 Jul 21). Interoperable Blockchains for Highly-Integrated Supply Chains in

Collaborative Manufacturing. *Sensors (Basel)*, 21(15), 4955. doi: 10.3390/s21154955. PMID: 34372191; PMCID: PMC8347888.

[57] Leng, J., Ye, S., Zhou, M., Zhao, J. L., Liu, Q., Guo, W., Cao, W., & Fu, L. (2021a). Blockchain-secured smart manufacturing in Industry 4.0: A survey. *IEEE Transactions on Systems, Man, and Cybernetics: Systems*, 51(1), 237–252. doi: 10.1109/TSMC.2020.3040789

[58] Ahram, T., Sargolzaei, A., Sargolzaei, S., Daniels, J., & Amaba, B. (June 2017). Blockchain technology innovations. In Proceedings of the IEEE Technology and Engineering Management Conference, Santa Clara, CA, USA. pp. 1–6.

[59] Yuan, Y., & Wang, F. (Sep. 2018). Blockchain and cryptocurrencies: Model, techniques, and applications. *IEEE Trans. Syst., Man, Cybern., Syst.*, 48(9), 1421–1428.

[60] Shahbazi, Z., & Byun, Y. C. (2021 Feb 20). Integration of blockchain, iot and machine learning for multistage quality control and enhancing security in smart manufacturing. *Sensors (Basel)*, 21(4), 1467. doi: 10.3390/s21041467. PMID: 33672464; PMCID: PMC7923442.

[61] Wan, P. K., Satybaldy, A., Huang, L., Holtskog, H., & Nowostawski, M. (2020 Oct 28). Reducing alert fatigue by sharing low-level alerts with patients and enhancing collaborative decision making using blockchain technology: Scoping review and proposed framework (MedAlert). *J Med. Internet Res.*, 22(10), e22013. doi: 10.2196/22013. PMID: 33112253; PMCID: PMC7657729.

[62] Alkhader, W., Salah, K., Sleptchenko, A., Jayaraman, R., Yaqoob, I., & Omar, M. (2021 Oct 5). Blockchain-based decentralized digital manufacturing and supply for COVID-19 medical devices and supplies. *IEEE Access*, 9, 137923–137940. doi: 10.1109/ACCESS.2021.3118085. PMID: 34812401; PMCID: PMC8545200.

[63] Steinwandter V., & Herwig C. (2019 Jul-Aug). Provable data integrity in the pharmaceutical industry based on version control systems and the blockchain. *PDA J Pharm. Sci. Technol.*, 73(4), 373–390. doi: 10.5731/pdajpst.2018.009407. Epub 2019 Feb 15. PMID: 30770485.

[64] Omidian, H., & Omidi, Y. (2022 Apr). Blockchain in pharmaceutical life cycle management. *Drug Discov. Today*, 27(4), 935–938. doi: 10.1016/j.drudis.2022.01.018. Epub 2022 Feb 3. PMID: 35124248.

[65] Pustišek, M., Chen, M., Kos, A., & Kos, A. (2022 Jan 3). Decentralized machine autonomy for manufacturing servitization. *Sensors (Basel)*, 22(1), 338. doi: 10.3390/s22010338. PMID: 35009881; PMCID: PMC8749772.

[66] Krundyshev, V., & Kalinin, M. (2020). Prevention of cyber attacks in smart manufacturing applying modern neural network methods. *IOP Conference Series: Materials Science and Engineering*, 940(1). doi: 10.1088/1757-899X/940/1/012011

[67] Fernández, S. (2007). An application of recurrent neural networks to discriminative keyword spotting. Proceedings of the 17th International Conference on Artificial Neural Networks (Springer-Verlag). pp. 220–229.

[68] Graves, A., & Schmidhuber, J. (2009). Offline handwriting recognition with multidimensional recurrent neural networks. Advances in Neural Information Processing Systems, 545–552.

[69] Montuschi, P. (2015). Spiking neural network architecture computer, in Computer, 48(i 10), 6, 10.1109/MC.2015.293.

[70] Govindarajan, M. &Chandrasekaran, R. (2014). A hybrid multilayer perceptron neural network for direct marketing, International Journal of Knowledge-Based Organizations, 2(3), 63-73.. doi: 10.4018/ijkbo.2012070104

[71] Gallant, S. (1990). Perceptron-based learning algorithms. *IEEE Transactions on Neural Networks*, 1(2), 179–191.

[72] Ramya, R., Anandanatarajan, R., et al. (2012). Applications of fuzzy logic and artificial neural network for solving real world problem. IEEE-International

Conference on Advances In Engineering, Science And Management (ICAESM – 2012). pp. 443–448.
[73] Ghalia, M., & Alouani, A. (1995). Artificial neural networks and fuzzy logic for system modeling and control: a comparative study. Proceedings of the 27th Southeastern Symposium on System Theory. pp. 258–262.
[74] Bakir, T., Boussaid, B., et al. (2016). Artificial neural network based on wavelet transform and feature extraction for a wind turbine diagnosis system. 2016 14th International Conference on Control, Automation, Robotics and Vision (ICARCV). pp. 1–6.
[75] Chao, L., Tao, J., et al. (2016). Long short term memory recurrent neural network based encoding method for emotion recognition in video. IEEE International Conference on Acoustics, Speech and Signal Processing (ICASSP).
[76] Dai, J., Liang, S., et al. (2016) Long short-term memory recurrent neural network based segment features for music genre classification 10th International Symposium on Chinese Spoken Language Processing (ISCSLP). pp. 1–5.
[77] Youmans, M., Spainhour, C., & Qiu, P. (2017) Long short-term memory recurrent neural networks for antibacterial peptide identification. IEEE International Conference on Bioinformatics and Biomedicine (BIBM). pp. 498–502.
[78] Radford, A., Metz, L., & Chintala, S. (2015). Unsupervised representation learning with deep convolutional generative adversarial networks. arXiv preprint arXiv:1511.06434.
[79] Yin, C., Zhu, Y., et al. (2018). An enhancing framework for botnet detection using generative adversarial networks Int. Conference on Artificial Intelligence and Big Data (ICAIBD). pp. 228–234.
[80] Sze, V., Chen, Y., et al. (2017). Efficient processing of deep neural networks: A tutorial and survey. in *Proc. of the IEEE*, 105(12), 2295–2329.
[81] Naik, D., & Mammone, R. (1992). Meta-neural networks that learn by learning. International Joint Conference on Neural Networks Baltimore, USA.
[82] Hans, A., & Udluft, S. (2010). Ensembles of neural networks for robust reinforcement learning. 2010 9th Int. Conference on Machine Learning and Applications. pp. 401–406.
[83] CIP008-5 Cyber Security – Incident Reporting and Response Planning. 2016 http://www.nerc.com/_layouts/PrintStandard.aspx?standardnumber=CIP-008-5.
[84] USC Chapter 1, Subchapter XVI: Chemical Facility Anti-Terrorism Standards. 2014 http://uscode.house.gov/view.xhtml?path=/prelim@title6/chapter1/subchapter16&edition=prelim.
[85] Line, M., Tondel, I., & Jaatun, M. (2014). Information security incident management: Planning for failure. 2014 Eighth International Conference on IT Security Incident Management IT Forensics (IMF). pp. 47–61. doi: 10.1109/IMF.2014.10
[86] Spyridopoulos, T., Tryfonas, T., & May, J. (2013). Incident analysis amp; digital forensics in SCADA and industrial control systems. 8th IET International System Safety Conference incorporating the Cyber Security Conference 2013. pp. 1–6. doi: 10.1049/cp.2013.1720

4 Integration of Circular Supply Chain and Industry 4.0 to Enhance Smart Manufacturing Adoption

Monika Vyas
L. D. College of Engineering, Ahmedabad, Gujarat, India

Gunjan Yadav
Swarnim Startup and Innovation University, Gandhinagar, India

CONTENTS

DOI: 10.1201/9781003333760-4

4.1 INTRODUCTION

Manufacturing industries capture a large worldwide market and play a major role in building a country's economy; they progressed with the advent of time and the industrial revolutions [2,3]. Incorporation of the latest technologies with real-time data in smart manufacturing and material management techniques upstream and downstream of a circular supply chain yields beneficial outputs with regards to resource availability and environmental sustainability [4]. Smart manufacturing is self-adaptive to real situations and yields optimized output, whereas circular supply chain practices on the other end save virgin materials [5]. Thus, adoptions of circular practices in smart manufacturing are of prime importance and also a point of study in this chapter.

4.1.1 Bakground and Motivation

Research in the field of circular supply chain management revealed the need to adopt possible circularity at every stage of the supply chain. This needs to incorporate a possible R framework throughout product life cycle management that drives into saving of earth resources, material recovery, value from waste with environmental concern, and many other things. Incorporation of Industry 4.0 technologies geared the CSC and thus smart manufacturing [6]. There are many challenges to overcome for developing countries during this major shift, and hence this has become the point of the study.

4.1.2 Study Objective

Most of the industrial sector deals with constant fluctuating demands, product varieties, and innovation, with on-time, effective, quality product delivery. Thus, there is a need to focus on the integration of CSC practices in light of the latest technologies that enhance smooth transitions in manufacturing. Incorporation of CSC practices leads to quality performance in smart manufacturing and results in firm growth. Here, the main study objective is:

- To identify different CSC practices and individual weight computations to derive their relative importance using the FAHP technique framework.

4.2 LITERATURE REVIEW

A literature review was carried out to identify prominently adopted CSC practices reported by various researchers. Different authors had reported individual domain work; for this article, integration of domains yielded results. Different methods and framework adoptions within manufacturing industries are also observed. For clarity, prominent CSC practices and various MCDM techniques adopted in different industries are briefly mentioned in the next two parts.

4.2.1 Circular Supply Chain Practices

Adoption of smart manufacturing (SM) and CSC are challenging tasks with many limitations and hurdles to overcome. Enablers and practices that promote the advancement of technology appeared as one major shift. Several practices are identified, may differ as per industries and their geographic location, and are listed and described briefly in Table 4.1.

TABLE 4.1
List of Circular Supply Chain Practices Integrated with I4.0

Sr. No.	Prominent Practices	Description	Reference
1	Value addition in products and services with a long term vision	Waste reduction, increased product life, less rework, and few other parameters at the design stage and manufacturing stage increase value within the supply chain	[7–9]
2	Design the right end to end architecture in supply chain configuration	Reliability and robustness of the system depend upon end-to-end architecture and framework of supply chain	
3	Intelligent decision-making techniques	Technologies empower decision makers with smart, quick, and most beneficial decisions	[6,10]
4	Smart tracking tracing system and resource efficiency	Smart logistics manages critical demand effectively with the help of technologies	
5	Predictive maintenance	An optimal condition with a possible minimum wait or downtime is an added advantage	[11,12]
6	Industrial IOT platform for interconnectivity and information sharing	Key supporting technology for smooth, large data collection and processing at different stages in SM, CC, and big data drives proper inventory with resource efficiency and manages the product life cycle in a broader sense	[13]
7	Cyber-physical system	Leads to changeable design and highly customized products its milestone for integrated encapsulation as an embedded system	[13,14]
8	Data capture by cloud computing and big data in CSC	Cloud computing and big data enhance the supply chain by recognizing patterns and improving production	
9	Innovative technologies: AI, ML, AR, robotics, smart manufacturing approaches	Smart technologies such as AR, which provides 3D information, and ML in the design and production stages with robotics enhance different actors in the supply chain	
10	Employee rewards and incentive implementation	Recognizing and rewarding imparts social enhancement that results in loyalty	
11	Key stakeholder collaboration	Helps to improve the supply chain, by managing inventory levels, providing flexibility, and improving response time	[15,16]

(Continued)

TABLE 4.1 (Continued)
List of Circular Supply Chain Practices Integrated with I4.0

Sr. No.	Prominent Practices	Description	Reference
12	Asset building infrastructure and technology investment	Infrastructure supports technology advancement, and it is thus a necessary investment that yields long-term benefit	
13	Smart supply chain and association with CSC experts	Supply chain adopts recycle, reuse, and reduce as major practices; this smart process creates additional value	[15,17]
14	Choose the right partner with shared beliefs	To locate different recoverable actions from product life cycle analysis, is important to involve domain experts for clarity	[18]
15	Digital transformation and modularity adoption in LCA	Life cycle analysis shows the value possesses and the worth it can create with proper techniques at the end of life	[19,20]
16	Social inclusion	Different cultures and beliefs observe slight variations in adoption, and techniques to include	[21,22]
17	Corporate social responsibility (CSR)	To practice prevailing laws, should be stringent policy as it delivers quality standards and CSR leads to sustainability	[23,24]
18	Customer cooperation and best practices	Active link for return of product, role in recovery, and adoption of secondary material with environmental awareness	[25]
19	Digital twin	Its digital model of a real component or virtual base saves costs and imparts visionary output	[5,26]
20	Green purchases and sustainable practices	"Go green" campaign is adopted in almost all fields to save unnecessary resources by e-commerce and logistics in the supply chain	[27]
21	Campaign and awareness	To drive programs that bring awareness to society and educate on the benefits	[8]
22	Adoption of renewable energy practices	Minimal energy utilization and saving natural resources by using renewable energy are two pillars for sustainability	[1,28]
23	Agile production	Process improves factory real performance	
24	Environmental concern and eco-design	Design and performance to be in relation with environmental concern from design to production stage	[29]

TABLE 4.2
Applications of MCDM Techniques in Various Industries

Sr. No.	Field of Application	Applied MCDM	Reference
1	Remanufacturing sector	AHP-TOPSIS	[30]
2	Supply chain with big data	Fuzzy ANP	[8]
3	Manufacturing sector–lean	AHP-DEMATEL	[31]
4	Lean manufacturing	Fuzzy-VIKOR	[32]
5	Manufacturing sector for lean Six Sigma	FAHP-PROMETHEE	[33]
6	Supplier selection	FAHP	[34]

4.2.2 MCDM Methods

The role of the MCDM technique is to select the best alternative when a decision is not straightforward as alternatives are composed of multiple objectives. Several researchers used single or hybrid methodologies in CSC but were limited in I4.0 and SM. Table 4.2 depicts different MCDM applications in various industries.

4.3 RESEARCH METHODOLOGY

Adopted research methodology is mentioned in Figure 4.1. Initially, a literature review was conducted, and a research gap was identified to incorporate CSC integration with Industry 4.0 to enhance SM. The steps and their formulation of fuzzy AHP [35] are described in detail.

4.4 WEIGHT DETERMINATION BY FUZZY AHP (FAHP) [35]

FAHP is one of the MCDM techniques that finds its major application for complex problems of decision making irrespective of sectors. It also overcomes the limitation of vagueness that is observed in the simple analytic hierarchy process by introducing fuzzy numbers. Therefore, decision makers have more flexibility and assurance for optimal decisions. FAHP allocates weight to each parameter and thus adds relative importance to the hierarchy of identified parameters. In this study, eight experts form a decision-making panel to incorporate triangular fuzzy numbers. Table 4.3 shows the triangular fuzzy number (Tfn) and corresponding membership function against each linguistic value; this is further adopted for a pair-wise matrix comparison [35]. The DM panel suggests the construction of five major groups that include all 24 identified CSC practices. Stated below, four steps are executed to derive the results.

Step I. To construct fuzzy pair wise comparison matrix (FPM- Z)

To check the relationships of systems, propose selected n factors $f = \{f_1, f_2, ..f_n\}$. Suppose that to obtain the influence that factor fi has on factor fj, and l experts considered in a decision group, $E = \{E_1, E_2..,E_n\}$ using a triangular fuzzy scale [36]. In other words, an n × n matrix was constructed, considering the effect of each row

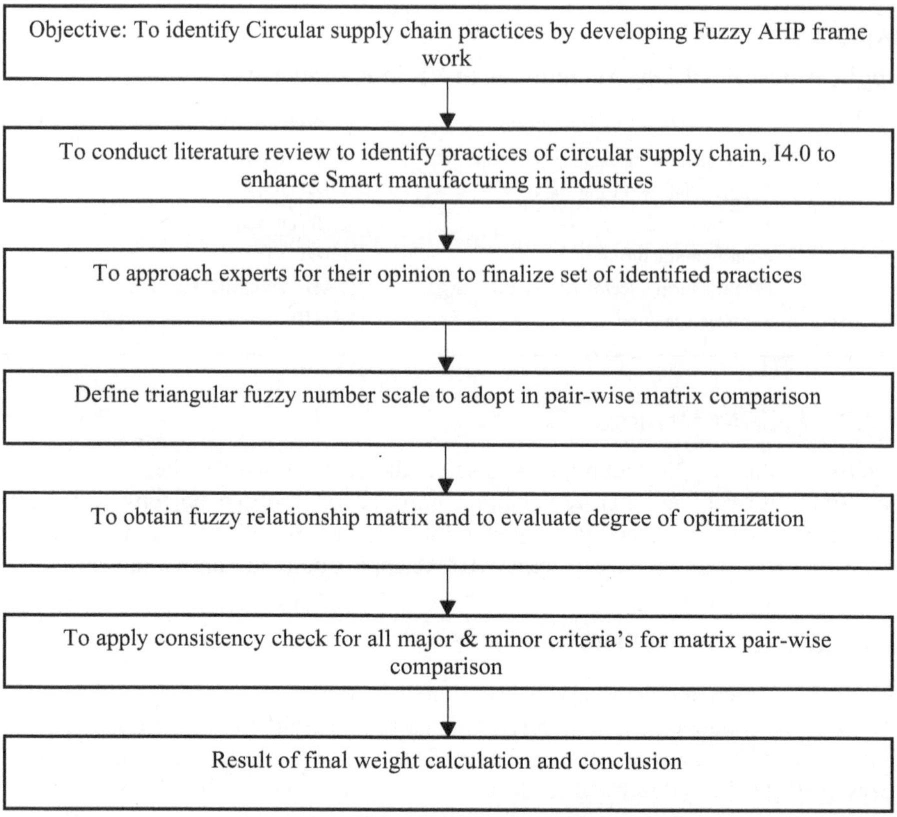

FIGURE 4.1 Flowchart of adopted research methodology.

TABLE 4.3
Scale of Relative Importance

Fuzzy Number	Membership Function	Membership Function
$\tilde{1}$	(1,1,3)	Equal importance
$\tilde{3}$	(1,3,5)	Weal importance
$\tilde{5}$	(3,5,7)	Strong importance
$\tilde{7}$	(5,7,9)	Very strong importance
$\tilde{9}$	(7,9,11)	Extremely strong importance

element on each column element by all experts individually with relative importance weight and this overcame any bias chance of qualitative judgment from Table 4.3.

if *Here,* $(\tilde{x}_{ij}) = 1$, *if i is similar to j and if i and j values are not similar than* $\tilde{x}_{ij} = \tilde{1}, \tilde{3}, \tilde{5}, \tilde{7}, \tilde{9}$ *and this highlights importance of i attribute over j attribute according to Tfn*

Step II. To compute fuzzy crisp matrix (FCM) from fuzzy pair wise comparison (FPM -Z)

Now there is a need for uniting all eight expert panel's assurance over judgment obtained previously by fuzzy pair-wise comparison. Adopt α cut method for ranking. Here, $\alpha = 0.5$ set (2, 3, and 4) is considered.

Now construct a pair-wise comparison in a similar way for α cut and adjust the degree of optimism as μ leads to calculating the satisfaction level. Were enhance μ leads to an increase degree of optimism where index of optimism equation as

$$\tilde{x}_{ij}^{\alpha} = \mu \tilde{x}_{ijl}^{\alpha} + (1 - \mu)\tilde{x}_{iju}^{\alpha} \quad (4.1)$$

were value of μ lies between 0 to 1

By substituting μ in Eq. 4.1, α cut fuzzy comparison matrix is converted into a crisp comparison.

Step III. Consistency check adoption

Panel decision and judgment are of paramount importance, further, to assess the results and compute the consistency ratio for optimal results. Thus, we need to determine the largest Eigen value for each matrix λ where value depends upon matrix composition say for 8*8 matrixes, resulting in eight attributes and eight values of λ; thus, whichever is highest can be considered λ_{max} and incorporated to calculate the consistency index. The consistency ratio is based on consistency and random index, as in Eq. 4.2:

CR is the consistency ratio.
CI is the consistency index.
RI is the random index.

$$CR = CI/RI \quad (4.2)$$

$$CI = \lambda_{\max} - n/(n - 1) \quad (4.3)$$

RI as in Table 4.4

If all pair-wise comparisons for major and sub-criteria CR ≤<_0.1 were not achieved, the expert panel would need to discuss and revise the matrix.

TABLE 4.4
Random Consistency Index

1	2	3	4	5	6	7	8
0	0	0.52	0.89	1.11	1.25	1.35	1.40

Step IV. Final weight calculation

Finally. either apply row or column normalization for final weight ascertain to each parameter. Here, parameters are practices that can be further adopted for ranking alternatives.

4.5 FRAMEWORK DEVELOPMENT

All identified practices were analyzed and finalized by the DM panel, with suggestions of allocation into five major theme groups. Thus, main practices are categorized into five major groups that consist of a total of 24 sub-criteria for further processing. They are categorized as organizational, managerial, technological, sociocultural, and sustainability-green practices [22]. Classification appears as in Table 4.2.

Figure 4.2 highlights adopted practices grouped from the DM view. The influence hierarchy in Table 4.3 was shared with DM, and thus they can make an initial pair-wise comparison matrix. Panel members take care to make a comparison and to assign a proper weight. Here in Table 4.4, one comparison is considered, and the

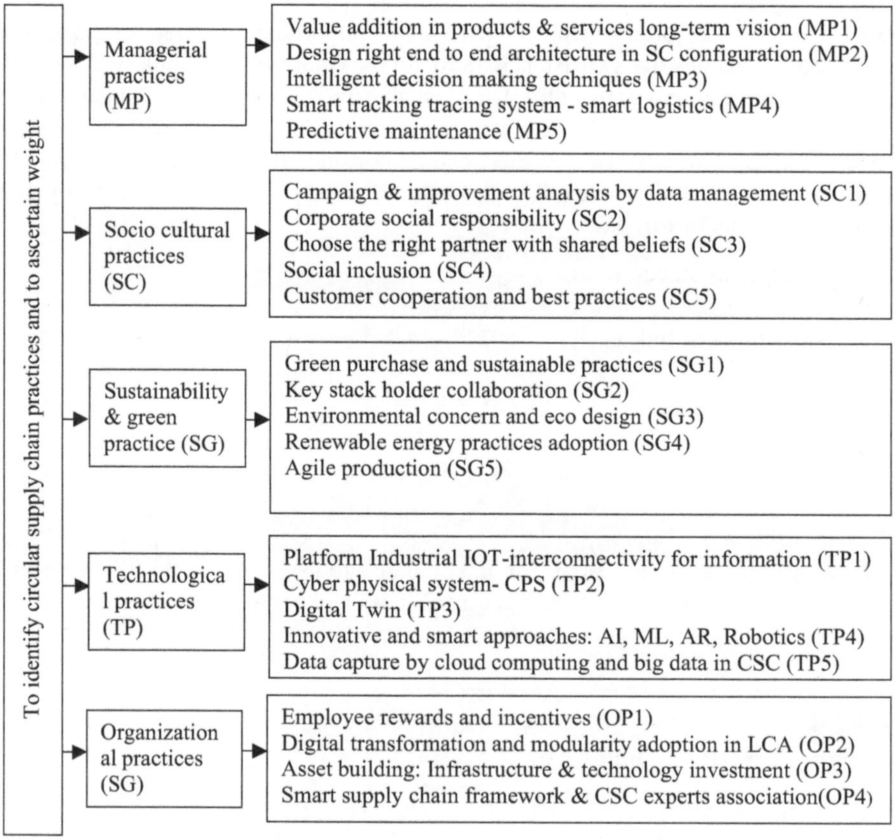

FIGURE 4.2 Research methodology framework.

TABLE 4.5
Computation of Weights for Practices

Sr. No.	Major Group	Major Criteria Weights	Sub-Group	Local Weight	Global Weight
1	Managerial practices (MP)	0.445	MP1	0.0386	0.0172
2			MP2	0.0671	0.0299
3			MP3	0.2575	0.1146
4			MP4	0.522	0.2323
5			MP5	0.112	0.0498
6	Technological practices (TP)	0.286	TP1	0.0393	0.0112
7			TP2	0.2669	0.0763
8			TP3	0.1391	0.0398
9			TP4	0.4845	0.1386
10			TP5	0.0705	0.0202
11	Sustainable and green practices (SG)	0.167	SG1	0.2672	0.0446
12			SG2	0.134	0.0224
13			SG3	0.4849	0.0810
14			SG4	0.0711	0.0119
15			SG5	0.0342	0.0057
16	Organizational practice (OP)	0.061	OP1	0.1104	0.0067
17			OP2	0.2506	0.0153
18			OP3	0.0474	0.0029
19			OP4	0.5914	0.0361
20	Socio-cultural practice (SC)	0.035	SC1	0.52	0.0067
21			SC2	0.0391	0.0182
22			SC3	0.27	0.0014
23			SC4	0.0681	0.0095
24			SC5	0.112	0.0024

same is repeated for the remaining major practice groups, too. Thus, the weights of major criteria and sub-criteria are mentioned in Table 4.5.

4.6 RESULT

The fuzzy AHP method is adopted as a tool for decision making that revealed important findings as listed. Major and minor criteria's associated weight shows their importance; higher value shows much impact comparatively. Thus, all major and sub-criteria are listed in Table 4.5. The preference of impact for most prominent are discussed and highlighted. The hierarchy turned into ***managerial practices*** is most important to ***drive technology.*** This drives to achieve ***sustainability and green practices*** as a third factor and is followed in line with ***organization vision*** that provides robustness with the ***sociocultural*** aspect in the last position, but still imparts strength to execution [37].

4.6.1 Managerial Practices

Managerial practices possess profound impacts on an identified group, based on trends and techniques working scenario yields different benefits in terms of customer satisfaction, quality products, worker progress and prosperity, team building, etc. that lead to benefits with harmony in a competitive era. Smart tracking and logistics govern the leading position, and intelligent and efficient decision making in daily activities adds value within the system. These are considered two important sub-criteria, followed in hierarchy order by predictive maintenance, the right architecture, and value addition processes.

4.6.2 Technological Practices

The technology subgroup holds the second-most significant group; it encompasses quality results in comparison to solely human practices in most cases. Innovative technology: AI, ML, AR, robotics, and smart manufacturing approaches appeared as a boon to manufacturers and formed the basis for adopted practices; all are integrated on a platform, and digital twin captures 3D models on augmented realities, and the Industrial Internet of Things platform all constitute the Industry 4.0 that enhance CSC as an active driver for other data management and manufacturing aspects.

4.6.3 Sustainability and Green Practices

Sustainable development goals' vision is in global demand; thus, this group possesses the unique sense necessary for environmental concern [15]. Two key prominent practices are eco-design adoption in CSC and concern with circularity in product imagination. End of life, green purchasing, and sustainable practices are highly focused, and vision and integration from all actors and stakeholder collaboration are vital parts of making SC more circular, along with a possible increase in the adoption of renewable resource, and last, with agile production adoption as a means of saving material. This comprises this major group, and all of this has its own importance based on type of industry.

4.6.4 Organizational Practices

This group leads and builds the base of each and every industry. Here it constitutes the fourth group with four sub-criteria in order of their weight. Also, well-executed plans by higher authorities close the loop with the adoption of identified goals. The utmost importance is to build a smart supply chain with flexibility and firmness adopted together; digital transformation changes the execution mode, followed by employee rewards to those who adopt and execute well its building block for propagation. This all comprises support for smart asset building in the form of different investments that are mandatory.

4.6.5 Sociocultural Practices

Humans and their cultural differences depend on demographic and geographic factors and other parameters that form social beliefs and culture. It should work to progress and not be a hindrance due to sociocultural effects [37]. Thus, an awareness campaign makes people aware of the actual benefits, followed by the right partnership with social norms in mind; customer linkage is the vital cental link in adoption with CSR by corporatations at the other end; and finally concern for social and cultural practices adoption in descending order compose sociocultural practice.

4.7 DISCUSSION

CSC practices will lead to beneficial results like processing time optimization, minimal virgin resource consumption, waste reduction, and cost efficiency in the long run [12,38]. Quality products in time result due to technology advancements, digitization, e-commerce, smart logistics, and a dynamic supply chain, i.e., equipped with blockchain [39,40]. Cyber-physical system (CPS) and real-time data play a basic role in supporting the system. Finally, a circular supply chain needs to practice big data analytics, digital modularity, distributed control, digital twins, artificial intelligence, machine learning with augmented reality and real-time data, right design of end-to-end architecture, and multidisciplinary cross-platform-based infrastructure for continuous production [16]. This all leads to enhanced execution of smart manufacturing practices along with the upgrading of the circular supply chain for sustainable development [41]. Overall, the most impactful practices are to enhance smart supply chains that logically cover action involvement from all actors; awareness campaigns yield impact because humans do not know the worth and, sometimes, in spite of knowing the importance, don't play an active role. A proper campaign is always beneficial. Environment and eco-design are of prime concern due to global climate change, ozone depletion, and to avoid harm. Innovative technologies such as these lead to enhanced CSC and thus smart manufacturing approaches in descending hierarchy.

4.8 CONCLUSION

Framework-based results are derived from the circularity of supply chain integration with Industry 4.0 adoption. From the hierarchy of all identified practices, a clear indication appears as managerial practices are the most prominent and pillared upon existing technology, and go green with sustainability on the third, followed by organizational practices and socioculture as the last group relatively based upon weights. Thus, the MCDM technique framework of FAHP suggests that circular practices lead to smart manufacturing practices resulting in sustainable development for developing countries in terms of achieving responsible production and consumption goals.

4.9 STUDY IMPLICATIONS

For practitioners, researchers, and managers, the conclusion derived sheds light to wisely adopt possible practices effectively, as many times trying to incorporate all of them turns up some limitations. Secondly, it is possible to know the impact value for practice corresponding to input and check its adoptability is feasibility or not; thus, most beneficial in terms of instant execution can be considered first.

It also provides an accurate and detailed framework with expert opinion taken into consideration for verifying the consistency of the framework. Additionally, a few unnoticed potential practices can account for this if witnessed; for example, as managerial practices are executed daily in all firms. Therefore, even minor attention can lead to huge potential benefits. Additionally, intensity also imparts significance, so detail-derived framework results fulfill the aim. This study can have various other approaches for further research and result in diverse investigations. This study provides a road map to researchers and visionary direction to policymakers and practitioners.

4.10 LIMITATIONS AND FUTURE SCOPE

Each and every study has its own limitations and supports future work options. Few over here are all three domains, namely CSC, Industry 4.0, and smart manufacturing, which are new, growing concepts that demand a lot of work in domain and integration for benefits. This study may have neglected a few important practices in identification and some important work may have been overlooked; different techniques other than a literature review can be adopted, such as a questionnaire or interview; results can be achieved by different methods. Adoption of a hybrid framework can be adopted. Here, experts play a key role, and they may have some individual limitations derived from matrix computation with expert field knowledge. Different search strings can be adopted for analysis; other methods can be adopted with modification in parameters. A study with comparative analysis from developed countries can highlight benefits for developing countries.

REFERENCES

[1] M. S. R. A. Hamid, N. R. Masrom, and N. A. B. Mazlan, "The key factors of the industrial revolution 4.0 in the Malaysian smart manufacturing context," *Int. J. Asian Bus. Inf. Manag.*, vol. 13, no. 2, pp. 1–19, 2022, 10.4018/IJABIM.20220701.oa6

[2] P. Lengyel *et al.*, "Development of the concept of circular supply chain management—A systematic review," *Processes*, vol. 9, no. 10, pp. 1–20, 2021, 10.3390/pr9101740

[3] T. S. Deepu and V. Ravi, "Supply chain digitalization: An integrated MCDM approach for inter-organizational information systems selection in an electronic supply chain," *Int. J. Inf. Manag. Data Insights*, vol. 1, no. 2, p. 100038, 2021, 10.1016/j.jjimei.2021.100038

[4] D. Huybrechts, A. Derden, L. Van den Abeele, S. Vander Aa, and T. Smets, "Best available techniques and the value chain perspective," *J. Clean. Prod.*, vol. 174, pp. 847–856, 2018, 10.1016/j.jclepro.2017.10.346

[5] J. Friederich, D. P. Francis, S. Lazarova-Molnar, and N. Mohamed, "A framework for data-driven digital twins for smart manufacturing," *Comput. Ind.*, vol. 136, p. 103586, 2022, 10.1016/j.compind.2021.103586

[6] T. D. Mastos *et al.*, "Introducing an application of an industry 4.0 solution for circular supply chain management," *J. Clean. Prod.*, vol. 300, 2021, 10.1016/j.jclepro.2021.126886

[7] J. Wang, Y. Ma, L. Zhang, R. X. Gao, and D. Wu, "Deep learning for smart manufacturing: Methods and applications," *J. Manuf. Syst.*, vol. 48, pp. 144–156, 2018, 10.1016/j.jmsy.2018.01.003

[8] N. K. Dev, R. Shankar, R. Gupta, and J. Dong, "Multi-criteria evaluation of real-time key performance indicators of supply chain with consideration of big data architecture," *Comput. Ind. Eng.*, vol. 128, pp. 1076–1087, 2019, 10.1016/j.cie.2018.04.012

[9] J. Reinhold, C. Koldewey, and R. Dumitrescu, "Value creation framework and roles for smart services," *Procedia CIRP*, vol. 109, pp. 413–418, 2022, 10.1016/j.procir.2022.05.271

[10] J. Valenzuela, M. Alfaro, G. Fuertes, M. Vargas, and C. Sáez-Navarrete, "Reverse logistics models for the collection of plastic waste: A literature review," *Waste Manag. Res.,* 39(9), pp.1116-1134, 2021, 10.1177/0734242X211003948

[11] J. C. Serrano-ruiz, J. Mula, and R. Poler, "Smart manufacturing scheduling: A literature review," *J. Manuf. Syst.*, vol. 61, no. September, pp. 265–287, 2021, 10.1016/j.jmsy.2021.09.011

[12] J. C. Serrano-ruiz, J. Mula, and R. Poler, "Development of a multidimensional conceptual model for job shop smart manufacturing scheduling from the Industry 4. 0 perspective," *J. Manuf. Syst.*, vol. 63, no. February, pp. 185–202, 2022, 10.1016/j.jmsy.2022.03.011

[13] B. Elahi and S. A. Tokaldany, *Application of Internet of Things-aided simulation and digital twin technology in smart manufacturing*. INC, 2021.

[14] S. Blömeke, J. Rickert, M. Mennenga, S. Thiede, T. S. Spengler, and C. Herrmann, "Recycling 4.0 – Mapping smart manufacturing solutions to remanufacturing and recycling operations," *Procedia CIRP*, vol. 90, pp. 600–605, 2020, 10.1016/j.procir.2020.02.045

[15] H. Gupta, A. Kumar, and P. Wasan, "Industry 4.0, cleaner production and circular economy: An integrative framework for evaluating ethical and sustainable business performance of manufacturing organizations," *J. Clean. Prod.*, vol. 295, p. 126253, 2021, 10.1016/j.jclepro.2021.126253

[16] Y. Riahi, T. Saikouk, A. Gunasekaran, and I. Badraoui, "Artificial intelligence applications in supply chain: A descriptive bibliometric analysis and future research directions," *Expert Syst. Appl.*, vol. 173, no. November 2020, p. 114702, 2021, 10.1016/j.eswa.2021.114702

[17] S. Luthra, M. Sharma, A. Kumar, S. Joshi, E. Collins, and S. Mangla, "Overcoming barriers to cross-sector collaboration in circular supply chain management: A multi-method approach," *Transp. Res. Part E Logist. Transp. Rev.*, vol. 157, no. May 2021, p. 102582, 2022, 10.1016/j.tre.2021.102582

[18] S. Mittal, M. A. Khan, D. Romero, and T. Wuest, "Building blocks for adopting smart manufacturing," *Procedia Manuf.*, vol. 34, pp. 978–985, 2019, 10.1016/j.promfg.2019.06.098

[19] S. Cisneros-Cabrera, G. Pishchulov, P. Sampaio, N. Mehandjiev, Z. Liu, and S. Kununka, "An approach and decision support tool for forming Industry 4.0 supply chain collaborations," *Comput. Ind.*, vol. 125, p. 103391, 2021, 10.1016/j.compind.2020.103391

[20] E. Cesur, M. R. Cesur, Y. Kayikci, and S. K. Mangla, "Optimal number of remanufacturing in a circular economy platform," *Int. J. Logist. Res. Appl.*, pp. 1–17, 2020, 10.1080/13675567.2020.1825656

[21] E. K. Zavadskas, J. Antucheviciene, and P. Chatterjee, "Multiple-criteria decision-making (MCDM) techniques for business processes information management," *Inf.*, vol. 10, no. 1, pp. 1–7, 2018, 10.3390/info10010004

[22] B. M. Bringsken, I. Loureiro, C. Ribeiro, C. Vilarinho, and J. Carvalho, "Community involvement towards a circular economy: A sociocultural assessment of projects and interventions to reduce undifferentiated waste," *Eur. J. Sustain. Dev.*, vol. 7, no. 4, pp. 496–506, 2018, 10.14207/ejsd.2018.v7n4p496

[23] T. Arauz, P. Chanfreut, and J. M. Maestre, "Cyber-security in networked and distributed model predictive control," *Annu. Rev. Control*, vol. 53, no. October 2021, pp. 338–355, 2022, 10.1016/j.arcontrol.2021.10.005

[24] I. Mezgár and J. Váncza, "From ethics to standards – A path via responsible AI to cyber-physical production systems," *Annu. Rev. Control*, vol. 53, no. April, pp. 391–404, 2022, 10.1016/j.arcontrol.2022.04.002

[25] S. Alonso-Muñoz, R. González-Sánchez, C. Siligardi, and F. E. García-Muiña, "Building exploitation routines in the circular supply chain to obtain radical innovations," *Resources*, vol. 10, no. 3, pp. 1–18, 2021, 10.3390/resources10030022
[26] S. Çetin, C. De Wolf, and N. Bocken, "Circular digital built environment: An emerging framework," *Sustain.*, vol. 13, no. 11, 2021, 10.3390/su13116348
[27] M. V. Hernandez Marquina, M. A. Le Dain, P. Zwolinski, and I. Joly, "Sustainable performance of circular supply chains: A literature review.," *Procedia CIRP*, vol. 105, pp. 607–612, 2022, 10.1016/j.procir.2022.02.101
[28] M. Despeisse *et al.*, "Unlocking value for a circular economy through 3D printing: A research agenda," *Technol. Forecast. Soc. Change*, vol. 115, pp. 75–84, 2017, 10.1016/j.techfore.2016.09.021
[29] C. Dalhammar, "Industry attitudes towards ecodesign standards for improved resource efficiency," *J. Clean. Prod.*, vol. 123, pp. 155–166, 2016, 10.1016/j.jclepro.2015.12.035
[30] X. Zhang, Z. Li, Y. Wang, and W. Yan, "An integrated multicriteria decision-making approach for collection modes selection in remanufacturing reverse logistics," *Processes*, vol. 9, no. 4, 2021, 10.3390/pr9040631
[31] G. Yadav, S. Luthra, D. Huisingh, S. K. Mangla, B. E. Narkhede, and Y. Liu, "Development of a lean manufacturing framework to enhance its adoption within manufacturing companies in developing economies," *J. Clean. Prod.*, vol. 245, p. 118726, 2020, 10.1016/j.jclepro.2019.118726
[32] R. K. A. Bhalaji, S. Bathrinath, and S. Saravanasankar, "A Fuzzy VIKOR method to analyze the risks in lean manufacturing implementation," *Mater. Today Proc.*, vol. 45, no. xxxx, pp. 1294–1299, 2021, 10.1016/j.matpr.2020.05.123
[33] G. Yadav, D. Seth, and T. N. Desai, "Application of hybrid framework to facilitate lean six sigma implementation: A manufacturing company case experience," *Prod. Plan. Control*, vol. 29, no. 3, pp. 185–201, 2018, 10.1080/09537287.2017.1402134
[34] M. V. C. Fagundes, B. Hellingrath, and F. G. M. Freires, "Supplier selection risk: A new computer-based decision-making system with fuzzy extended AHP," *Logistics*, vol. 5, no. 1, p. 13, 2021, 10.3390/logistics5010013
[35] G. Yadav, D. Seth, and T. N. Desai, "Prioritising solutions for Lean Six Sigma adoption barriers through fuzzy AHP-modified TOPSIS framework," *Int. J. Lean Six Sigma*, vol. 9, no. 3, pp. 270–300, 2018, 10.1108/IJLSS-06-2016-0023
[36] G. Yadav, S. K. Mangla, A. Bhattacharya, and S. Luthra, "Exploring indicators of circular economy adoption framework through a hybrid decision support approach," *J. Clean. Prod.*, vol. 277, p. 124186, 2020, 10.1016/j.jclepro.2020.124186
[37] P. Gerli, J. Clement, G. Esposito, L. Mora, and N. Crutzen, "The hidden power of emotions: How psychological factors influence skill development in smart technology adoption," *Technol. Forecast. Soc. Change*, vol. 180, no. April, p. 121721, 2022, 10.1016/j.techfore.2022.121721
[38] S. Lahane, R. Kant, and R. Shankar, "Circular supply chain management: A state-of-art review and future opportunities," *J. Clean. Prod.*, vol. 258, p. 120859, 2020, 10.1016/j.jclepro.2020.120859
[39] C. Garrido-Hidalgo, F. J. Ramirez, T. Olivares, and L. Roda-Sanchez, "The adoption of internet of things in a circular supply chain framework for the recovery of WEEE: The case of Lithium-ion electric vehicle battery packs," *Waste Manag.*, vol. 103, pp. 32–44, 2020, 10.1016/j.wasman.2019.09.045
[40] J. A. C. Bokhorst, W. Knol, J. Slomp, and T. Bortolotti, "International journal of production economics assessing to what extent smart manufacturing builds on lean principles," *Int. J. Prod. Econ.*, vol. 253, no. September 2021, p. 108599, 2022, 10.1016/j.ijpe.2022.108599
[41] A. S. Homrich, G. Galvão, L. G. Abadia, and M. M. Carvalho, "The circular economy umbrella: Trends and gaps on integrating pathways," *J. Clean. Prod.*, vol. 175, pp. 525–543, 2018, 10.1016/j.jclepro.2017.11.064

5 Artificial Intelligence with Additive Manufacturing

Devarajan Balaji
Department of Mechanical Engineering, KPR Institute of Engineering and Technology, Coimbatore, India

M. Priyadharshini
School of computer science and engineering, Vellore Institute of Technology – AP University, India

B. Arulmurugan and V. Bhuvaneswari
Department of Mechanical Engineering, KPR Institute of Engineering and Technology, Coimbatore, India

S. Rajkumar
School of Mechanical and Electrochemical Engineering, Institute of Technology, Hawassa University, Hawassa, Ethiopia

CONTENTS

DOI: 10.1201/9781003333760-5

5.1 INTRODUCTION

Smart 3D printing has the potential to revolutionize industrial logistics as well as inventory management. Manufacturers must adopt new digital techniques that encourage machines to interact with devices and humans to interact with machine interaction in the virtual environment. This is because the manufacturing environment is becoming increasingly digital and automated [1]. Advanced digitization of manufacturing processes, such as the incorporation of the web and intelligent things, is possible through the use of digital technologies such as artificial intelligence (AI) (machines and products) [2]. Regarding manufacturing, Industry 4.0 is the new standard, allowing faster production and more incredible customizability than ever. Modern electronic technologies and industry approaches are speeding up a manufacturing transformation that is reshaping business practices as well as the framework of the industry [3]. Digitizing manufacturing supply chain implementations, inventory control, distribution networks (SCs), and value chains also significantly impact this modern electronic industrial paradigm [4]. Researchers are constantly looking for new ways to use AM to speed up digitalization in the production segment. Using AM, businesses can interact with their customers more digitally, print locally quickly, personalize large-scale production, and move to rapid manufacturing while saving money and time and SCs can also be simplified by using it [5]. SAM, which also integrates AM with intelligent technology, devices, and systems, has emerged in this context. There are numerous benefits and risks to consider as SAM is integrated into DSCs and eventually into DVCs because of the rapid growth and development of Industry 4.0. The relationship between evolving crucial allowing innovations of Industry 4.0 (e.g., SAM) and manufacturing supply chain and stock control, DSCs, and DVCs is one of the more hopeful research areas [6]. AM technology has a significant impact on manufacturing processes like supply chain management (SCM), inventory management (IM), and manufacturing supply chains (VCs). The above survey provides an in-depth assessment of the most recent research and developments on the impact of SAM on the production industry.

Machine learning algorithms can analyze human behavior in an intelligent home environment by analyzing sensor data sets that track user activity and behavior. Sensor data from smart homes can be used to identify the activities of older people, one of the most crucial research areas. Intelligent home sensor data sets can track activities like sleeping, eating, cooking, and watching television. In the first module of our system, which consists of two modules, the activities are recognized by sensors that monitor the use of home equipment [7]. However, large-scale AM processes can print parts up to 8 feet long along the lengthiest axis and, therefore, can take days or weeks to complete. Regulation of characteristics and effectiveness is required before AM can be used to manufacture critical structural elements. Thermally stimulated part distortion and retained stress management are two of the most challenging problems in large-scale AM [8–14]. Diffraction methods use Bragg's law to measure elastic strain and Hooke's law to calculate stress. It is possible to determine the current surface stress state using X-ray diffraction. However, this is only useful if it is paired with contrasting internal surface stresses, which requires cutting the part to determine. A vital advantage of the ND is its ability to measure residual stress in significant, intact components due to the neutrons' deep penetration into materials. Residual stress measurements and mapping are critical for evaluating the reliability of parts and identifying the root causes of failure [15,16]. Large-scale, long-term printing experiments can be avoided using numerical deformations and fatigue strength simulations. Because of their colossal diffusion capacity, temporary heating element and resource communication, and nonlinear part deformations, part-scale AM simulations are still expensive. Based on welding mechanics, the underlying deformation technique [16,17] can quickly predict deformations and residual stresses in a piece of metal. A single-line scan was used to test the modified method on a simple wall geometry. There needs to be more research into the impact of intricate tooling and component geometries on the intrinsic strain method [10]. It's possible to simulate heat transfer using thermo-mechanical simulation as well. Sequential or fully coupled heat and structural modeling can be carried out.

Fully coupled simulations better approximate physical phenomena, but the computational cost is enormous [9]. New numerical methods have made it possible to probe optimization and strain rate predictions at a lesser expense and faster processing time in recent years. Using Octree coarsening as the basis for an adaptive mesh, Li et al. [18] performed a thermo-mechanical simulation that was weakly coupled. Two stages of mesh coarsening resulted in the most remarkable algorithmically influential prototype, with little less than 5% of exactness damage. An adaptive coarsening of the mesh was used in conjunction with a quiet, inactive approach to creating large-scale AM parts by Denlinger et al. [19]. Until the deposition process activates an element, it is unutilized in the simulation. Reduced computation costs are achieved by coarsening two different strands underneath the energetic concentration surface.

The model was able to predict the deformation of a 3,180 mm larger piece with a total deviation of 29% using this approach. Physically based modeling was used by Stockman and colleagues [20] to coarsen meshes in two very different periods and storage depending on device control factors. Huang et al. [9] demonstrated that the GPU-based technique reduced the computational burden effectively. Models that are simple and fast to compute often have a significant contract with exploratory data. Because of this, the models are still in their infancy regarding practical applications [21–23] or the validation of yield stress in huge AM parts. Amplitude modeling (AM) simulations can be more affordable using surrogate models (SMs). A wide range of statistical models, including regression trees [24], Gaussian processes [19], polynomial chaos expansions [20], and machine learning techniques [25], can be used to analyze data in the SM.

Manufacturing systems are being replaced or redesigned by many inventors, scientists, entrepreneurs, and businesses [26]. They can also have unique properties in terms of their linear actuator, temperature, and directional characteristics. Industry 4.0 processes could benefit significantly from this scenario [27]. Developing computer algorithms that mimic human visual behavior was an early goal of computer vision research in the 1960s [28,29]. After that, we can expect to see significant progress where algorithms try to mimic human vision and improve on it as they deal with new data. Using this principle, new AI applications for computer vision were discussed. Magnetic resonance imaging is a standard imaging method used in the medical field to assess a patient's health and make diagnoses [30,31]. For example, computer vision can be used in manufacturing to help automate and inspect processes [32,33]. Another well-known use of computer vision is in autonomous vehicles [34]. Real-time people counting [35] and human-computer interaction [36] are examples of how computer vision and AI have been used in commercial monitoring. Today, extrusion-based bioprinters are widely used in biomedical implementations due to their ability to print cells and biomaterials with a broad target area of flowability [37–41]. Bioinks or biomaterials with or without living cells are printed to create scaffolds or structures. Therefore, biomaterials must have the following properties: biocompatibility, biological mimicry, mechanical integrity, degradability, and printability [42]. The printed scaffolds for tissue engineering are expected to imitate specific organs' constructions, characteristics, and features thanks to 3D printing. Printing of implants and their subsequent culture into functional structures are the keys here. Before printing, it is essential to consider the biomaterial's printability to avoid discrepancies between printed and designed forms [43,44].

5.2 SMART AM

SAM impacts digital supply chains (DSCs), corresponding digital value chains (DVCs), industrial production, logistics, and inventory management. The authors are supplemented by their own experience in the field. According to the findings, manufacturers' business models are affected by the digital transformation of manufacturing. Industry 4.0's successful introduction of intelligent manufacturing is one of its most outstanding achievements. Additive manufacturing (AM) plays a pivotal role in the digital transformation of manufacturing. Using SAM, AM can be integrated into an intelligent factory. SAM simplifies DSCs and allows for greater adaptability in logistics and inventory control. Mass customization and decentralization manufacturing have also grown and become more popular. Other positive effects include increased product design flexibility, improved product manufacturing efficiency, and resource conservation have also been realized as a result of this development. All five of Fletcher's stages in DVCs are affected by SAM technology.

5.3 AI FOR CONTROLLING

The ongoing global outbreak of COVID-19 has detailed research. Strategies powered by AI may be helpful in accurately forecasting the epidemic's parameters, risks, and effects while keeping costs low. AI has a lot of trouble due to the limited data and the unknown essence of the illness. In this piece, we explore how AI, with the aid of deep learning, can identify COVID-19-afflicted breast X-rays and whether or not this can be used to predict the infection outbreak. In this paper, we've tried to convey that AI can help us recognize specific characteristics of the illness epidemic that might be essential in protecting the human population from this deadly disease, which poses enormous economic, ecological, and societal challenges on a global scale. To create three-dimensional objects, additive manufacturers build them up layer by layer. There is still work to be done before it can be used widely in the industry, including addressing issues like low-quality output and high overhead. Post-production audits for fault identification are an example of an offline action that is not only expensive and inefficient but also unable to issue remedial activity signals during the building's life span. However, there are alternatives to online anomaly detection and process control that can be implemented with the help of in-situ tracking and ideal regulation techniques. However, conventional and parametric approaches to control and monitoring processes perform poorly due to the assumptions' complexity and the data collected. This work focuses on two subsets of additive manufacturing—large-scale additive manufacturing (LSAM) through a material extrusion and laser powder bed fusion—particularly vulnerable due to their more advanced technologies. Heat dissipation in LSAM can be affected by the morphology of huge portions, resulting in significant temperature differences between distant surface locations. The dynamics of surface cooling are formulated by taking an infrared thermal camera's reading of the surface's temperature profile and fitting it to a nonlinear multivariate regression. The method for predicting surface temperatures is then integrated into a probabilistic constraint optimization model to control layers' thickness in real time. The process can be observed in real time through on-axis optical high-speed cameras, which can record streams of melt pool images showing the laser's and powder's interaction. Model-agnostic deep learning techniques provide more leeway than one's intimate and linear extrapolation modeling counterparts when dealing with such unstructured big data. We propose a setup for a convolutional long-short term memory auto-encoder to gain knowledge of a deep spatio-temporal depiction from action scenes of melt pool images gathered from experimental builds. By further mining the unrolled bottleneck tensors, we can build an abnormality sensing and monitoring system with greater precision and a low false alarm rate [45].

5.4 AI TO PREDICT THE ANOMALIES ABNORMAL ACTIVITY RECOGNITION ALGORITHM

The Internet of Things (IoT) uses intelligent sensors in various contexts, from extreme conditions to smart homes and remote sensing. One application of this is in the care of the elderly, where sensors in the house help us understand the patterns of human behavior. There are advantages and disadvantages to using machine learning and AI algorithms for action acknowledgment in the IoT, even though many studies are currently devoted to the topic. Anomaly activity in an intelligent home setting is anything outside the norm that signals a shift in the frequency with which certain events occur or how people typically behave. The isolated forest algorithm (IFA) is an outlier detection algorithm used to predict humans' actions based on the score values of a subset of the features in a given data set. An abnormal activity recognition (AAR) system was developed to spot outliers in intelligent home sensor data sets. Predicted values of IFA and AAR are used to determine performance metrics for each task, with recall, precision, and accuracy ranging from 98% to 100%. This paper also analyzes the data set's activity distribution, groups the data set's activity values into bins, evaluates each activity's distribution, and calculates the similarity among any two tasks as well as the frequency of those activities using incredibly rapid and Gaussian distributions to help us spot anomalous behavior [46].

5.5 AI FOR COMPONENT SCALE

The wire-arc process-based additive manufacturing known as metal big area additive manufacturing (MBAAM) is progressing from the prototyping stage to full-scale production. MBAAM printing can produce parts up to 8 feet tall and can continue for several days. Management of deformation and tensile stress presents significant difficulties in making such large printed structures. Very small-time increments in transient thermo-mechanical simulations have led to accurate process predictions for tiny pieces, as time screen sizes are not algorithmically possible for huge parts. Therefore, assessing the modeling precision and simulation feasibility of the moment boost-up in thermo-mechanical experiments of building bodies is essential. In this work, MBAAM was used to print two thin walls, and data on thermal and performance requirements was collected for later use in verifying and calibrating the model. With the help of nucleon beam refraction in the high flux isotope reactor (HFIR), the dislocations and residual stresses of the parts were evaluated before and after the stress relaxation process. The data were compared to what was expected from the simulations. In this study, we looked into the impacts of time advancement amplitude on the accuracy and efficiency of the model when simulating large-scale MBAAM. The overall distortion of the part was captured by the model with a rough duration of 20 s, but the model failed to capture the growth of yield stress in the platen. The research concluded that the best supercomputing efficacy and precision for residual stress prediction could be achieved by utilizing a mix of fine and coarse time increments [47].

5.6 QUANTIFY THE MATERIAL DISTORTION USING AI

An AI-dubbed approach is a "material deformation finding algorithm" based on artificial intelligence. In AM, the core of modern 3D-printing technology, this method has direct relevance and is highly demanded in determining and designing the thermal remuneration setup of a 3D-printed product. Suggested AI continuum/material deflection discovering method can precisely find the fixed thermal deflection setup for a complicated 3D-printed structure component. Allowing us to recognize the heat-balanced design configuration needed to reduce the effect of temperature fluctuations on such parts [48]. The study created a platform that facilitates technical workflow for CNC milling and 3D printing. A system with these capabilities would act similarly to human experts in their fields, gathering and analyzing data and drawing conclusions to find solutions to problems. To accomplish this, such systems may employ AI techniques. The research confirmed the value of AI techniques and demonstrated their significant efficiency in bolstering technological process planning. This article aims to establish an intelligent machining and 3D-printing system incorporating company-specific knowledge, models, and procedures. However, there is a shortage of works utilizing a hybrid processing approach. However, these two processes now blend into a single one:

hybrid processing. A system of this scope is essential. Rulesets, rules, and facts represent the system's in-depth understanding. The system is demonstrated with the help of a case study of a functioning business. The intelligent expert system was created for engineers who either lack extensive expertise in technical planning or are new to a specific manufacturing facility and its equipment [49].

5.7 AI FOR REMOTE DEFECT DETECTION

An innovative approach to real-time quality control in fused filament fabrication (FFF) 3D printing uses artificial intelligence (AI) and computer vision. In the context of 3D printing, neural networks are specifically designed to analyze printed-out video in search of flaws. Stringing is one defect that can occur in 3D-printed objects; it typically has some relationship to the printing variables or the geometry of the printed object. This stringing of defects can cover a wide area and is generally found in areas of the thing given in the camera. A deep convolutional neural network, a type of AI, was deployed in real-world conditions to start detections and forecasts on a video camera feed after being trained on images exhibiting the stringing problem. Specifically, we outline a process for creating and launching depth neural networks for stringing recognition. The prepared model can be effectively applied to a real-world setting after the necessary components, like a camera and processors, have been assembled. Then, stringing can be recognized linearly, allowing for rapid speed and precise classification. The printing process can be modified with further development of this method. This allows the proposed approach to either stop the printing process or adjusts the parameters responsible for the flaw [50].

Human visual drain inspection is time consuming, dangerous, and requires much effort. In this paper, we introduce Raptor, a reconfigurable robot that can inspect and map drains remotely with the help of AI. The four-layer IoRT connects humans and robots so that data can be exchanged without hiccups. An IoRT framework remote defect detection task was used to train the Faster RCNN ResNet50, Faster RCNN ResNet101, and the Faster RCNN Inception-ResNet-v2 deep learning frameworks using a transfer learning scheme and six classes representing common types of concrete defects. Using the simultaneous localization and mapping (SLAM) method, the effectiveness of the trained CNN algorithm and the sink investigation robot Raptor was assessed through a series of real-time sink investigation field trials. According to the data collected during the experiments, the robot exhibited stable maneuverability, and its modeling and localization were accurate across various drain types. Finally, the lidar-SLAM map defect detection results were fused to produce the SLAM-based defect map, a vital tool for efficient drain maintenance [51].

5.8 AI FOR BIOPRINTING

Share our predictions for the future of bioprinting, specifically how incorporating AI and robot-assisted equipment will help expedite the process of automating, standardizing, and translating bioprinted tissues into clinical practice [52]. An emerging technology called extrusion bioprinting is used for tissue engineering applications to precisely apply biomaterials with living cells (called bioink) layer after layer to generate three-dimensional (3D) operational structures. The printability of a construct and the viability of the cells it contains have been shown to be sensitive to design choices, bioinks, and bioprinting method variables. This paper provides a brief literature review to identify the relevant parameters and highlight the methods or tactics for meticulously deciding or optimizing those parameters to enhance printability and cell viability. The experimental, computational, and machine learning (ML) perspectives are emphasized in this paper because of their potential in this area. It is hoped that machine learning will be a game-changer in the field of bioprinting for tissue engineering [53]. Since many variables are involved in bioprinting, ML technology can be a useful tool for fine-tuning these settings. For the most part, this paper is devoted to discussing how machine learning can be applied to 3D printing and bioprinting to fine-tune various settings and processes [54]. Through integrating scaffolds, cells, and biomolecules, tissue engineering has developed from biomaterials science to create artificial functional tissues. The primary goal of tissue engineering is the creation of functional, artificial tissues and organs in 3D to enhance, repair, or replace native tissue injured or destroyed by the disease. Tissue engineering and studies of how the body can repair itself using either its own or foreign materials are part of the broader field of regenerative medicine [55]. Tissue engineering and regenerative medicine are two names for the same field of study, which seeks to restore lost functions due to age, illness, injury, or congenital disabilities. However, personalized medicine is an emerging field that tailors patients' care to their specific needs [56]. It is not advisable to treat everyone with the same medication [57] because everyone has a unique medical history and disease. As time passes, more people choose personalized medicine because of its low immune rejection rate. Precision medicine, also known as patient-specific therapy, differs from personalized medicine [58]. In the field of medicine known as "precision medicine," doctors develop treatments tailored to the unique needs of individual patients. Personalized therapy is tailoring medical care to each patient's specific needs by taking into account their unique set of symptoms and genetic makeup. 3D bioprinting is an additive manufacturing method propelled forward by the field of tissue engineering. It prints 3D structures out of biologically compatible materials, such as living cells (or bioinks) [59]. When a biodegradable polymer, life forms, and hydrocolloids are combined, the result is a bioink; the most popular bioink because it provides an extremely moisturized atmosphere for cell proliferation. Hydrogels from natural ingredients like keratin, hyaluronic, or glycolate make up most matrix bioinks. Bioink is typically made from mucilage sodium polyacrylate, keratin, polyethylene glycol, pluronic, glycolate, and its properties allow extracellular matrix-based materials. Heterostructures that cannot be made with conventional production methodologies can be made with 3D bioprinting, helping to solve the body part shortage issue in transplantation [60]. In this context, 3D bioprinting, which can refer to several different technologies, is particularly broad. The most common method of bioprinting is based on an extrusion process. Scaffolds are manufactured without potentially toxic organic solvents, and this method also allows for precise fabrication thanks to the use of controlled pressure on the biomaterials [61]. Printing different types of cells in a precise location is a strength of inkjet bioprinting [62]. The liquid photocurable resin in stereolithography bioprinting is crosslinked using ultraviolet light [63]. Pre-bioprinting, biomaterials, and post-bioprinting are the three main phases of the bioprinting process [64]. In the preprinting stage, various software packages utilize computer-aided design (CAD). After the CAD design is complete, it will go through a slicing process to become a standard tessellation language (STL) file and, ultimately, G-code. G-code is a computer language that can do just that to send a command to a printer. The number of tissues or organs that can receive the necessary nutrients and oxygen is a key consideration in determining whether or not a design is adequate. The printing process is where real-time research into flaw identification of printable patterns would be conducted. Post-printing, methods of optimal storage, and constant quality monitoring are explored to ensure optimal conditions for cell growth. There are thus many obstacles to overcome when trying to establish an appropriate printing setup for 3D bioprinting or 3D printing. This problem prompted the introduction of ML methods to the world of 3D printing [65]. ML algorithms can automatically learn from data, predict outcomes, and enhance their own performance. Instead of running a set of predetermined instructions, these algorithms create a theoretical framework for making inferences and taking action. Synergy can be achieved by integrating these computing technologies and state-of-the-art medical technologies. When referred to the physics-based model, the ML-based model performs significantly better [66]. Nonlinearities, variable parameters, and uncertainties present a problem for the standard physics-based model. As a counterargument, ML is extremely malleable, uses flexible methods, and is a superb predictor. An important issue is deciding between a physics-based prototype and a data-driven ML model. Optimization allows a wide range of printing parameters to be adjusted to get the best possible print and design environment within a given set of prioritized criteria or constraints. The ML's ability to synthesize components with optimized printing parameters has led to its widespread use in additive manufacturing and traditional 3D printing [67]. Multiple aspects of the 3D-printing process (such as tungsten filament shape, high temperature, printing speed, printing layer thickness, high density, and injector size) have been optimized with the help of ML technology [68]. Despite its potential, 3D bioprinting has seen fewer studies involving ML than traditional 3D printing.

5.9 PATENT LANDSCAPE FOR ADDITIVE MANUFACTURING

In the future, the general scope will be assessed based on the research articles, industry bulletins, forums of scientific communities, and patent landscape. This patent landscape has been chosen due to its immediate updating over another database. (Table 5.1).

Patent landscape analysis is carried out for the keywords "artificial intelligence" and "additive manufacturing" in the "English All" category; the total count pops out at 1,160 (as on 12 August 2022).

The above table reveals the USA dominates in this technology. A researcher who wants to work in this domain and looking for research collaboration can look to these countries. The principal applicant is "General Electric Company" with a filing

TABLE 5.1
AI with AM Based on Country, IPC Code, and Year [69]

S. No.	Countries	Count	IPC code	Count	Year	Count
1	USA	636	B33Y	254	2013	17
2	Patent Cooperation Treaty	313	G06F	242	2014	21
3	European Patent Office	82	B29C	219	2015	25
4	Australia	47	G06Q	156	2016	27
5	Canada	32	G05B	153	2017	45
6	India	28	G06N	146	2018	77
7	China	8	A61B	127	2019	140
8	United Kingdom	7	A61F	99	2020	253
9	Singapore	5	G06T	99	2021	270
10	Israel	1	H04L	85	2022	211

count of 22 patents; none of the other players are able to compete. As an individual player, "Mr. Eric J. Varady" has a patent filing count of 13. Researchers aim for this company, and this collaborator might have good scope for their research. The intentional patent classification (IPC) code plays a pivotal role for the researchers to converge their search; G06F, B33Y, and B29C are dominant. Growth in this technology is depicted by the number of patents filed over the years, showing a progressive rise. In the future, it will be a promising technology for versatile applications (Table 5.2).

A patent search was carried out for the keywords "artificial intelligence" and "3D printing" in the "English All" category. The total count popped out at 1,469 (as on 12 August 2022). Single-family member option is enabled. The same patent application filed in multiple countries is considered a single application, and the count is taken accordingly.

TABLE 5.2
AI with 3D Printing Based on the Country, IPC Code, and Year [69]

S. No.	Countries	Count	IPC code	Count	Year	Count
1	USA	720	G06F	358	2013	3
2	Patent Cooperation Treaty	622	G06T	260	2014	10
3	India	40	A61B	240	2015	27
4	China	33	G06N	238	2016	26
5	European Patent Office	26	G06Q	172	2017	44
6	United Kingdom	12	B33Y	168	2018	107
7	Canada	7	G06K	154	2019	219
8	Australia	5	B29C	120	2020	307
9	Republic of Korea	3	G16H	106	2021	407
10	Sweden	1	G01N	89	2022	309

The above table reveals the USA dominates in this technology. A researcher who wants to work in this domain and looking for research collaboration can look to these countries. The principal applicant is "Nvidia Co." with a filing count of 119 patents; none of the other players are able to compete. As an individual player, "Mr. Jason E. Duff" has a patent filing count of 19. Researchers aiming for this company and this collaborator might have good scope for their research. The intentional patent classification (IPC) code plays a pivotal role for the researchers to converge their search; G06F, G06T, A61B, and G06N are dominant. Growth in this technology is depicted by the number of patents filed over the years, showing a progressive rise. In the future, it will be a promising technology for versatile applications.

5.10 CONCLUSION

In this chapter, the role of artificial intelligence in additive manufacturing is consolidated in the manner in which how innovative additive manufacturing plays a role in today's industry. Further, artificial intelligence for controlling, and predicting anomalies, scaling the component, and quantifying material distortion, lead to a new revolution in the industry forum. Artificial intelligence is further used to identify defects even from far away. Specifically, bioprinting requires more accuracy, so AI is mainly needed in this role. So, it is also discussed. In addition, the growth of the technology with the aid of AI is predicted by the patent landscape. It will be found to be one of the promising techniques shortly.

REFERENCES

[1] Napoleone, A., Macchi, M., Pozzetti, A. A review on the characteristics of cyber-physical systems for the future smart factories, *J. Manuf. Syst.*, 54 (2020), 305–335.

[2] Grabowska, S. Smart factories in the age of industry 4.0, *Manag. Syst. Prod. Eng.*, 28 (2020), 90–96.

[3] Kagermann, H., Wahlster, W., Helbig, J. *Recommendations for Implementing the Strategic Initiative Industrie 4.0: Final Report of the Industrie 4.0 Working Group*, Acatec–National Academy of Science and Engineering: Munich, Germany, (2011).

[4] Tu, M., Lim, M., Yang, M. IoT-based production logistics and supply chain, *Ind. Manag. Data Syst.*, 118 (2018), 96–125. [Google Scholar] [CrossRef][Green Version].

[5] Khajavi, S. H., Ituarte, I. F., Jaribion, A., An, J., Kai, C. C., Holmstrom, J. Impact of additive manufacturing on supply chain complexity. In Proceedings of the 53rd Hawaii International Conference on System, Maui, HI, USA, 7–10, January (2020).

[6] Devi, A., Mathiyazhagan, K., Kumar, H. Additive manufacturing in supply chain management: A systematic review. In *Advances in Manufacturing and Industrial Engineering*. Lecture Notes in Mechanical Engineering, Springer: Singapore, (2021), pp. 455–464.

[7] Gaddam, A., Wilkin, T., Angelova, M., Gaddam, J. Detecting sensor faults, anomalies and outliers in the internet of things: A survey on the challenges and solution, *Electronics*, 9 (2020), 511, 10.3390/electronics9030511www.mdpi.com/journal/electronics

[8] Savolainen, J., Collan, M. How additive manufacturing technology changes business models? –Review of literature, *Addit. Manuf.*, 32 (2020), 101070.

[9] Huang, H., Ma, N., Chen, J., Feng, Z., Murakawa, H. Toward large-scale simulation of residual stress and distortion in wire and arc additive manufacturing, *Addit. Manuf., 34 (C)*, (2020) 101248.

[10] Liang, X., Cheng, L., Chen, Q., Yang, Q., To, A. C., A modified method for estimating inherent strains from detailed process simulation for fast residual distortion prediction of single-walled structures fabricated by directed energy deposition, *Addit. Manuf.*, 23 (2018) 471–486.

[11] Jimenez, X., Dong, W., Paul, S., Klecka, M. A., To, A. C. Residual stress modeling with phase transformation for wire arc additive manufacturing of B91 steel, *JOM, 72*, (2020) 4178–4186.

[12] Lee, Y., Bandari, Y., Nandwana, P., Gibson, B., Richardson, B., Simunovic, S. Effect of interlayer cooling time, constraint and tool path strategy on deformation of large components made by laser metal deposition with wire, *Appl. Sci.*, 9 (23) (2019), 5115.

[13] Nycz, A., Noakes, M., Richardson, B., Messing, A., Post, B., Paul, J., Flamm, J., Love, L. Challenges in making complex metal large-scale parts for additive manufacturing: A case study based on the additive manufacturing excavator, Proceedings of the 28th Annual International Solid Freeform Fabrication Symposium—An Additive Manufacturing Conference, (2017).

[14] Song, X., Feih, S., Zhai, W., Sun, C.-N., Li, F., Maiti, R., Wei, J., Yang, Y., Oancea, V., Brandt, L. R. Advances in additive manufacturing process simulation: Residual stresses and distortion predictions in complex metallic components, *Mater. Des.*, 193 (2020), 108779.

[15] Walker, D., Residual stress measurement techniques, *Adv. Mater. Process.*, 159 (8) (2001), 30–33.

[16] Ueda, Y., Fukuda, K., Tanigawa, M. New measuring method of three dimensional residual stresses based on theory of inherent strain (welding mechanics, strength & design), *Trans. JWRI*, 8 (2) (1979), 249–256.

[17] Deng, D., Murakawa, H., Liang, W. Numerical simulation of welding distortion in large structures, *Comput. Methods Appl. Mech. Eng.*, 196 (45–48) (2007), 4613–4627.

[18] Li, C., Denlinger, E. R., Gouge, M. F., Irwin, J. E., Michaleris, P. Numerical verification of an Octree mesh coarsening strategy for simulating additive manufacturing processes, *Addit. Manuf.*, 30 (2019), 100903.

[19] Denlinger, E. R., Irwin, J., Michaleris, P. Thermomechanical modeling of additive manufacturing large parts, *J. Manuf. Sci. Eng.*, 136 (6) (2014).

[20] Stockman, T., Schneider, J. A., Walker, B., Carpenter, J. S. A 3d finite difference thermal model tailored for additive manufacturing, *JOM*, 71 (3) (2019) 1117–1126.

[21] Ding, D., Pan, Z., Cuiuri, D., Li, H. A practical path planning methodology for wire and arc additive manufacturing of thin-walled structures, *Rob. Comput. Integer. Manuf.*, 34 (2015), 8–19.

[22] Hu, X., Nycz, A., Lee, Y., Shassere, B., Simunovic, S., Noakes, M., Ren, Y., Sun, X. Towards an integrated experimental and computational framework for large-scale metal additive manufacturing, *Mater. Sci. Eng.*, A 761 (2019), 138057.

[23] Rodrigues, T. A., Duarte, V., Miranda, R., Santos, T. G., Oliveira, J. Current status and perspectives on wire and arc additive manufacturing (WAAM), *Materials*, 12 (7) (2019), 1121.

[24] Paul, A., Mozaffar, M., Yang, Z., Liao, W.-K., Choudhary, A., Cao, J., Agrawal, A. A real-time iterative machine learning approach for temperature profile prediction in additive manufacturing processes, In 2019 IEEE International Conference on Data Science and Advanced Analytics (DSAA), IEEE, (2019), pp. 541–550.

[25] Tapia, G., Khairallah, S., Matthews, M., King, W. E., Elway, A. Gaussian process-based surrogate modeling framework for process planning in laser powder bed fusion additive manufacturing of 316L stainless steel, *Int. J. Adv. Manuf. Technol.*, 94 (9–12) (2018), 3591–3603.

[26] Pereira, T., Kennedy, J. V., Potgieter, J. A comparison of traditional manufacturing vs additive manufacturing, the best method for the job, *Procedia Manuf.*, 30 (2019), 11–18.

[27] Rojek, I., Mikołajewski, D., Kotlarz, P., Macko, M., Kopowski, J. Intelligent system supporting technological process planning for machining*In MATEC Web of Conferences, vol. 357, p. 04001,*EDP Sciences, 2022.

[28] Grzesik, W. Hybrid machining processes. Definitions, generation rules and real industrial importance, *Mechanik*, 5–6 (2018), 338–342.

[29] Abouna, G. M. Organ shortage crisis: Problems and possible solutions, *Transplant. Proc.*, 40 (2008), 34–38.

[30] Mandrycky, C., Wang, Z., Kim, K., Kim, D. H. 3D bioprinting for engineering complex tissues, *Biotechnol. Adv.*, 34 (2016), 422–434.

[31] Khademhosseini, A., Langer, R., Borenstein, J., Vacanti, J. P. Microscale technologies for tissue engineering and biology, *Proc. Natl. Acad. Sci. USA*, 103 (2006), 2480–2487.

[32] Chen, X. B.. *Extrusion Bioprinting of Scaffolds for Tissue Engineering*. Springer Nature Switzerland AG: Cham, Switzerland, (2019).

[33] Kozior, T., Bochnia, J., Gogolewski, D., Zmarzły, P., Rudnik, M., Szot, W., Szczygieł, P., Musiałek, M. Analysis of metrological quality and mechanical properties of models manufactured with photo-curing polyjet matrix technology for medical applications, *Polymers*, 14 (2022), 408.

[34] Betancourt, N., Chen, X. Review of extrusion-based multi-material bioprinting processes, *Bioprinting*, 25 (2022), e00189.

[35] You, F., Wu, X., Kelly, M., Chen, X. Bioprinting and in vitro characterization of alginate dialdehyde–Gelatin hydrogel bio-ink, *Bio-Des. Manuf.*, 3 (2020), 48–59.

[36] Delkash, Y., Gouin, M., Rimbeault, T., Mohabatpour, F., Papagerakis, P., Maw, S., Chen, X. Bioprinting and in vitro characterization of an eggwhite-based cell-laden patch for endothelialized tissue engineering applications, *J. Funct. Biomater.*, 12 (2021), 45.

[37] Sadeghianmaryan, A., Naghieh, S., Alizadeh Sardroud, H., Yazdanpanah, Z., Afzal Soltani, Y., Sernaglia, J., Chen, X. Extrusionbased printing of chitosan scaffolds and their in vitro characterization for cartilage tissue engineering, *Int. J. Biol. Macromol.*, 164 (2020), 3179–3192.

[38] Ning, L., Zhu, N., Smith, A., Rajaram, A., Hou, H., Srinivasan, S., Mohabatpour, F., He, L., McLnnes, A., Serpooshan, V., et al. Noninvasive three-dimensional in situ and in vivo characterization of bioprinted hydrogel scaffolds using the X-ray propagation based imaging technique, *ACS Appl. Mater. Interfaces*, 13 (2021), 25611–25623.

[39] Naghieh, S., Sarker, M. D., Sharma, N. K., Barhoumi, Z., Chen, X. Printability of 3D printed hydrogel scaffolds: Influence of hydrogel composition and printing parameters, *Appl. Sci.*, 10 (2020), 292.

[40] Zimmerling, A., Yazdanpanah, Z., Cooper, D. M. L., Johnston, J. D., Chen, X. 3D printing PCL/nHA bone scaffolds: Exploring the influence of material synthesis techniques, *Biomater. Res.*, 25 (2021), 3.

[41] Suntornnond, R., Tan, E. Y. S., An, J., Chua, C. K. A mathematical model on the resolution of extrusion bioprinting for the development of new bioinks, *Materials*, 9 (2016), 756.

[42] Zhang, Z., Jin, Y., Yin, J., Xu, C., Xiong, R., Christensen, K., Ringeisen, B. R., Chrisey, D. B., Huang, Y. Evaluation of bioink printability for bioprinting applications, *Appl. Phys. Rev.*, 5 (2018), 041304.

[43] Müller, M., Becher, J., Schnabelrauch, M., Zenobi-Wong, M. Nanostructured pluronic hydrogels as bioinks for 3D bioprinting, *Biofabrication*, 7 (2015), 035006. [CrossRef].

[44] Kyle, S., Jessop, Z. M., Al-Sabah, A., Whitaker, I. S. 'Printability' of candidate biomaterials for extrusion based 3D printing: State-of-the-art, *Adv. Healthc. Mater.*, 6 (2017), 1700264.

[45] Fathizadan, S. Real-time monitoring and optimal control for smart additive manufacturing (Doctoral dissertation, Arizona State University), (2022).

[46] Rani, R. M., Kavitha, R., Deeptha, R. Analysis of human activities in smart home using abnormal activity recognition algorithm (AAR) and visualization techniques, International Journal of Early Childhood Special Education, vol. 14, issue 5, pp. 749-761, Turkey, 2020.

[47] Nycz, A., Lee, Y., Noakes, M., Ankit, D., Masuo, C., Simunovic, S., Bunn, J. et al. Effective residual stress prediction validated with neutron diffraction method for metal large-scale additive manufacturing, *Materials & Design*, 205 (2021), 109751.

[48] Wang, C., Li, S., Zeng, D., Zhu, X. An Artificial-intelligence/statistics solution to quantify material distortion for thermal compensation in additive manufacturing, *arXiv preprint arXiv:2005.09084* (2020).

[49] Rojek, I., Mikołajewski, D., Kotlarz, P., Macko, M., Kopowski, J. Intelligent system supporting technological process planning for machining and 3D printing, *Bulletin of the Polish Academy of Sciences. Technical Sciences*, 69(2), pp. 1–8, (2021).

[50] Paraskevoudis, K., Karayannis, P., Koumoulos, E. P. Real-time 3D printing remote defect detection (stringing) with computer vision and artificial intelligence, *Processes*, 8(11) (2020), 1464.

[51] Palanisamy, P., Mohan, R. E., Semwal, A., Melivin, L. M. J., Félix Gómez, B., Balakrishnan, S., Elangovan, K., Ramalingam, B., Ng Terntzer, D. Drain structural defect detection and mapping using AI-enabled reconfigurable robot raptor and IoRT framework, *Sensors*, 21(21) (2021), 7287.

[52] Jo, Y., Hwang, D. G., Kim, M., Yong, U., Jang, J. Bioprinting-assisted tissue assembly to generate organ substitutes at scale, *Trends in Biotechnology, vol. 41, issue 1, pp. 93–105*, (2022).

[53] Malekpour, A., Chen, X. Printability and cell viability in extrusion-based bioprinting from experimental, computational, and machine learning views, *Journal of Functional Biomaterials*, 13(2) (2022), 40.

[54] Shin, J., Lee, Y., Li, Z., Hu, J., Park, S. S., Kim, K. Optimized 3D bioprinting technology based on machine learning: A review of recent trends and advances, *Micromachines*, 13(3) (2022), 363.

[55] Mao, A. S., Mooney, D. J. Regenerative medicine: Current therapies and future directions, *Proc. Natl. Acad. Sci. USA*, 112 (2015), 14452.

[56] Skardal, A., Shupe, T., Atala, A. Body-on-a-chip: Regenerative medicine for personalized medicine, *Atala, A., Lanza, R., Mikos, T., & Nerem, R. (Eds.), Principles of regenerative medicine. Academic press*, (2019), 769–786, 10.1016/B978-0-12-809880-6.00044-8.

[57] Risse, G. B., Warner, J. H. Reconstructing clinical activities: Patient records in medical history, *Soc. Hist. Med.*, 5 (1992), 183–205.

[58] Hamburg, M. A., Collins, F. S. The path to personalized medicine, *N. Engl. J. Med.*, 363 (2010), 301–304.

[59] Jovic, T. H., Combellack, E. J., Jessop, Z. M., Whitaker, I. S. 3D bioprinting and the future of surgery, *Front. Surg., vol. 7, article 609836*, (2020).

[60] Mironov, V., Kasyanov, V., Markwald, R. R. Organ printing: From bioprinter to organ biofabrication line, *Curr. Opin. Biotechnol.*, 22 (2011), 667–673.

[61] Ramesh, S., Harrysson, O. L. A., Rao, P. K., Tamayol, A., Cormier, D. R., Zhang, Y., Rivero, I. V. Extrusion bioprinting: Recent progress, challenges, and future opportunities, *Bioprinting*, 21 (2021), e00116.

[62] Li, X., Liu, B., Pei, B., Chen, J., Zhou, D., Peng, J., Zhang, X., Jia, W., Xu, T. Inkjet Bioprinting of biomaterials, *Chem. Rev.*, 120 (2020), 10793–10833.

[63] Wang, Z., Abdulla, R., Parker, B., Samanipour, R., Ghosh, S., Kim, K. A simple and high-resolution stereolithography-based 3D bioprinting system using visible light crosslinkable bioinks, *Biofabrication*, 7 (2015), 045009.

[64] Nair, K., Gandhi, M., Khalil, S., Yan, K. C., Marcolongo, M., Barbee, K., Sun, W. Characterization of cell viability during bioprinting processes, *Biotechnol. J.*, 4 (2009), 1168–1177.

[65] Delli, U., Chang, S. Automated process monitoring in 3D printing using supervised machine learning, *Procedia Manuf.*, 26 (2018), 865–870.

[66] Ji, C., Mandania, R., Liu, J., Liret, A., Kern, M. Incorporating risk in field services operational planning process. In *International Conference on Innovative Techniques and Applications of Artificial Intelligence*, Springer: Cham, Switzerland; Cambridge, UK, (2018), Volume 11311 LNAI, ISBN 9783030041908.

[67] Goh, G. D., Sing, S. L., Yeong, W. Y. A review on machine learning in 3D printing: Applications, potential, and challenges, *Artif. Intell. Rev.*, 54 (2021), 63–94.

[68] Menon, A., Póczos, B., Feinberg, A. W., Washburn, N. R. Optimization of silicone 3D printing with hierarchical machine learning, *3D Print. Addit. Manuf.*, 6 (2019), 181–189.

[69] https://patentscope.wipo.int/search/en/result.jsf?_vid=P20-L6Q3E6–52733 – Accessed 12 August 2022.

6 Robotic Additive Manufacturing Vision towards Smart Manufacturing and Envisage the Trend with Patent Landscape

V. Bhuvaneswari, Devarajan Balaji, and B. Arulmurugan
Department of Mechanical Engineering, KPR Institute of Engineering and Technology, Coimbatore, Tamil Nadu, India

S. Rajkumar
School of Mechanical and Electrochemical Engineering, Institute of Technology, Hawassa University, Hawassa, Ethiopia

CONTENTS

DOI: 10.1201/9781003333760-6

6.1 INTRODUCTION

Typical 3D printing uses a gantry system, which has hardly three degrees of freedom and is consequently limited in what it can produce. Therefore, printing can occur only along this axis (z-direction). Mechanical characteristics are anisotropic in unidirectional 3D-printed parts. In unidirectional 3D printing, support structures are required for sophisticated 3D objects with cantilever features. This lengthens the time necessary to complete the project and the number of materials needed [1]. To lessen the need for support structures, researchers can create a 3D-printing method that allows for printing in multiple directions simultaneously. The number of joints in an articulated robot determines how many degrees of freedom the robot has and is termed "degrees of freedom." Industrial robots come in various sizes, have a six-degrees-of-freedom (DoF) articulated configuration, and have cost advantages. They are widely used in today's production automation, making them ideal for robotic additive manufacturing. Virtual models can be transformed into physical objects using a process referred to as additive manufacturing or 3D printing [2]. Various materials, including plastics, metals, and composites, are used in a layer-by-layer construction process. In the last two decades, the use of 3D printers has skyrocketed, and now you can find them in many classrooms, offices, and even people's homes. For robotic additive manufacturing (RAM), a material extruder is affixed to a robot arm, expanding the capabilities of traditional 3D printing. By combining robotics with 3D printing, RAM opens up previously unthinkable avenues for exploration and development, including free-form and nonplanar printing. In conventional additive manufacturing, an object is constructed layer upon layer of material, beginning at its base and progressing upward. This concept is flipped on its head with free-form printing, which enables the creation of objects that can be oriented in any path and at every angle, even in a downward direction. Free-form printing allows for creating things that would typically necessitate a great deal of discarded support material to be formed. Free-form printing also produces sharp, smooth edges, like those used in airfoils [3]. Table 6.1 lists the difference between traditional additive manufacturing and robotic additive manufacturing.

TABLE 6.1
Traditional AM vs. RAM

S. No.	Traditional Additive Manufacturing	Robotic Additive Manufacturing
1	It uses planar layers	Robots enable the deposition of nonplanar layers
2	3D object converted to 2D layers	Minimize the number of layers
3	Constrained by deposition orientation and geometric irregularity	Reduce support material
4	Building curved geometries take a long time	Reduces cost, time, and material waste
5	Strength is compromised	Improve strength
6		Ability to exploit composite material
7		The capability of material deposition on irregular surfaces and platforms

Much progress has been made in terms of accuracy, increased flexibility, and decreased production time due to technological innovations of this type. The stair-step effect [3] occurs naturally due to the additive manufacturing process of slicing the geometric digital model with a level plane and fabricating the component by stacking the flat layer of material; this is depicted. In contrast to three-dimensional printing of PLA or selective laser sintering, where the layer thickness is only 0.1–0.5 mm, wire and arc additive manufacturing can have a layer thickness of 2.0–4.0 mm. Since the stair-step consequence is so much more severe, the forming appearance and the tensile strength of the complex-surfaced component will suffer. Scientists have devised various techniques, from slicing but also path-planning methods to process controlling strategies, to enhance the forming aesthetic of additive manufacturing, which is degraded by the stair-step effect. These techniques include the blended path planning algorithm, which uses contour paths along the boundary and zig-zag trajectories to fill the internal area. The responsive slicing algorithm uses a variant layer thickness by the surface curvature [4] and the multi-directional process control strategy. Since the design optimization is still sectioned along a plane. These techniques cannot eliminate the stair-step effect caused by an uneven surface finish brought on by the accumulation of flat layers of material.

A few researchers have suggested a contoured layer fusion deposition modeling technology (CLFDM). In this method, the digital model is sliced along a nonplanar surface, as well as the material accumulation trajectory follows the same path. This is because the material is distributed following the complex surface's changing regular and curvature, eliminating the stair-step effect and allowing for a seamless, high-quality finish. For robotic 3D printers of PLA wire substance, Zhao et al. [5] created a nonplanar slicing and path generation method for constructing overhanging structures and curved parts.

6.2 LARGE-AREA ADDITIVE MANUFACTURING

By trying to remove the simple geometric constrictions of commercial production as well as shortening the lead time for components, especially for large-scale parts, additive manufacturing (AM) does have the potential to alter the manufacturing landscape radically. Using direct energy deposition (DED), 3D-printing techniques in conjunction with robotic systems enables support-free printing of components with a size range from sub- to multi-meters [6].

Due to their low construct volume as well as high installation cost, traditional 3D printers cannot be used for production on a large scale. When using a conventional 3D printer to create a large part, the machine envelope needs to be large enough to hold the entire region. The deposition head's actuators must keep consistent precision across the complete build platform as they move the head. Because of the need for a robust motion system as well as accurate calibration, AM machines tend to be bulky and expensive. The need for a large build plate, along with its aids, controllers, calibration system, motion, etc., causes machines to have overall dimensions that are much larger than their build volume. This gantry-style machine is also relatively cumbersome, making it challenging to move around. Permanently remove printed components from the build platform before continuing. Using robots allows us to increase productivity in terms of (i) constructing volume size, (ii) development time, as well as (iii) cost, all of which are limitations of the gantry-type print machine [7]. An extended robot arm's capabilities include moving tools around a workspace and increasing build volumes. The use of robots for AM on a massive scale also has the added benefit of reducing production time. If you print a component where it will ultimately be used, you won't have to worry about taking it off the build plate, moving it, and then putting it together [8]. The time savings from not having to construct supporting structures or perform post-processing are also substantial. Due to the significant size, erecting and dismantling the necessary support structures can consume considerable time. As a base, a 6 DOF robot utilizes an already existing item. This allows the robot to print a substantial component directly onto the existing structure. Due to the potential for collision, this is impossible with a standard gantry-based 3D printer. Eliminating the time and effort needed to move the component or assembly can be a significant cost saver. Compared to a hoisting system 3D printer, a robotic system seems more space-efficient, leading to lower upfront and ongoing costs. When they are not being used during 3D printing, industrial robots can do a variety of other jobs. Additionally, savings may result from this. Using 3D printing, as an illustration, an industrial robot could create a part and paint or finish it [9].

Recently, systems with large-scale robotic manipulators carrying the welding torch have been suggested as a method for achieving large-scale printing volumes of less than one cubic meter. A common addition to these robotic systems is a positioning system that allows the part designed and manufactured to be tilted and rotated as it is being made. Overhanging attributes of the component could be printed in orientation with the gravity vector thanks to multi-directional deposition, which is made possible by a positioning system that reorients the part during manufacturing. This eradicates the requirement for structural components [10].

6.3 MULTI-DEGREES OF FREEDOM

Several obstacles arise when applying robotic manipulators to cutting-edge AM techniques. Getting the component into the robot's working area is the first obstacle to overcome. Workspace and configuration space mapping is complicated for articulated manipulators. Finding the optimal part location requires meeting both the limitations of the robot and the AM process. Attached to the manipulator's end-effector, the AM tools have greater flexibility and can perform a wider variety of tasks than is possible with a traditional setup. This leads to the second difficulty: planning tool paths efficiently enough to complete all of the functions, prevent tool collisions, and produce an excellent final product. The robot trajectories must be generated in the manipulator's configuration space before the part can be built in the manipulator's workspace [7]. In addition, the robot trajectories must meet several constraints, including the reachability of the singularity, collision, continuity, manipulator, velocity, and much more. Articulated manipulators are less precise than a standard AM setup when performing AM. The fourth difficulty is constructing exact AM components by compensating for and enhancing manipulator accuracy. Multiple robots are required for large-scale AM to ensure the part is built quickly as well as efficiently. The fifth difficulty arises from dismantling and installing the AM component.

To improve the mechanical and physical characteristics of the parts, conventional AM methods permit the manufacturing of multi-material components. However, this is limited by the size of the tool, which is, in turn, determined by how many materials it must be able to switch between. Because of the limited degree of freedom (DOF) of the setup as well as the size of the tool, multi-material conventional AM limits the flexibility with which materials can be swapped out and used during the fabrication process [11]. However, high-DOF robots can provide significantly more maneuverability for a variety of tools. By combining a variety of materials, robots are capable of quickly constructing intricate shapes. Robots are equipped with tool-switching mechanisms to accommodate the use of a wide variety of materials in AM methods.

In contrast to traditional AM [12], a 6 DOF machine can employ a sophisticated three-nozzle extruder tool to construct multi-material parts. Multiple robots equipped with unique materials could be used to build multi-material features to have ten productions. In robot-based multi-material AM, the setup design is a significant challenge. Because of the complexity of coordinating multiple tools by a single robot or by multiple robots sharing a single workspace, effective task planning, scheduling, and motion planning seems essential for the robotic fabrication of multi-material components.

6.4 RAM FOR TOOLING

Traditional AM faces difficulties in tool size limitations due to restricted machine build space. This issue was solved by segmenting the production of the necessary tools, which added complexities and labor hours to the project. RAM could address this issue because it has longer reach distances and larger build spaces than traditional cartesian machines. As a result of AM, conventional subtractive manufacturing techniques can be abandoned in favor of more efficient and versatile additive ones [13].

6.4.1 Tool Path Planning

There are more factors to consider when planning a robot's trajectory in 3D printing than when using a traditional printer. A robot's planned course is deemed invalid if it confronts a collision but rather a joint limit across its path; in such a case, the system must resurrect a safe way for the robot to follow. If you're having trouble finding a trajectory that works, hatching patterns, or joint angle velocities, try experimenting with different material deposition directions or even the physical location of the part printing in the workspace [14]. Trajectory planning, as well as optimization, is the subject of a great deal of study for use with industrial robots in applications as diverse as welding, painting, cleaning, and finishing [15–18]. Ding et al. [19] introduced medial axis transformation for use in responsive trajectory tracking for wire-feed AM in robotic additive manufacturing. There were improvements in material efficiency, void-free deposition, and boundary accuracy after the proposed algorithm was put into practice by a robotic wire and arc AM system. Our work will fill in the gaps in this research by addressing robotic trajectory planning in addition to presenting the improved additive manufacturing algorithm used in this study.

The cylinder surface is sliced from a mesh model to determine the axis as well as radius using a cylinder surface slicing technique. The cylindrical mesh model is flattened out into a planar one. The model has been sliced just after the transformation, and a tool path is generated for it in the same way it would for planar slicing using the conventional methods. The tool paths have been inverted to correspond to the contoured layer tool paths in a three-dimensional environment. An AM system is built using robots to validate the tool paths that were pre-simulated and pre-programmed offline by a robot simulation system validate validate validate the tool paths that have been pre-simulated and pre-programmed offline by a robot modeling system; an AM system is built using robots. The optimization techniques are illustrated in two instances [20].

The field of additive manufacturing must have advanced over the course of several decades, as well as it seems to have recently attracted a lot of attention from academics. When representing geographic information in 3D printing, the standard tessellation language (.stl) structure is typically used. There are some inherent problems only with .stl format files that are becoming increasingly apparent to the public. Furthermore, using the straightforward polygonal facet representation with the five-axis strategic approach is extremely difficult. A five-axis 3D printer is therefore understudied and underutilized, despite their widespread use in subtractive machining. The authors suggest a feature-based five-axis approach to improve and broaden the applicability of additive manufacturing. We define and classify the features of 3D printing as either 5D AM features or free-form AM features. The suggested technique for feature extraction could indeed instantaneously identify the 3D-printer features from the input framework. To produce tool paths for the free-form AM feature, a five-axis trajectory scheduling technique is proposed as well as broken down into three distinct steps: (1) offsetting the source surface, (2) spatial and temporal slicing of the freeform layers, as well as (3) generating the toolpaths for every free-form layer. The suggested technique operates as an advanced secondary plug-in within the CATIA and can create reliable tool paths for the five axes used in the 3D printer. Using the simulation and off-programming results, the suggested algorithm produces five-axis 3D-printer tasks that are then post-processed by a RAM system. Some illustrations are provided to demonstrate the proposed technique's practicality and effectiveness [3].

6.4.2 Process Planning

Traditional 3D printers have reshaped the manufacturing sector by allowing for the creation of one-of-a-kind items with no need for specialized equipment. In other words, they allow for the realization of complex geometries that could be impossible to construct using traditional production methods. Traditional AM, however, has drawbacks like lengthy construction times, unfavorable mechanical properties, a subpar final product, and wasteful support structures. Due to the high flexibility of RAM, these restrictions can be overcome. The constraints of traditional AM can only be overcome with several technological advances, all of which are necessary for implementing robotic additive manufacturing. The following are four significant contributions made by this dissertation toward the goal of automating AM with robots: Four distinct algorithms are needed for efficient manufacturing: (1) a part placement algorithm to maximize the precision with which the robot's trajectories are carried out; (2) a tool-path planning technique to make full use of the tool's versatility; (3) a robot path compensation strategy to enhance the precision with which the robotic manipulators operate; as well as (4) a robot placement algorithm to facilitate multi-robot manufacturing and reduce the total time required for manufacturing. The dissertation illustrates the underlying advancements using the following four examples of robotic additive manufacturing. To begin, the robot's accuracy is elevated by the compensation scheme to carry out supportless material extrusion 3D printing. Second, we show how multi-resolution material extrusion 3D printing can benefit from flexible tool-path planning. For the third step, the conductive wire arc AM process uses the part placement algorithm. Finally, a multi-robot cell is designed to increase the deposition rate in multi-axis wire arc manufacturing. Compared to a large-scale gantry setup, robotic manipulators seem more cost effective but have a larger workspace for the same footprint. In turn, this drastically lowers the initial investment required for AM deployment on a massive scale [21].

6.4.3 Optimization Techniques

There is a new method of manufacturing called AM. Current additive manufacturing techniques use a three-dimensional printing mechanism in which material is added primarily in a vertical (z) plane. This ultimately restricts creative control over the printed product. Overhanging objects can't be printed without support structures. In the long run, print efficiency suffers when these supports are removed. We suggest a framework control system for robot-assisted AM, which allows for multi-directional printing without needing support structures. Stewart platform manipulator with 6 DOF is developed to replace the printer's build plate due to its high stiffness and the payload-to-weight ratio [22,23]. It is designed to show how the manipulator's kinematics and dynamics work. After that, a sliding mode controller with an extended proportional derivation is developed for trajectory tracking. The optimal controller parameters are determined using a customized grey wolf optimization algorithm version. In this method, we use integral absolute error (IAE) to lead the way to the cost function and stop iterating when the IAE reaches its minimum value (75,100). With the aid of MATLAB, the simulation of something like the analytical model is operated for 10 seconds. The findings demonstrate that after 3.5 seconds, the manipulator's six legs have reached their requested length pathways. The analytical model's efficacy is tested using computer-assisted dynamic analysis of just a mechanical system. The ability of 3D printing to construct complicated forms has made it a popular topic of study. It uses slicing and material accumulation to go from design files to finished products. For the most part, the AM industry employs the planar slicing tactic when transforming CAD models into successive layers. However, it requires support structures and produces a significant quantity of planar layers, increasing manufacturing time, when constructing overhang structures and curved parts. This work discusses two nonplanar slicing methodologies to accomplish these goals: a decomposition-based contoured surface slicing strategy and a transformation, depending on the cylinder surface slicing technique. Both are implemented in a similar way, but while the former uses STEP models, the latter can slice mesh models. By printing two components with something like a robotic fused deposition modeling system, the methodologies proposed here are practical [17,20].

Wire arc additive manufacturing (WAAM), depending on gas metal arc welding (GMAW), is a promising technique for manufacturing large-scale metallic structures. By depositing metal in thin layers and welding them together, it's possible to create nearly net-shaped parts. A hybrid robotic additive/subtractive assembly system is described. Uneven layer thickness is a serious problem when making tall, thin buildings. It is impossible to sustain depositing multi-layer materials caused by the accumulation of folds with poor flatness, which generates distinct changes in the altitudes of various layers' roles [24]. The study's overarching goal is to find the best ways to use a robotic additive and a subtractive manufacturing system to create massive, tall metal structures with thin walls. Arc igniting and extinguishing control, accumulation with weaving, and a local measuring and milling strategy are the three suggested optimization strategies for achieving flat layers. Various experiments have been set up to examine the outcomes of employing these methods. Experimental consequences demonstrate that combining these methods can decrease layer height differences and increase layer flatness. Using these methods, a large and highly thin-walled component is produced, showcasing the capability of the robotic subtractive and AM system to make large metallic parts rapidly. Additionally, regression analysis is applied to determine the significance of process parameters on bead topography in a deposition during weaving [25,26].

6.5 LATEST TECHNOLOGIES LIKE AI, ML, AND DEEP LEARNING FOR RAM

WAAM is a production method that uses welding methods to deposit metal layers upon layers to create 3D parts. Many studies have focused on the width, height, and penetration of the weld bead as indicators of successful WAAM. Layer roughness, however, is crucial due to its impact on machining expenses and the mechanical characteristics of finished products. Less machining time and the material will be saved if the roughness of something like a deposited layer could be decreased. Better adhesion between layers is another benefit of reducing layer roughness. As a result, it's crucial to pay close attention to the process of depositing weld beads to ensure the least amount of roughness possible. Few researchers have explored roughness in WAAM, and those have focused on straight pathways for the deposition process. This is despite the weaving path's excellent ability to lower layer roughness, which has received little attention. Contribution is its successful implementation of two machine learning methods, random forest as well as multilayer perceptron (MLP), as well as artificial neural network (ANN), to appropriately model a surface finish in WAAM utilizing a weaving path. Random forest outperformed MLP in terms of accuracy and computational time when modeling and predicting layer roughness for a particular set of robotic WAAM parameters and utilizes a weaving path to explore the effect of layer roughness on a robotic WAAM [27]. The layer roughness was predicted using a combination of a random forest and a multi-layer perceptron model fed with data from a robot's WAAM. Both methodologies seem to be effective in modeling the connection among robotic WAAM parameters and layer roughness, according to the experimental results; nevertheless, the modeling approach utilizing random forest achieves better outcomes than that using MLP. The produced random forest model predicted a weld bead roughness out from input parameters, including a typical error rate below 6% during both validations and testing. Due to its low computational cost, random forest could be easily adapted for use in online roughness modeling in quality control [28,29].

Using the strengths of combining the tools, standards, and algorithms of both fields, we introduce a distributed control and conversation architecture for agent training and strategy implementation. Specifically, it's designed for use in CDRF-based industrial settings. Using one such structure, a robotic agent is instructed to independently plan and construct facilities in two case studies (robotic block stacking and sensor-adaptive 3D printing) utilizing two model-free DRL algorithms (TD3, SAC). This first study shows how computational design environments can be used for DRL training as well as how effective the employed algorithms are in comparison to one another. The second case study demonstrates the advantages of our setup in the areas of tool trajectory tracking, geometric state restructuring, the integration of fabrication constraints, and the evaluation of actions as components of the training and execution processes. The study benefits from using real-time physics modeling in CAD, industrial-grade hardware control, and geometric scripting to supplement distinct actions. The most open-source codebase is made available [30,31].

Learning many different process plans with ML allows you to optimize your material usage and decrease your production time besides assessing the relative costs of the various options. Structure for 3D printing, method enhancement, and in-situ monitoring are just a few of the many uses of ML in the 3D-printing industry. ML has also been proven to be an effective tool for data-driven simulation analysis, feature guidelines in design, real-time intrusion detection, and cyber security [32,33]. Results show that ANN has become the most popular and effective ML technique for method enhancement. You can get an accuracy of 98% with just a three-layer ANN. It has been determined that CNN's capability to acquire spatial features makes it more effective than ANN when working with 2D images as well as 3D models. Thus, in-situ monitoring, CNN has been used for intrusion detection systems [34,35], functionality suggestions in 3D object design, and feature recognition. Using only a tiny number of monitored samples of low quality (including noisy and blurry images), Li et al. [36] presented a deep learning–based reliability authentication mechanism for the metal AM technique with encouraging results. The efficiency of the suggested approach, which provides a hopeful tool for real industry sectors, is verified by implementing experimentations on an actual metal AM data set.

6.6 AUTOMATIC INSPECTION

Davtalab et al. [37] suggest a mechanical layer deficiency detection structure for 3D printing in construction. Starting with images as input, a step-by-step procedure is used to build the deep convolutional neural network capable of semantic pixel-wise segmentation, which is then used to separate concrete layers from several other objects in the image. The CNN model is trained and evaluated using 1M synthetically labeled images created with data enhancement techniques. The images are processed through a convolutional neural network (CNN) model, and then a fault diagnostic configuration is created that can identify distortions in the printable concrete layers. Accuracy, F1 score, as well as miss rate findings, show that the advanced system performs satisfactorily [38].

6.6.1 Tool-Path Strategies

Required to test physical hardware with something like a model designed to simulate the actual underlying process is known as "hardware in the loop" (HiL). HiL testing can be cost-effective before testing hardware in a production environment. This research will focus on the physical process of WAAM because it is a state-of-the-art AM technique for depositing successive layers of a metal-based material. This study mimics the robot's actions while the real robot carries out the procedure (additive manufacturing). By using slicer software to break down a CAD model into individual layers, a robot could now perform 3D printing. The robot must move under these layers to construct a three-dimensional object. There are discrepancies between the heights depicted in CAD models and the measurements of the completed buildings. Since the printed layers will have different sizes, a robot's route must be adjusted after each. This is achieved by substituting a mathematical formula for the CAD model in the creation of the printed structure (model). This study updates the mathematical model after each layer is produced thanks to real-time feedback from sensors monitoring the additive manufacturing process. One definition of real-time simulation is the process of continuously running as well as modifying a mathematical model (RTS). In this study, we developed a HiL-based real-time simulation system to predict the optimal printing layer height as well as the total number of layers. A cyber-physical framework was created by merging hardware and software to ease the transition from facilitated to fully autonomous robotics and to further the aims of Industry 4.0 [39–41].

AM has experienced an increment in the use of industrial robots in recent decades as a replacement for traditional three-axis CNC cartesian robots to address drawbacks such as the staircase effect (hull quality), the need for a support structure in the case of overhanging features (scrap creation), and the limitation of AM to producing only small and medium-sized components. The article explains how to convert G-code written for CNC equipment to a format that can be read by an industrial robot with minimal change to the final product. Particular attention is paid to machines where the programmer cannot specify the trajectory through each cycle time, resulting in significant discrepancies between the intended as well as the produced product. In order to accomplish the desired acceleration-deceleration portfolio, the G-code is subjected to a post-processing routine that redefines the tool path by using the controller's native primitives. Experiments were conducted on an Epson T3 SCARA by attaching an FDM extruder and having the robot give instructions for the extruder's speed in real time based on the robot's tool center point (TCP).

6.6.2 Spherical

To avoid the stair-stepping effect and boost structural integrity and tensile strength [42], spherical introduces a spherical slicing technique for 3D printing that places additive layers on a circular exterior. A tool path is needed to direct the extruder in the desired directions in a 3D space. Slicing algorithms are used to calculate these tool paths. Planar slicing is a tried-and-true technique in the AM business. This process generates 2D tool trajectories from the 3D computer-aided design approach by slicing it along planes. Novel nonplanar slicing techniques are being developed in response to the advent of multi-axis AM machines. A new kind of nonplanar slicing, based on a sphere, is presented here. The model is cut into rings, much like an onion, using the spherical slicing methodology developed. These sphere slices are then employed to generate tool paths for directing a robotic manipulator throughout a 3D space, along with the appropriate configuration, to place the spherical layer upon layer atop the traditional extended planar base. Fused deposition modeling with a six-axis robotic sequential manipulator is utilized to create a model and evaluate the outcomes [43].

6.7 PATENT LANDSCAPE ROBOTIC ADDITIVE MANUFACTURING

The patent landscape is one of the better databases to determine the future trend than any other database. Herein, the keywords taken are "robots" and "additive manufacturing" and searched in all English categories, which reveals the patent count of 5,890 along with the option "single-family"; it means that the same patent can be filed in multiple countries, but it is considered as a patent (Table 6.2).

Herein, the key words taken are "robotic additive manufacturing" and searched in all English categories, which reveals the patent count of 39 along with the option "single-family," which means that the same patent can be filed in multiple countries, but it is considered a patent (Table 6.3).

TABLE 6.2
Patent Landscape for Robots with Additive Manufacturing [44]

S. No.	Country	Major Applicants	Year Wise Count
1	USA – 3,025	The Boeing Company – 166	2013 – 95
2	PCT – 2,291	General Electric Company – 141	2014 – 167
3	China – 213	Desktop metal inc – 108	2015 – 222
4	European Patent Office – 209	Continuous Composites Inc – 67	2016 – 399
5	United Kingdom – 58	Velo3d Inc – 67	2017 – 592
6	India – 44	Divergent Tech Inc – 64	2018 – 696
7	Canada – 26	Markforged Inc – 62	2019 – 838
8	Australia – 9	Arevo Inc – 61	2020 – 915
9	Netherlands – 4	CC3D LLC – 60	2021 – 888
10	Singapore – 3	SEURAT TECH INC – 60	2022 – 597

* Patent Cooperation Treaty – PCT.

TABLE 6.3
Patent Landscape for Robotic Additive Manufacturing [44]

S. No.	Country	Major Applicants	Year Wise Count
1	PCT – 16	Nanjing University of Science And Tech – 5	2014 – 1
2	United States of America – 13	Stratasys Inc – 5	2015 – 1
3	China – 6	Carbon Inc – 3	2016 – 3
4	United Kingdom – 2	Grale tech – 3	2017 – 1
5	European Patent Office – 1	British Telecomm – 2	2018 – 5
6	India – 1	Clark fixture tech inc – 2	2019 – 7
7		Lincoln global inc – 2	2020 – 9
8		XR downhole llc – 2	2021 – 8
9		ABB Schweiz AG – 1	2022 – 4
10		Board of Regents The University of Texas System – 1	2014 – 1

6.8 CONCLUSION

Recent years have seen a proliferation of studies that back the idea of using robots to fix the problems with conventional AM. This article showed how robots could reduce construction time, boost quality, quicken the procedure, and sometimes even produce large parts. In this study, typical AM robot setups, as well as the features they have created, are analyzed. This article draws from a body of research demonstrating the great potential of robotic AM technologies. However, at the moment, this technology is still in its infancy. We have discovered how AM-based robots can be utilized to develop novel capabilities. We conclude with some recommendations for where this field could go next. To that end, we anticipate that robots will play a significant role in the next era of the additive production system.

REFERENCES

[1] C. K. Chua, K. F. Leong, *3D Printing and Additive Manufacturing: Principles and Applications*, 5th ed., SG: World Scientific, (2017).

[2] D. Bengs, M. Z. Cordero, Technobothnia has a new robotic additive manufacturing environment (2021). Vaasa Insider 30 December 2021.

[3] Z. Hu, L. Hua, X. Qin, M. Ni, Z. Liu, C. Liang, Region-based path planning method with all horizontal welding position for robotic curved layer wire and arc additive manufacturing, *Robotics and Computer-Integrated Manufacturing* 74 (2022), 102286.

[4] J. C. S. McCaw et al. Curved-layered additive manufacturing of nonplanar, parametric lattice structures, *Mater. Des., vol. 160, pp. 949–963,* (2018).

[5] Z.-Y. Liao, J.-R. Li, H.-L. Xie, Q.-H. Wang, X.-F. Zhou, Region-based toolpath generation for robotic milling of freeform surfaces with stiffness optimization, *Robotics, and Computer-Integrated Manufacturing* 64 (2020), 101953.

[6] T. Lehmann, D. Rose, E. Ranjbar, M. Ghasri-Khouzani, M. Tavakoli, H. Henein, T. Wolfe & A. J. Qureshi. Large-scale metal additive manufacturing: A holistic review of the state of the art and challenges, *International Materials Reviews* 67 (4) (2022), 410–459. 10.1080/09506608.2021.1971427

[7] P. M. Bhatt, R. K. Malhan, A. V. Shembekar, Y. J. Yoon, S. K. Gupta, Expanding capabilities of additive manufacturing through use of robotics technologies: A survey, *Additive Manufacturing* 31 (2020), 100933.

[8] N. Hack, W. Lauer, S. Langenberg, F. Gramazio, M. Kohler, Overcoming 845 repetition: Robotic fabrication processes at a large scale, *International Journal of Architectural Computing* 11 (3) (2013), 285–299.

[9] B. G. de Soto, I. Agust-Juan, J. Hunhevicz, S. Joss, K. Graser, G. Habert, B. T. Adey, Productivity of digital fabrication in construction: Cost and time analysis of a robotically built wall, *Automation in Construction* 850 (92) (2018), 297–311.

[10] T. Lehmann, A. Jain, Y. Jain, H. Stainer, T. Wolfe, H. Henein, A. J. Qureshi, Concurrent geometry- and material-based process identification and optimization for robotic CMT-based wire arc additive manufacturing, *Materials & Design* 194 (2020), 108841.

[11] J. Duro-Royo, L. Mogas-Soldevila, N. Oxman, Flow-based Fabrication: An integrated computational workflow for design and digital additive manufacturing of multi-functional heterogeneously structured objects, *Computer-Aided Design* 69 (2015), 143–154.

[12] Y. J. Yoon, M. Yon, S. E. Jung, S. K. Gupta, Development of threenozzle extrusion system for conformal multi-resolution 3D printing with a robotic manipulator, in ASME International Design Engineering Technical Conferences and Computers and Information in Engineering Conference, American Society of Mechanical Engineers, vol. 59179, p. V001T02A024, Anaheim, California, (2019).

[13] I. E. Yigit, S. A. Khan, I. L. Lazoglu, Robotic additive manufacturing of tooling for composite structures, in The 18th International Conference on Machine Design and Production (UMTIK), Eskişehir, Turkey, (July 2018).

[14] A. V. Shembekar, Y. J. Yoon, A. Kanyuck, S. K. Gupta, Generating robot trajectories for conformal 3D printing using non-planar layers, *Journal of Computing and Information Science in Engineering, vol. 19, issue 3, pp. 031011* (2019). 10.1115/1.4043013

[15] L. B. L. Huo, The joint-limits and singularity avoidance in robotic welding, *Industrial Robot: The International Journal of Robotics Research and Application* 35 (5) (2008), 456–464.

[16] H. Chen, T. Fuhlbrigge, X. Li, Automated industrial robot path planning for spray painting process: A review, in 2008 IEEE International Conference on Automation Science and Engineering, (August 2008), pp. 522–527.

[17] A. M. Kabir, K. N. Kaipa, J. Marvel, S. K. Gupta, Automated planning for robotic cleaning using multiple setups and oscillatory tool motions, *IEEE Transactions on Automation Science and Engineering* 14 (3) (July 2017), 1364–1377.

[18] A. M. Kabir, B. C. Shah, S. K. Gupta, Trajectory planning for manipulators operating in confined workspaces, in 2018 IEEE International Conference on Automation Science and Engineering (CASE), Munich, Germany, (August 2018).

[19] D. Ding, Z. Pan, D. Cuiuri, H. Li, N. Larkin, Adaptive path planning for wire-feed additive manufacturing using medial axis transformation, *Journal of Cleaner Production* 133 (2016), 942–952.

[20] G. Zhao, G. Ma, J. Feng, W. Xiao, Nonplanar slicing and path generation methods for robotic additive manufacturing, *The International Journal of Advanced Manufacturing Technology* 96 (9–12) (2018), 3149–3159. 10.1007/s00170-018-1772-9

[21] P. M. Bhatt, Process Planning for Robotic Additive Manufacturing, Ph.D Thesis, (2022), https://www.proquest.com/openview/1a58998e9b5ca1e430ff59258bda2333/1?pq-origsite=gscholar&cbl=18750&diss=y

[22] T. S. Tamir, G. Xiong, X. Dong et al., Design and optimization of a control framework for robot-assisted additive manufacturing based on the stewart platform, *Int. J. Control Autom. Syst.* 20 (2022), 968–982. 10.1007/s12555-021-0058-4

[23] G. Ma, G. Zhao, Z. Li et al., Optimization strategies for robotic additive and subtractive manufacturing of large and high thin-walled aluminum structures, *Int J Adv Manuf Technol* 101 (2019), 1275–1292. 10.1007/s00170-018-3009-3

[24] D. Ding, Z. Pan, D. Cuiuri, H. Li, A multi-bead overlapping model for robotic wire and arc additive manufacturing (WAAM). *Robot Comput Integr Manuf* 31 (2015), 101–110.

[25] Y. Li, Y. Sun, Q. Han, G. Zhang, I. Horvath, Enhanced beads' overlapping model for wire and arc additive manufacturing of multi-layer multi-bead metallic parts, *J Mater Process Technol* 252 (2018), 838–848.

[26] Y. Chen, Y. He, H. Chen, H. Zhang, S. Chen, Effect of weave frequency and amplitude on temperature field in weaving welding process. *Int J Adv Manuf Technol* 75 (5–8) (2014), 803–813.

[27] A. Yaseer, H. Chen, Machine learning based layer roughness modeling in robotic additive manufacturing, *Journal of Manufacturing Processes* 70 (2021), 543–552.

[28] Chen, H., A. Yaseer, Y. Zhang, Top Surface Roughness Modeling for Robotic Wire Arc Additive Manufacturing, *Journal of Manufacturing and Materials Processing* 6 (2) (2022), 39. 10.3390/jmmp6020039

[29] D. Ding, C. Shen, Z. Pan, D. Cuiuri, H. Li, N. Larkin, S. van Duin, Towards an automated robotic arc-welding-based additive manufacturing system from CAD to finished part, *Computer-Aided Design* 73 (2016), 66–75. 10.1016/j.cad.2015.12.003

[30] B. Fabric, T. Schork, A. Menges Autonomous robotic additive manufacturing through distributed model-free deep reinforcement learning in computational design environments. *Constr Robot* 6 (2022), 15–37. 10.1007/s41693-022-00069-0

[31] S. Amarjyoti Deep reinforcement learning for robotic manipulation—The state of the art, (2017). http://arxiv.org/pdf/1701.08878v1

[32] G. D. Goh, S. L. Sing, W. Y. Yeong, A review on machine learning in 3D printing: applications, potential, and challenges. *Artif Intell Rev* 54 (1) (2021), 63–94. 10.1007/s10462-020-09876-9

[33] Z. Li, Z. Zhang, J. Shi, D. Wu Prediction of surface roughness in extrusion-based additive manufacturing with machine learning. *Robot Comput Integr Manuf* 57 (2019), 488–495.

[34] A. Koeppe, C. A. Hernandez Padilla, M. Voshage, J. H. Schleifenbaum, B. Markert, Efficient numerical modeling of 3D-printed lattice-cell structures using neural networks, *Manuf Lett* 15 (2018), 147–150.

[35] G. Williams, N. A. Meisel, T. W. Simpson, C. McComb, Design repository effectiveness for 3D convolutional neural networks: Application to additive manufacturing, *J Mech Des* 141 (11) (2019), e4044199.

[36] X. Li, X. Jia, Q. Yang, J. Lee (2020). Quality analysis in metal additive manufacturing with deep learning. *Journal of Intell Manuf, 31, 2003–2017* (2020).

[37] O. Davtalab, A. Kazemian, X. Yuan et al., Automated inspection in robotic additive manufacturing using deep learning for layer deformation detection. *Journal of Intell Manuf* 33 (2022), 771–784. 10.1007/s10845-020-01684-w

[38] H. Lin, B. Li, Automated defect inspection of LED chip using deep convolutional neural network. *Journal of Intell Manuf* 30 (2019), 2525–2534.

[39] K. Castelli, A. M. A. Zaki, A. Y. Balakrishnappa, M. Carnevale, H. Giberti, A path planning method for robotic additive manufacturing, in 2021 3rd International Congress on Human-Computer Interaction, Optimization and Robotic Applications (HORA), (2021), pp. 1–5. 10.1109/HORA52670.2021.9461380

[40] S.-H. Suh, S. K. Kang, D.-H. Chung, I. Stroud, *Theory and Design of CNC Systems*, London: Springer, (2008).

[41] F. M. Ribeiro, J. N. Pires, A. S. Azar, Implementation of a robot control architecture for additive manufacturing applications, *Industrial Robot* 46 (1) (2019), 73–82.

[42] Y. Ding, R. Dwivedi, R. Kovacevic, Process planning for 8-axis robotized laser-based direct metal deposition system: A case on building revolved part, *Robot Comput Integr Manuf* 44 (2017), 67–76. 10.1016/j.rcim.2016.08.008

[43] I. E. Yigit, I. Lazoglu Spherical slicing method and its application on robotic additive manufacturing. *Prog Addit Manuf* 5 (2020), 387–394. 10.1007/s40964-020-00135-5

[44] https://patentscope.wipo.int/search/en/result.jsf?_vid=P12-L6WFLZ-18738. Accessed on 13 August 2021.

7 Smart Materials for Smart Manufacturing

Bhavna, Aryan Boora, Supriya Sehrawat, Priya, and Surender Duhan

Advanced Sensors Lab, Department of Physics, Deen-Bandhu Chhotu Ram University of Science and Technology, Murthal, Sonipat, Haryana, (INDIA)

CONTENTS

DOI: 10.1201/9781003333760-7

7.1 INTRODUCTION

Nanoparticles and other passive nanostructures were the main focus of the first generation of nanomaterials. The second generation of nanomaterials brought about the creation of active nanostructures. However, the third generation was suggested by nano-systems such as hierarchical 3D structures [1]. The development of intelligent materials and buildings is possible with smart materials. It is still difficult to create nanomaterials with appropriate functional characteristics for biomedical applications, specifically to improve treatments. A system of multifunctional components that can perform **sensing, control, and actuation** is referred to as a smart structure. In the past, these "smart materials" were frequently described as ones that could react quickly to their environment. The notion of the novel smart materials has since been expanded to include any material that may respond to stimuli from outside sources and provide a new class of useful characteristics. Additional categories of stimuli include light, temperature, electromagnetic fields, stress, pressure, pH, and others. Smart materials are particularly intriguing to be used in a variety of medical applications, such as release of the drug in a controlled manner, the treatment of various diseases, biosensors, etc. because of their controlled capacities [2].

The limits of contemporary medicine have been greatly expanded by the development of smart biomaterials. Due to being rich in their distinct properties such as good response to the unobservable changes occurring in the environment, such class of materials related to medical fields can function as an **"on-off"** switch for several applications lying in the range of nanometer to macrometer scale. For instance, smart biomaterials have several uses in tissue engineering because they can direct stem cells to improve tissue regeneration [3]. Additionally, the use of biomaterials that are responsive to the stimulus in conjunction with additional manufacturing structures results in the introduction of intelligent structures that resembles closely to the realistic tissues [4]. Therefore, advancements in smart and novel biomaterials have the potential to significantly impact biomedical engineering.

Historical manufacturing companies based on vertically optimised businesses, practices, market share, and competitiveness are being replaced by companies that are far more responsible across the entire value chain to customer demands for customized products and dynamic markets, as well as highly sustainable environment, reduced energy consumption, and free of incidents. These companies are also accelerating the adoption of new technologies and products [5]. Agile production and innovation paired with dramatically higher productivity serve as engines for competitiveness and reinvestment rather than just cost reduction [6]. Opportunities that go far beyond lowering market volatility are created when priority is over the issues such as lower energy consumption, agility, ease of productivity, and environmental sustainability. Innovation, time to market, and a quicker more thorough investigation of the trade sector are all directly impacted by agility. The emergence of new information technology infrastructure for the synthesis of smart materials are driven by these developments.

7.2 SHAPE MEMORY MATERIALS

When a certain stimulus is given, shape memory materials (SMMs) have the ability to regain their previous shape and size after suffering a huge and seemingly plastic distortion. This phenomenon is known as the **shape memory effect** (SME). Under specific circumstances, superelasticity or viscoelasticity are also frequently seen in polymers or metals [7]. The SME can be used in a variety of industries, including medicine and aerospace engineering (for example, in deployable structures and morphing wings) (e.g., in stents and filters). The purpose of this chapter is to review the most recent developments in SMMs (including shape memory-metals, polymers, ceramics, and composites). These newly emerging SMM types will be discussed in detail in a later section.

7.2.1 Shape Memory Alloys

The SME was discovered in an Au-Cd alloy in 1932, but it wasn't until 1971 that the Naval Ordnance Laboratories in the USA2 saw a strain in the nickel-titanium alloy that was recoverable significantly and thus the appeal of this phenomenon became clear. Today, a vast variety of SMAs have been synthesized such as solid-state-based films, and even the foam-based shapes are also being produced.

But strictly three of them, such as nickel-titanium-based, copper-based (Cu-Al-Ni and Cu-Zn-Al), and iron-based alloy systems, are currently of greater commercial significance. NiTi, CuAlNi, and CuZnAl SMAs have been systematically compared in terms of a number of performance indices relevant to engineering applications [8]. The present trend toward micro- and even nano-electro-mechanical systems (MEMS) has led to the production of thin-film SMAs, primarily based on NiTi.

7.2.2 Shape Memory Polymers

Engineering-wise, it is far simpler to tune the material properties of polymers than it is for metals and alloys. Additionally, polymers have historically had substantially cheaper costs (both for the materials and for processing) [9]. Numerous SMPs with prosperous properties have been developed and are listed in the well-documented list of the literature; yet, newly explored ones continue to be created on a weekly, if not daily, basis. Along with the advantages already described, SMPs also have the advantages of being substantially lighter, having recoverable strain that is at least an order larger than that of SMAs, and being able to be active for the shape recovery phase of various stimuli, including multiple stimuli being applied at a particular time [10]. Different than that from heat, two such stimuli include light (ultraviolet and infrared region) and chemical (moisture, solvent, and pH change). A lot of SMPs are also naturally biocompatible.

7.2.3 Shape Memory Composites

As we've seen, SMPs can be prepared to acquire the specific characteristics that may be useful for a particular application. However, it takes practice and a solid foundation (professional knowledge and experience). If the characteristics of SMA/SMP are familiar to us, we may handle shape memory composites (SMCs), which incorporate at least one type of SMM, either shape memory alloy or shape memory polymer, as one of the elements, with ease [11]. More phenomena and newly discovered properties can be achieved in SMCs through careful design and the incorporation of one or more additional mechanisms (such as elastic buckling70).

7.2.4 Shape Memory Hybrids

Even for those with only a basic knowledge in science or engineering, shape memory hybrid (SMH) is a more approachable and adaptable method. Conventional materials are used to construct SMHs (properties are well-known and/or simple to locate, but all without the SME as an individual) [12]. As a result, an SMM can be designed by our own selves in order to accomplish the necessary function(s) in a specific application. SMPs are also similar to a great extent as are SMHs and similarly built on a duality of the domain system, where the first domain (the elastic part domain) is constantly elastic, whereas the second (the transition domain) has the ability to dramatically modify its rigidity in response to the correct stimuli. However, any potential or chemical linkage between the elastic domain and transition domains must be considered while choosing the domains for SMHs [13].

7.3 PIEZOELECTRIC, CHROMO-ACTIVE, PHOTOACTIVE, MAGNETO-RHEOLOGICAL, ELECTROSTRICTIVE, AND SELF-HEALING SMART MATERIALS

The majority of smart materials mimic natural biological materials, which can change one or more properties in response to external stimuli (such as mechanical, thermal, electrical, photic, auditory, magnetic, and chemical ones) in order to adapt to their surroundings. Smart materials that are inspired by biological systems are currently a significant area of study in the field of material science. Since it is challenging for a homogeneous material to perform various functions, it is typically combined in a specific way with materials that perform the functions of sensing, actuation, control, etc. to create a unique composite with a variety of properties. The developed smart material is multileveled with several domains; each domain bearing its unique properties and their microstructures, and there is a coupling effect observed between various domains that causes the smart material to respond to external stimuli in a complex manner.

7.3.1 Piezoelectric and Electrostrictive Materials

In 1880, two brothers named Jacques (1856–1941) and Pierre (1859–1906) Curie discovered the piezoelectric effect. A piezoelectric material shape gets changed as an electric current passes through them, and when their shape is altered, they also produce an electric current [14].

Due to their knowledge with pyroelectricity, a phenomenon that has existed for many millennia, the Curie brothers were intrigued by the study of piezoelectricity. Pyroelectricity is the propensity for some materials to produce the flow of current whenever the heating effect is produced in them. As depicted in Figure 7.1, applying pressure in one direction to some crystals results in developing a scalar electric potential (voltage) at the opposite face of that crystal that is normal (perpendicular) to the direction of the applied pressure. Numerous substances, including quartz crystals, topaz, tourmaline, and Rochelle salts, are discovered to show this

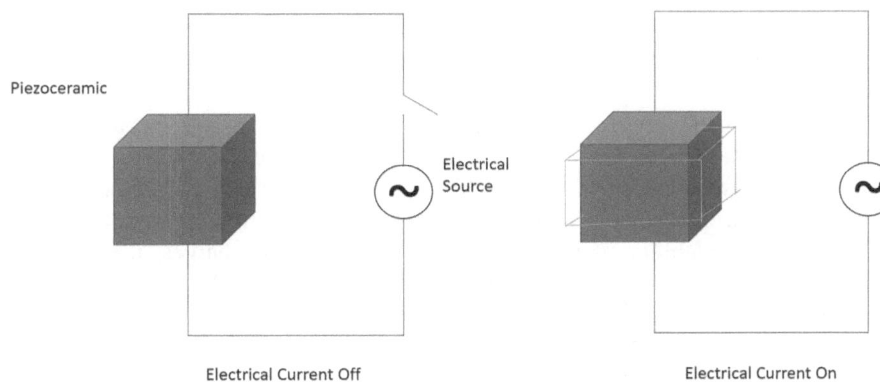

FIGURE 7.1 Piezoelectric effect.

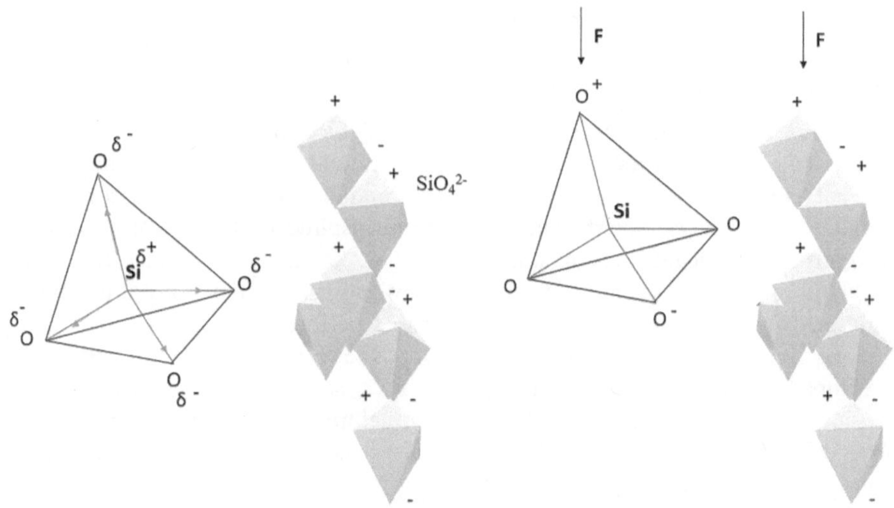

FIGURE 7.2 Atomic mechanism of the piezoelectric effect. Variance of charge distribution in a quartz molecule after the applied force per unit area.

effect (sodium-potassium tartrate tetrahydrate). Currently, it is known that roughly 20 kinds of crystals exhibit piezoelectricity, among which also lie a variety of organic substances like bone, tissue, and collagen.

Take quartz, for instance, which is made up of extended chains of silicate (SiO_4^{2-}) groups organized in a helical manner (see schematic below, Figure 7.2).

The silicon atom is located in the core of each silicate group, whereas the four vertices of a tetrahedron are occupied by the oxygen atom. The tetrahedron is non-polar overall, with its outer surface being uniformly more negative in contrast to the inside surface, despite each silicon-oxygen link being polar. As shown in Figure 7.2, pressure exerted on a single tetrahedral unit separates the charges within it, making one side of the tetrahedron more positive and the other more negative. Whenever the polarization process of tetrahedron starts to occur, adjacent units arrange themselves in a complementary pattern with positive and negative ends facing in the opposite directions, even if adjacent tetrahedra are not attached to one another (as they are in some silicates).

These silica and oxygen anions exhibit no general separation of electrical charges, and pressure has no electrical effect on such crystals. However, in quartz, the position of adjacent tetrahedral units on the helix is fixed. The positive and negative ends of every tetrahedron become polarized in the same way, pointing in the same direction. Due to the availability of mobile positive and negative charges, the crystal can conduct an electric current in this conformation.

Twentyof the 32 crystal classes that are now understood by science display the piezoelectric action. These crystal classes share an asymmetric shape that allows electrical charges to separate when pressure is applied. However, the effect's strength varies significantly among distinct crystal kinds; therefore, piezoelectric effects only have practical relevance in a limited subset of all scenarios.

For instance, in 1942 and 1944, researchers discovered that although the substance barium titanate ($BaTiO_3$) was ceramic but it still exhibits a potential to piezoelectric effects. Barium titanate is a material that can be used to create piezoelectric sensors with a piezoelectric constant that is nearly hundred times greater than that of quartz [15]. Such sensors were used for the first time in 1947 during the production of phonograph pickups (needles). Later, they were employed in a variety of other gadgets that include detectors for vibrations, sonar systems, ignition systems, microphones usable under water (hydrophones), tiny sensitive microphones, and relays.

The ceramic substance lead zirconate titanate ($PbTiZrO_3$), sometimes known as PZT, exhibits piezoelectric properties, which were first identified in 1954 by American chemist Bernard Jaffe (1916–1986) [16]. PZT is the most widely used piezoelectric material right now and is currently commercially available in a number of forms. Barium titanate among the class of ceramics is now predominantly utilized in the production of electrical capacitors.

There are several uses for piezoelectric materials in the modern world. Systems including intruder alarms, remote-controlled devices, record players, mirrors with deformable surfaces, ultrasonic motors, "smart skis," waste gas generators, and hydrophones are a few of these. The use of sensors in auto airbag systems is one of the most widespread applications. A front-end collision with another item may cause the release of an airbag from the dashboard. Whenever the chemical compound consisting of sodium azide (NaN_3) breaks down into metal ion of sodium and the nitrogen gas further inflates the airbag:

$$2NaN_3(S) \rightarrow 2Na(s) + 3N_2(g)$$

The airbag's ability to inflate hinges on a relatively complicated procedure [17]. A set of sensors positioned throughout the front of the car in various locations serve as the initial step in the procedure. These sensors give an electrical signal to the processing unit when they identify a collision with enough force, and the processing unit subsequently instructs the airbag to inflate. In most cases, this entire process lasts no longer than 50 milliseconds.

7.3.2 Chromoactive Materials

Chromoactive materials change color in response to an outside stimulation.

- **Photochromic materials**. These have color changes reversibly when exposed to sunshine or other light with a high UV component. In the dark, these materials are invisible. Advertising posters, T-shirts, shoes, laces, bags, brochures, invisible ink security, and document detection are some of their applications.
- **Thermochromic materials**. These have hue changes in response to temperature. They provide a very broad range of applications because they let you choose the color and temperature range. Moreover, depending on the pigment used, they may change color temporarily or permanently as

the temperature rises. Typically, they are made of semiconductor-related substances [18]. Its applications include labeling, signaling, temperature-cold chain control, safety pipelines and conduits, hazardous materials, and everyday items like microwave containers, pans, hot plates, glasses as well as artworks, textiles, and buildings.

- **Electroactive materials.** These have oxidation states that vary as a result of the application of an external potential difference, altering their absorption spectra and, typically, their color. Automotive anti-glare rearview mirrors, smart windows, and screens are just a few of their uses.

7.3.3 Photoactive Materials

Unlike the earlier ones, photoactive materials can release light in response to various outside stimuli. We differentiate the following kinds within this group:

- **Electroluminescent materials**. These materials are electrically connected to produce light emission. This makes it possible to pretty precisely control how it operates using electronics.
- **Photoluminescent materials.** When exposed to a specific wavelength, typically in the ultraviolet (UV) range, the photoluminescent materials in this instance are capable of emitting light.

7.3.4 Magnetorheological Materials

Magnetorheological (MR) materials can be viewed as a sort of bioinspired smart material due to how easily an external magnetic field may be used to affect their viscoelastic properties. Magnetorheological materials can be generally categorized into gels, fluids, and elastomers depending on the class of carrier matrix used and their physical state when the applied magnetic field is turned off [19]. Practical MR fluid-based devices primarily include of dampers, buffers, clutches, artificial muscles, and other similar components. Several dampers that use MR fluid as their working medium can be found to use MR fluid in a variety of applications. In addition, MR fluid has a wide range of potential applications in the biomedical, precision machining, thermal conduction, and sound transmission domains.

7.3.4.1 Magnetorheological Material Development

The first magneto-sensitive smart material to be created was magnetorheological (MR) fluid, which is a particulate suspension created by combining additives, non-magnetic fluid, and ferromagnetic fillers that are smaller than a micrometer. The MR fluid swiftly transforms from a fluid with a Newtonian-like behavior to a semi-solid state with the application of an external magnetic field (within several microseconds). Through magnetic interaction, the randomly distributed magnetic fillers are reorganized to create chain-like homogeneously ordered microstructures [20]. Additionally, the degree of order in the microstructure has an impact on the magnetic field intensity. In other words, a larger magnetic field will result in a microstructure that is more organized and in phase with the direction of the applied external magnetic field.

7.3.5 Self-Healing Materials

A type of intelligent material known as self-healing material has the capacity to repair structural flaws on its own. This power may fix harm that a machine sustained from repeated use.

The ability of self-healing materials to detect and "autonomously" repair damage has drawn the attention of the research community in recent years [21]. Every year, numerous initiatives are made with the goal of creating various self-healing systems and integrating them into large-scale manufacturing with the best possible property-cost relationship.

7.4 SMART BIOMATERIALS

7.4.1 Smart Hydrogels

Scientists have worked hard over many years to build hydrogels by modifying their chemical and physical properties, dubbed "smart" hydrogels. Depending on the water supply and their surroundings, conventional hydrogels can change the way they swell and deflate. However, in reaction to external stimuli, stimuli-responsive hydrogels can significantly alter their swelling performance, mechanical strength, sol-gel transition, permeability, and network topology. What's more intriguing is that a number of smart hydrogels can react to specific molecules, such as antigens and enzymes or even cells or tissues, without the need for an artificial stimulus [22]. There are numerous possible biological uses for smart hydrogels. They can be crucial in tissue engineering, biomedical applications (like smart valves), injectable systems, and biomedical applications (like surfaces that respond to stimuli to control the adherence of cells).

7.4.2 Smart Nanomaterials

Numerous nanomaterials have been created since the advent of nanotechnology and have been employed successfully in a variety of applications. Nanoparticles and other passive nanostructures were the focus of the first generation of nanomaterials. Active nanostructures were introduced in the second generation of nanomaterials [23,24]. Nanosystems, such as hierarchical 3D structures, on the other hand, signaled the third generation. The fourth generation of nanomaterials is thought to be represented by molecular nanosystems. The application of innovative stimulus-responsive nanomaterials has rapidly increased in recent years in a variety of medical sectors, including engineering of tissues, biological sensors, antimicrobials, bioimaging, various types of cancer therapies, controlled delivery of drug and genes. Spherical, core-shell, and hollow nanoparticles are examples of isotropic shape nano-objects that have an ability to change their shapes, sizes, and colous consistently.

7.4.3 Smart Bioconjugates

Combining peptides, proteins, cells, and DNA with smart polymers is a cutting-edge method for giving inert polymers new value, surprising capabilities, and inventive features. In fact, when they are coupled to certain biomolecular enzymes, liposomes, plasmid vectors, and polymers based on stimuli-response can provide switchable solubilization/precipitation for these biomolecules. These new smart biomaterials show behaviors and characteristics that can be changed by adding outside inputs. As a result, the development of smart conjugation quickly found use in the field of medicine for a variety of purposes, including molecular switching, drug/gene delivery, bio-separations, and immunoassays [25]. Although it may be difficult to synthesize well-defined conjugates, the development of contemporary controlled/"living" polymerization techniques, which are tolerant to a wide range of functional groups, have resulted in the widespread use of protein-reactive initiators for creating well-defined polymers. This method is quicker and makes polymers that can be coupled to proteins without further transformation possible. Additionally, the combination of non-natural protein engineering and controlled radical polymerization can lead to exceptional control over polymer conjugation, producing intriguing bioconjugates for a variety of applications [26].

7.5 BICOMPONENT FIBERS

A unique category of synthetic fibers/filaments comprised of two distinct polymers is known as bicomponent fibers. Several configurations are possible, but out of them the most taken into consideration are side-by-side, core-sheath, mixed polymer, and islands in the sea effect [27]. A bicomponent fiber is created by extruding from a single spinneret two or more polymers of various physical (e.g., average molecular weight, crystallinity) and/or chemical (e.g., composition, additives) natures into a single fiber. This kind of fabric is popular in the carpet industry since each polymer shrinks at a different temperature.

It is feasible to create fibers with many polymeric components using a spinneret design. The polymers in this configuration remain as distinct regions within the fiber rather than being mixed together. There are many other bicomponent designs, but the most common ones are discussed below (Figure 7.3). Microfibers with exceptional elasticity and softness are frequently produced by fibrillating the segmented pie and islands-in-the-sea designs. The 4DG fiber is a high surface area fiber made for applications like particle capture and wicking.

When compared to typical fibers, bicomponent fibers can have unusual physical and aesthetically pleasing characteristics, making them a high-value commodity. It is

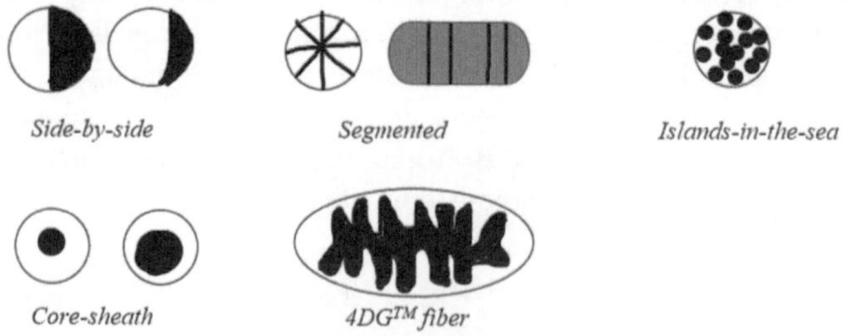

FIGURE 7.3 Common types of bicomponent fiber.

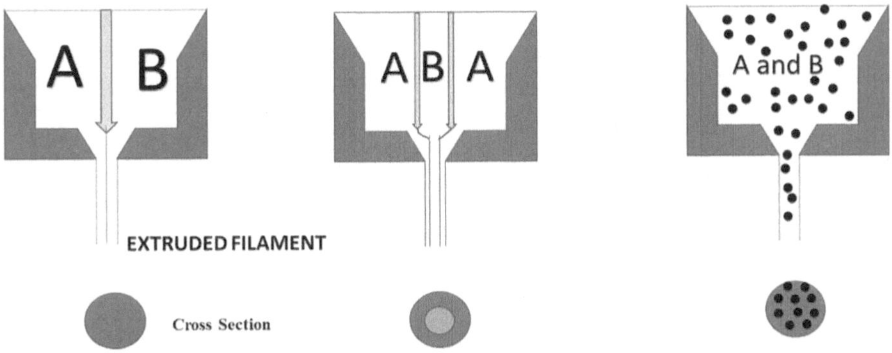

FIGURE 7.4 The cross-section and the bicomponent spinning principles.

frequently done to mix the properties of polymers or to make use of variations in a feature, like a melting point. For instance, bicomponent astroturf made of a polyamide core and a polyethylene sheath maintains the core's resilience while lowering the likelihood of friction burn through the polyethylene. Although not common, the process for creating such bicomponent materials is still young.

Bicomponent fibers are those filaments that can be extruded from two or more components made of different polymers.

Two components are split into two more or less different regions along the length of a bicomponent fiber.

Figure 7.4 shows a few different cross-sectional forms of bicomponent fibers.

7.5.1 Types of Bicomponent Fibers

The terms "composite," "conjugate," and "hetro" are sometimes used to describe bicomponent fibers. Each type of bicomponent fiber has unique characteristics and uses. According to the component distribution within the fiber cross-section area, these can be categorized into the following groups:

1. Core-sheath (C/S),
2. Side-by-side (S/S),
3. Segmented-pie (orange),
4. Islands in-the-sea (I/S) and
5. Polymer blends

These are the bicomponent fiber varieties that are used most frequently.

7.5.1.1 Production Methods for Bicomponent Fibers

The most popular technique for producing commercial synthetic fibers is melt spinning. One of the most intriguing advances in the world of synthetic fibers is the change of fiber morphology by bicomponent (conjugated) spinning, a trend in polymer melt spinning. Bicomponent fibers undergo standard melt-drawing procedures in a manner similar to that of traditional synthetic fibers. The primary goal of bicomponent melt spinning is to make use of characteristics that are not present in either polymer alone. By combining the beneficial mechanical, physical, or chemical qualities of two polymers into one fiber, the potential applications are increased.

7.5.1.2 There Are Three Categories under Which Side-by-Side Component Fiber Production Can Be Divided

1. In the first group, the two materials—solutions or melts—are delivered directly into the spinneret orifices where they combine to form bicomponent fibers at or close to the orifices.
2. In the second group, the two fiber bicomponents are molded into a multi-layered structure (either flat sheets or concentric cylinders) and supplied to the spinneret without turbulence, with the rows of orifices in the spinneret positioned to intersect the interfaces of the various layers of polymer.
3. In the third group, the two components are likewise combined into a non-turbulent layer structure (such as a mixed stream) and fed to the spinnerets; however, no attempt is made to line up the rows of orifices on the spinnerets with the component interfaces, leading to bicomponent spinnerets.

These bicomponent fibers can be used to create bonded fabrics, upholstery fabrics, floor coverings, high crease-resistant fabrics, etc.

7.5.1.3 Aftertreatment of Bicomponent Fibers

In order to boost strength, texturize yarns, or crimp and cut fibers for a staple fiber or wet-laid process, fibers typically go through a variety of processing procedures. To prevent shrinking, heat setting is frequently used to crystallize the fibers. It should be clear that hot air flow at a low temperature and slow speed must be used to attain the desired temperature instead of a hot contacting surface.

A segmented-pie technique is typically used to create splittable fibers. It is preferred that splitting be simple, and hollow segmented-pie fibers may help with this. However, the fibers must not split during the fiber melt spinning, drawing, crimping, or carding processes since this would seriously hinder these operations. Although there shouldn't be much intermolecular diffusion at the interface, the polymer melts should be compatible. In order to allow deformation during the crimping process with the least amount of force, the polymers should likewise exhibit identical drawing behavior and extensibility. Steam aids in crimping but can also lead to splitting and shrinking.

Fabrics made from splittable fibers can be knit, woven, or nonwoven. The mechanical force used to brush, needle, or treat woven and knitted textiles with water jets separates the segments. The fabric's binding of the filaments permits a rigorous treatment while preventing the mechanically caused splitting from spreading.

If the two fiber components have different shrinkage behaviors, heat treatment using air or infrared heating are alternative ways to separate the material. Hot water, sodium hydroxide, caustic soda, and benzyl alcohol solutions are used to split chemicals with the help of ultrasonic force at the end. Hot drawing and heat setting of bicomponent fiber are more difficult than single-component fiber.

7.5.2 Application/Uses of Bicomponent Fibers

Bicomponent technology has changed numerous fiber applications. Products now weigh less, are more durable, and are easier to use. In order to address issues and satisfy consumer expectations, this kind of fiber is anticipated to become a cutting-edge material for a variety of uses, including hygiene, textiles, automotive, home furnishings, and many others. Bicomponent staple fibers have historically been one of the most important fibers in nonwovens. The key applications for bicomponent fibers are listed below.

7.5.2.1 Fibers Used in Nonwovens as Bonding Components

A fiber web contains thermo-bondable core-sheath bicomponent fibers which are treated by passing hot air through it. Bicomponent fiber is used to create nonwoven fabrics with a delicate touch that can be used to make diapers and other hygiene items.

7.5.2.2 Microfibers

The synthetic leather known as Alcantara® is one of the original, prosperous, and most useful component in microfiber products. The fiber used to make it is a bicomponent. Applying mechanical force to segregate the various segments of segmented-pie fibers is a solvent-free option to producing microfibers.

7.5.2.3 Fibers with Special Cross Sections

The logotype fiber is one of a kind where the islands polymer has a special hue or has a different dyeability from the matrix, and is a unique property of the islands-in-the-sea technology (e.g., PA vs. PET).

7.5.2.4 High-Performance Fibers

Fiber reinforcing of concrete can be a more cost-effective option than traditional steel bar reinforcement. Polyolefin-based bicomponent fibers have been utilized successfully in order to improve the mechanical qualities of concrete because they are rich in tensile strength and bear a greater modulus of rigidity inside the core. Their nanoparticles and other additives inside the sheath include a well-structured fiber surface.

7.5.2.5 Functional Surface Fibers

Core-sheath fibers provide the opportunity to alter the surface while maintaining the bulk intact. Most commercially available bicomponent fibers have a sheath that melts at a low temperature and are binder fibers.

7.5.2.6 Fibers for Fully Thermoplastic Fiber-Reinforced Composites

Fully thermoplastic fiber-reinforced composites can be made with islands-in-the-sea and core-sheath bicomponent fibers.

7.5.2.7 Shape Memory Fibers

A composite material with shape memory properties can be created by combining two polymers with differing phase transition temperatures. Bicomponent spinning offers an advantageous engineering foundation for shape memory composites or blends.

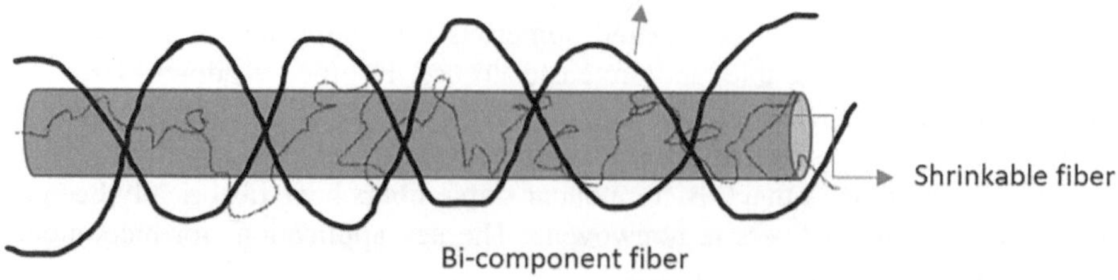

FIGURE 7.5 Shrinkable, non-shrinkable, and bicomponent fibers are used to make bulky acrylic yarn.

7.5.2.8 Polymer Optical Fibers

For textile applications, cyclic olefin polymer (COP)-based bicomponent melt-spun fibers with a tetrafluoroethylene-hexafluoropropylene-vinylidene fluoride outer layer are employed (THV).

7.5.2.9 Use of Bicomponent Fiber in Manufacturing Bulk Yarn for Knitting

Similar to what is seen in the case of wool, steam soaking and drying of acrylic bicomponent fibers causes them to crimp in both directions. In actuality, the morphological structure of wool fiber was studied in order to generate acrylic bicomponent fibers. Due to the presence of two components, namely ortho and para-cortex units in the cross section, the fiber exhibits a bilateral structural asymmetry, which is largely responsible for the characteristic hand and bulkiness of wool fibers [28]. The typical method for creating acrylic bicomponent fiber is to spin fibers from acrylic copolymers with various longitudinal shrinkage properties.

When bicomponent fiber is heated, a three-dimensional crimp is created as a result of differential shrinkage that develops in the fiber itself. As a result of the fiber's three-dimensional crimp, the yarn has better dimensional stability, a very excellent feel, and high elasticity and resilience. Figure 7.5 depicts the crimp arrangement of acrylic bicomponent and monocomponent fiber used to create bulky hand knitting yarns.

7.6 FUNCTIONALLY GRADED MATERIALS

The concept of structural gradients was introduced in 1972. In order to mimic the design and behavior of naturalistic materials like trees of bamboo, bones, teeth, etc., functionally graded materials (FGM) were first proposed for composites and polymeric materials. FGM was initially used in Japan in 1984 when a space shuttle was being developed. The goal was to gradually alter compositions to survive a severe temperature gradient of 1,000°C in order to produce a body from a material with superior mechanical and thermal capabilities [29]. The historical progression from purest form of metal to functionally upgraded metals is shown in Figure 7.6. When compared to traditional alloys and composite materials, FGMs have a number of advantages. FGMs introduce tools for regulating how a material reacts to

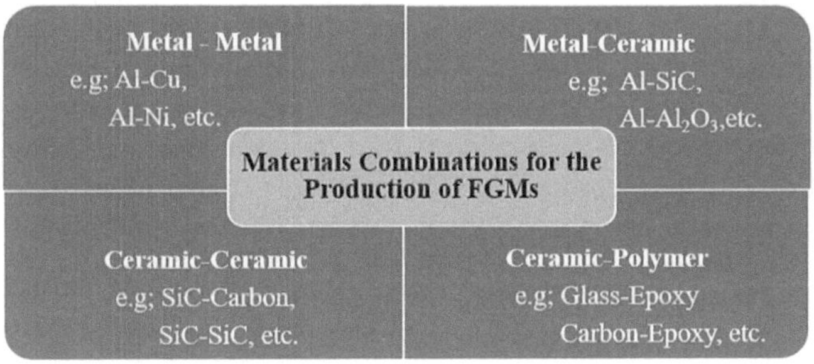

FIGURE 7.6 Material combinations for the production of FGMs.

deformation, dynamic loading, corrosion, and wear. Additionally, they offer the chance to profit from various material systems, such as metals and ceramics. Additionally, some FGMs are more suitable as bone replacements due to their biocompatibility. FGMs are also capable of acting as a thermal barrier, as well as coatings with good scratch resistance and low residual stress. FGMs can also be utilized to glue two incompatible materials together with great strength. In contrast to typical composites, a single scattered ingredient or phase can be used to create a single FGM, whereas a double FGM requires more than one constituent or phase [30]. Depending upon the change observed in the distribution density between the active constituents and various phases of the matrix, the continuous gradient is always obtained. Researchers initially categorized FGMs as conventional composite materials based on the constituent combinations they used. It is feasible to create FGMs using a wide variety of material combinations. Different classifications emerged over time as a result of the expansion of applications and technology used to generate FGMs at various scales. Six traditional classification criteria—state during processing, FGM structure, FGM type, nature of FGM gradient, key dimensions, and field of FGM categorize the FGMs. These characteristics or classifications enable the constant description of the manufactured FGM. These classifications, however, are not very helpful to FGM industrial producers in choosing the right fabrication technology that satisfies both the technical (e.g., shape complexity, accuracy, and minimum residual stress) and economic needs of the sector (e.g., productivity, minimum energy, minimum cost, lower environmental impact).

The classifications that will be presented in this study are intended to offer some guidance for industrial manufacturers looking for the best fabrication technique that satisfies both technical and financial requirements.

7.6.1 Production Method for FGMs

FGM production techniques involve centrifugal casting, powder metallurgy, plasma coating, chemical and physical vapor deposition (CVD/PVD), lamination, and infiltration procedures in addition to the family of solid freeform fabrication (SFF) or additive manufacturing (AM) and its subcategories. Modern AM methods can use a variety of materials, including biological material for inkjet printing and micro extrusion, polymer material for FDM and SLA, and metallic material for

LENS and DMD. There are a lot of publications that describe the specifics of the various manufacturing processes and talk about their technicalities, benefits, limitations, applications, and research trends. It is obvious that experimental mechanical characterization (especially tensile and hardness), wear rate prediction, or thermal characteristics evaluation comprise the majority of study activity [30]. Numerical simulation of FGMs is being considered by very few research organizations. The high level of complexity associated with the modeling of the various components and their characteristics, the modeling of interfaces, and the progressive change in structure may be the cause of this.

7.7 SMART NANOMATERIALS

Smart materials can adjust their characteristics in response to the precise stimulant. It is said that certain stimuli can cause certain materials to alter their form, denseness, appearance, color, elasticity, stiffness, and strength when needed. These "smart materials" were once widely described as having fast environmental responses.

Smart materials definition has been broadened to include those materials that can respond to stimuli from outside sources and display a new class of functional characteristics. In Figure 7.7, several nanomaterials are depicted schematically along with the relevant stimulating agents (Figure 7.7). Temperature variations, different light wavelengths, force, tension, an electromagnetic field, amount of substance, etc. can all act as specific stimuli, and the results can include color, heat, hyperthermia, magnetic deformation, etc. These materials can alter their own characteristics in response to particular stimuli, including change in surface area, volume, permeability, solubility, amount, shape, optical, and mechanical properties between other nanostructured materials. In general, the majority of smart materials show five distinguishing traits: directness, self-activation, immediateness, transience, and selection. When a stimulus first appears, a material is said to be immediate if it can react fast. However, a material can also be transient if it can adapt to various environments and have different characteristics relative to each environment. Among these materials, some have unique internal features that they can develop or induce on their own and are known as self-actuating. Directness refers to the output being created at the location of the input, indicating that this reaction is localized. As selectiveness is a repeatable and

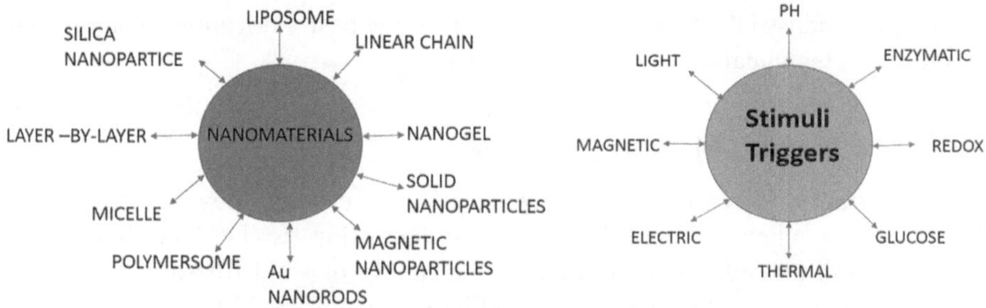

FIGURE 7.7 Distinct nanomaterial classes. Particular stimulating agents.

predictable aspect of the reaction, an individual environmental condition will result in a singular and consistent material's response.

7.8 SMART METALS, POLYMERS, CERAMICS, AND COMPOSITES

7.8.1 Smart Metals: Alloys with Brains

Metals in the history have always been employed extensively due to their high structural strength and associated low weight. Their appeal is also greatly influenced by their capacity to be bent into various shapes. A different class of metals, known as "smart" metals, go one step further by having the ability to "remember" a shape and return to it. What makes these metals so special when we have all bent paperclips and then straightened them out? Imagine bending the paperclip, and then watching it automatically bend back without any assistance. These shape memory alloys, commonly referred to as smart metals, are able to perform this. In a science fiction universe, this could be the discovery that people have long feared: the one that paves the way for the emergence of cruel self-repairing robots and may even be the beginning of the end for mankind. Our society, however, is considerably more open-minded about the technology and has accepted the numerous uses for its special qualities.

7.8.1.1 History of Smart Metals

Shape memory alloys, commonly referred to as smart metals, were the subject of the first known research study in 1932. Scientist Lander found the alloy to have what appeared to be elastic qualities while researching an alloy, or mixture of distinct metallic elements, made of gold and cadmium. Later in 1938, scientists Greninger and Mooradian studying a copper-zinc alloy discovered that they could change the metal's malleability by lowering and raising its temperature. In the 1940s and 1950s, when several studies were published describing the elastic and thermal properties of the alloys, smart metals became more and more popular. In 1961, scientist Buehler found that an alloy made of half nickel and half titanium, subsequently known as Nitinol, has shape memory qualities while conducting research for the U.S. Navy. Unintentionally, a bent portion of the Nitinol alloy was heated with a pipe lighter during a lab meeting, which is how this discovery was made. Everyone was shocked when the piece spontaneously returned to its former shape. Shape memory alloys first saw widespread use in 1971 when they were employed to seal the joint between lengths of pipe. Since then, shape memory alloys have gained popularity and are frequently employed for their distinctive qualities, and new uses are constantly being found.

7.8.1.2 How They Work

Pseudoelasticity and the shape memory effect are two critical characteristics of shape memory alloys that contribute to their unique behavior. An object's elastic properties refer to its capacity to stretch and adapt under stress. Shape memory alloys can change shapes elastically, but they are different from other elastic materials in that they do not maintain their shape over time. They are described as pseudo-elastic because the shape changes brought on by stress are reversible, preventing them from being truly elastic. Additionally, these alloys have the ability to recall their prior structural configurations and can go back to them under the right circumstances.

Phase changes are how these alloys produce their pseudo-elastic characteristic. People frequently visualize the gas, liquid, or solid phases while thinking of phase shifts. But shape memory alloys alternate between many types of solid phases. This is conceivable because, despite the fact that metals remain solids throughout, their internal particles are rearranged into various configurations between the solid phases. The two natural phases of all smart metals are **austenite** (rigid and homogeneous) and **martensite** (malleable) [31].

Metals are more malleable, or more easily bent and shaped, during the martensite phase. We shall use the definition of a metal's internal structure as being composed of several layers of molecules to demonstrate how this phase functions. Both martensite+ and martensite- type layers make up the molecular structure during the martensite phase. Although these two types are comparable, they differ significantly in that their molecules are organized in opposing directions. The metals are in a phase known as twinned martensite when they are composed of an equal number of layers of martensite+ and martensite- (see Figure 7.8b). The individual layers will adopt either a martensite+ or martensite- type, depending on the direction and orientation of the applied stress, if the metals are stretched or bent during this phase. The metals go through a phase change known as deformed martensite due to this stress during the twinned martensite phase (see Figure 7.8c). Thus, the metals in the

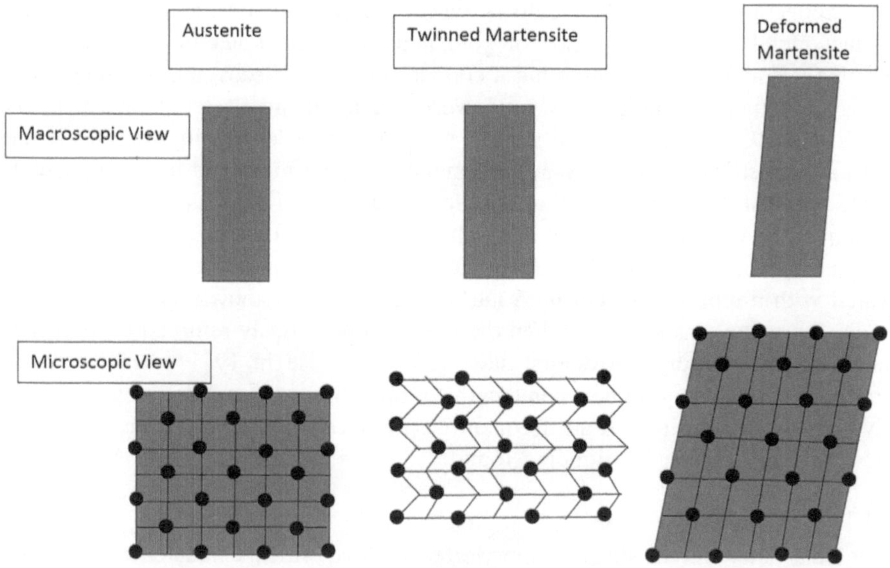

FIGURE 7.8 Images of the metal structure at different magnifications. (a) The cubic-structured austenite phase. (b) The phase of twinned martensite before any stress. consists of a mixture of martensite+ and martensite-. (c) Deformed martensite phase following application of a single stress.

martensite phase are able to achieve supposedly elastic qualities as a result of the metal's ability to divide between two types of layers.

Compared to the martensite phase, the austenite phase is harder and more stable, and its particle arrangement is particularly well organized and cubic (See Figure 7.8a). These are the "remembered" forms of the metals at this stage. The metals can correct any deformations they may have acquired during their softer martensite phase because of how strongly they are attracted to uniformity in this phase when they transition there. The metals can "remember" the shape during the austenite phase thanks to their capacity to rearrange their molecules into a rigid cubic crystal structure during this phase, which results in the shape memory phenomenon.

The temperature mostly determines the form the metals assume. The phases of the metals are austenite at higher temperatures and martensite at lower temperatures. The needed temperature differential for the majority of alloys is not very great, and for some alloys, it can even be as low as 10°C. Metals distorted in the martensite phase will regain their shape after heat treatment in the austenite phase. The physical characteristics of the metals appear to remain unchanged after cooling. However, they again become soft and flexible due to a rearrangement of their molecular structures into the twinned martensite phase. Shape memory alloys differ from conventional metals due to the transition between having discrete bending layers in the martensite phase and having a rigid cubic crystal structure in the austenite phase.

7.8.2 Smart Polymers

Smart polymers are high-efficiency materials that adjust themselves in accordance to the various environmental changes. These types of materials may react in different ways, including altering their colour or transparency, turning conductive or water-permeable, or altering their structure, or being responsive to a range of conditions, for example, a chemical compound's pH, chemical composition, light's wavelength or intensity, or an electromagnetic field (shape memory polymers). Typically, little environmental changes are enough to cause significant changes in the characteristics of the polymer.

7.8.2.1 Applications

Both highly specialized applications and commonplace items use smart polymers. They are used as sensors and actuators in the development of hydrogels, environment-friendly packaging, artificial muscles, and to a considerable extent, bioengineering. One illustration is a polymer that can be utilized for medication administration and changes conformation in response to pH variation. Another is a humidity-sensitive material that is used in dressings for self-adaptive injuries that autonomously control the balance of moisture within and outside the injury.

Smart polymers are special and powerful because of their nonlinear reaction. A very modest stimulation has the power to significantly alter form and properties. After the change takes place, there is no more change, resulting in an expected all-or-nothing reaction that is uniformly distributed throughout the polymer. Smart polymers may modify their structure, rheology, or water retention capabilities in accordance to minute changes in pH, surface charge, temperature, or other stimuli [31].

The performance of smart polymers is also influenced by the intrinsic features of polymers in general. The combined effect of individual monomer unit modifications, which would be insufficient on their own, determines how strongly each molecule reacts to stimuli. And if there are large number of these ineffectual reactions, they produce a sizable force that propels organic activities. The development of the polymer is directly tied to the pharmacy sector. This sector is dominated by polymers, whose improvements benefit entire populations throughout the globe. The human body operates as a complex device responding to chemical messengers. Polymers are used in the technology of medication delivery to regulate the release of medicinal substances in timed doses. Polymers can identify molecules and regulate subcellular distribution.

7.8.2.2 Stimuli

A lower analytical solution temperature phase transformation occurs in several polymer systems in response to temperature. With a transition temperature of about 33°C, poly(N-isopropylacrylamide) is one of these polymers that has received more research. There are a number of homologous N-alkyl acrylamides that exhibit LCST behavior as well, with the hydrophobic side chain's length influencing the transition temperature. In water, these polymers turn hydrophobic once they are above their transition temperature. This behavior is thought to be driven by entropy.

7.8.2.3 Classification and Chemistry

Targeted drug delivery is currently smart polymers' most popular use in biomedicine. Since the introduction of medications with timed release, scientists have had to figure out how to transport drugs to specific locations inside the body besides having them first break down in the very acidic nature of the stomach. The avoidance of damaging effects on good bone and tissue is another crucial factor. Smart polymers can be used to regulate drug release up until the delivery system reaches the intended target, according to research. Depending on the stimulus, it could be chemical or physiological.

Different features of linear and matrix smart polymers are dependent on side chains and chemical functional groups. These groups may respond to changes in light, electric or magnetic fields, temperature, pH, or ionic strength. Non-covalent linkages that are cross-linked reversibly in some polymers allow them to break and regenerate in response to environmental factors. Some nanoparticle polymers, such as dendrimers and fullerenes, that have been used for medication delivery were developed with the help of nanotechnology. Lactic acid polymers have been used in traditional drug encapsulation. Lattice-like matrices that carry the drug of relevance incorporated or trapped between the polymer filaments have been created in more recent study.

Drugs are released from smart polymer matrices through chemical or physiological reactions that change their structure; frequently, a hydrolysis event causes bonds to be severed, releasing the medication as the matrix disintegrates into biodegradable parts. The usage of synthetic polymers such polyanhydrides, polyesters, polyacrylic acids, poly (methyl methacrylates), and polyurethanes has replaced the use of natural polymers. The polymers that disintegrate the fastest are those that are hydrophilic, amorphous, low in molecular weight, and include heteroatoms (atoms other than carbon). By changing these qualities, scientists can modify the pace of breakdown and hence manage the rate of drug distribution.

Two distinct polymers are grafted together to form a graft-and-block copolymer. For various combinations of polymers with different reactive groups, there are already a number of patents. A smart polymer structure gains a new dimension from the product's dual component qualities, which may be advantageous in some applications. When hydrophobic and hydrophilic polymers are cross-linked, micelle-like structures are created that can protect drugs throughout delivery in an aqueous medium until environmental factors at the target area cause both polymers to break down simultaneously.

A graft-and-block strategy could be helpful for resolving issues with the use of polyacrylic acid, a popular bioadhesive polymer (PAAc). Although PAAc sticks to mucosal surfaces, it swells and degrades quickly at pH 7.4, which causes the medications that are trapped in its matrix to be released quickly. It is possible to lengthen the residence period and decrease the release of the medication by combining PAAc with a different polymer that is less susceptible to pH alterations at neutral levels. This will boost bioavailability and effectiveness.

Polymer networks called hydrogels inflate or collapse in shifting aqueous conditions but do not dissolve in water. The fact that they are recyclable or reusable makes them ideal for phase separation in biotechnology. Hydrogels are being used in research to find new techniques to control the flow or catch and release of target substances. The distribution and release of medications into particular tissues has been made possible by the development of highly tailored hydrogels. Because of their exceptional absorbency and bioadhesive qualities, hydrogels composed of PAAc are very popular.

The procedure of immobilising enzymes in hydrogels is comparatively well-established. Similar techniques can be used with hydrogels and reversibly cross-linked polymer networks in biological systems where the target molecule itself initiates the reaction and drug release. As an alternative, the end result of an enzyme reaction might activate or inhibit the response. This is frequently accomplished by adding an enzyme, receptor, or antibody that binds to the target molecule to the hydrogel. Once bonded, a chemical process occurs that causes the hydrogel to respond. Oxygen, recognized by oxidoreductase enzymes, or a pH-sensing reaction can all serve as the trigger. Entrapment of both insulin and glucose oxidase in a hydrogel that responds to pH is an illustration of the latter. When there is glucose present, the enzyme produces gluconic acid, which causes the hydrogel to release insulin.

Enzyme stability and swift kinetics are two requirements for this technology to function well (rapid reaction to the trigger and quick recovery once the trigger has been removed). The usage of same kind of smart polymers that can sense changes in blood glucose levels and activate the generation or release of insulin has been tested in a number of type 1 diabetes research strategies [32]. Similar hydrogels could also be used in a variety of other ailments and diseases as medication delivery systems.

7.8.2.4 Other Applications

Drug delivery is just one application for smart polymers. They are ideally suited for bioseparations due to their characteristics [33]. For many years, immunoassays and physical and affinity separations have both used conjugated systems. Precipitate formation, which results from microscopic changes in the polymer structure, can be employed to help separate trapped proteins from solution. These systems function by forming a bioconjugate between a protein or other molecule that needs to be isolated from a mixture and a polymer, which precipitates together when the environment changes. The desired conjugate component is isolated from the remaining mixture by removing the precipitate from the medium [34]. Hydrogels are extremely helpful in these operations since the elimination of this component from the conjugate relies on the recovery of the polymer and a return to its initial condition.

Recombinant proteins with integrated polymer binding sites close to ligand or cell binding sites are a different method for using smart polymers to influence biological events. Based on a number of stimuli, such as temperature and light, this technology has been utilized to modulate ligand and cell binding activity.

Self-adaptive wound dressing technology relies heavily on smart polymers. The dressing design uses patented, highly absorbent synthetic smart polymers that are immobilized in a matrix of three-dimensional fibers. Hydrogel is incorporated into the material's core to provide further hydration capabilities.

Due to the ability of polymers to detect and adapt to changes in humidity and water level in all parts of the injury simultaneously, the dressing functions by autonomously and rapidly transitioning from absorption to hydration. The smart polymer action ensures the active coordinated responsiveness of the dressing substance to changes in all directions of the wound in order to maintain the ideal moist healing conditions always.

7.8.2.5 Future Applications

It's been proposed that learning and self-correcting polymers could be created in the future. Even if this might be a very remote possibility, other, more likely applications seem to be on the horizon. One of these is the notion of smart lavatories that analyze urine and assist in the diagnosis of health issues. Intelligent irrigation systems have also been suggested in environmental biotechnology. A system that regulates fertilizer dependent on moisture in the soil, alkalinity, and nutritional composition would be tremendously helpful. There are also numerous inventive methods for developing self-regulating targeted drug delivery systems that take into account their particular cellular environment.

It is obvious that there are several potential problems with the usage of smart polymers in biomedicine. The most worrisome aspect of manufactured chemicals is their propensity to be poisonous or incompatible with the body, especially metabolites and breakdown products. Smart polymers have great potential in biotech and medical applications if these difficulties can be solved.

7.8.3 Smart Ceramics

Materials made from ultrafine particles are known as smart ceramics. They are inorganic, non-metallic materials that are shaped and then heated to a high temperature to form nitride or carbide materials, such as clay, that are hard, brittle, heat-resistant, and corrosion-resistant [35]. The preparation, inspection, and evaluation of ceramic microstructures is known as ceramicography. Any ceramic substance's crystalline structure and chemical makeup determine its physical qualities. In structural and building materials as well as textile fabrics, mechanical qualities play a crucial role.

Ceramics that are "smart" can work as actuators and sensors simultaneously. Materials that are "passively smart" adapt to changing environments in a positive way on their own, in contrast to actively smart materials, which incorporate a feedback mechanism that enables them to both notice the change and launch a proper reaction through an actuator circuit [36].

"Biomimetics," the mimicking of biological functions in engineered materials, is one method utilized to instill intelligence into materials. Composite ferroelectrics designed after the lateral line and swim bladder of fish are used to illustrate the idea. "Very smart" materials have the ability to "learn" by altering their feature coefficients in response to the environment, in addition to sensing and responding.

From a single smallest unit cell to the overall complex organism, the human body, life is considered to involve motion. The ability to move, alter, and adapt is what distinguishes living things from inanimate objects. The idea behind investigating smart materials is to develop higher-order novel materials and their nano-structures by giving them the mandatory life functions of sensing, actuating, controlling, and intelligence. Functional materials for a range of engineering applications, smart materials are a component of smart systems. Smart diabetes treatment systems that dispense insulin and monitor blood sugar change its shape with response to the air pressure and flight speed. Smart toilets that use urine analysis as a health problem early warning system. Intelligent buildings in space use vibration cancellation technologies to account for the lack of gravity and avoid metal fatigue. A smart toy like "Altered Beast" is a character awakens from the dead and must learn to survive in a harsh new age residence with electrochromic windows that can adjust the amount of heat and light coming in and out in response to environmental changes and human activity. Rackets used in the games such as tennis have quick internal adjustments for delicate drop shots and overhead smashes. Smart dental braces manufactured from shape-memory metals and smart muscle implants created from rubbery gels have a good response to the provided electric fields. Navy ships and submarines with intelligent hulls and propulsion systems that can identify flow noise, reduce turbulence, and avoid discovery. methods for intelligently detecting and removing harmful pollutants from water. For autos, a number of intelligent systems have already been created, but there will be many more. Many contemporary materials have been expertly crafted to perform beneficial tasks, and in some situations, we can label them "smart" with justification.

7.8.3.1 Passive Smartness

A material with passive intelligence can react to its surroundings in a helpful way. In contrast to actively smart materials, passively smart materials do not use external fields, pressures, or feedback mechanisms to improve their behavior. Many passively smart materials include built-in self-healing capabilities or stand-by phenomena that allow them to tolerate abrupt changes in their environment. An excellent example are the crack-suppressing processes seen in partly stabilised zirconia. Here, the main processes that can produce compressive stresses at the fracture tip are the tetragonal-monoclinic low phase change and ferro elastic twin wall motion. Similar to how structural composites used in aeroplanes or machinable glass-ceramics improve toughness by fiber pull-out or repeated crack-branching. Other passively intelligent materials include PTC thermistors and ceramic varistors. A zinc oxide varistor loses the majority of its electrical resistance when hit by high-voltage lightning, allowing the current to flow directly to ground. The resistance shift serves as a stand-by protection phenomena and is reversible. The extremely nonlinear I-V relationship of varistors can also be repaired by a self-repair technique that involves repeatedly applying voltage pulses to the device. The ferroelectric phase transition near 130 0C causes the electrical resistance of barium titanate PTC thermistors to significantly increase. The thermistor can stop current surges thanks to the increase in resistance, serving once more as a safety measure. The extremely nonlinear behavior of the varistor's R(V) behavior and the PTC thermistor's R(T) behavior, which function as standby protection phenomena, make ceramics intelligent in a passive state.

7.8.3.2 Active Smartness

In an analogy to the human body, sensing and actuating capabilities can also be used to describe a smart ceramic. A smart material made of ceramic reacts to a change in the environment by detecting it and employing a feedback system. It serves as both an actuator and a sensor. Examples include modern suspension systems in the cars that are electrically controlled and use piezoelectric ceramic-based sensors and actuators and vibration dampening systems for space platforms. The basic working principle of an actively smart material is demonstrated by the piezoelectric pachinko machine. In Japan, pachinko parlors with a large number of vertical pinball machines are particularly common. The PZT multilayer stacks used in the Nippon Denso engineers' Piezoelectric Pachinko game serve as both sensors and actuators. The force of impact caused when a ball strikes the stack produces a piezoelectric voltage. The voltage pulse causes the actuator stack to react by way of a feedback loop. A series of similar actions causes the stack to quickly expand, tossing the ball out of the hole, and causes the ball to go up a spiral ramp. It eventually drops into a hole and resumes its spiral ascent.

Similar logic guides the operation of a tape head positioner for video created by Piezoelectric Products, Inc. The sensing and actuation operations of the positioner are divided by a segmented electrode pattern in a bilaminate bender manufactured of tape-cast PZT ceramic (Figure 7.9). Through the feedback mechanism, the positioning electrode voltage is changed from the voltage across the sensing electrode. After that, the cantilevered bimorph curved toward the video tape track. Near the tape head, articulating electrode placement and sensing running at 450 Hz aid in maintaining the head's perpendicularity to the track.

7.8.4 Smart Composite

Any new composite smart material's ultimate goal is to combine two or more single smart materials in order to leverage synergistically the greatest qualities of each of

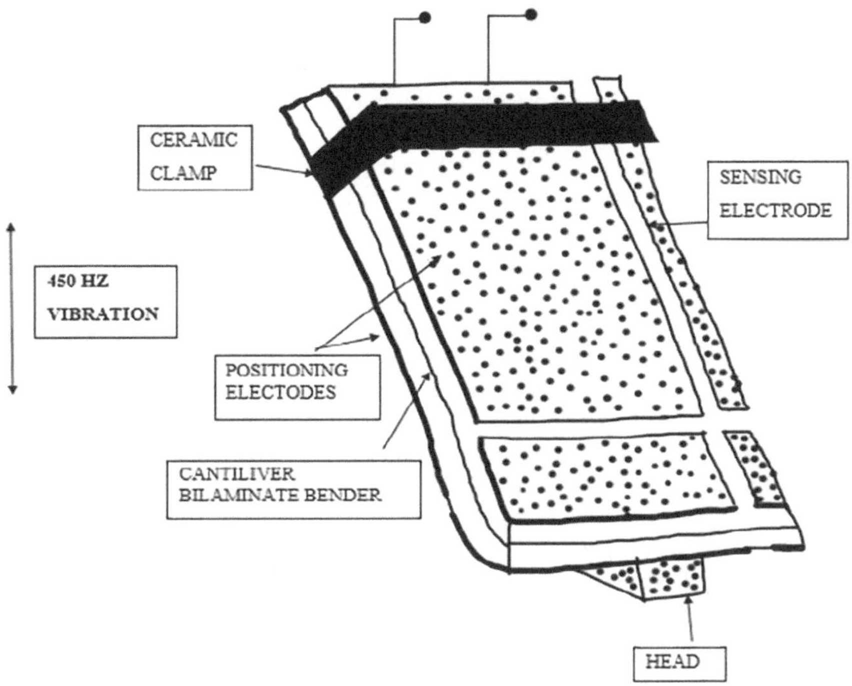

FIGURE 7.9 PZT bimorph video tape head positioner with separated electrodes for the sensor and actuator.

their unique components [32]. Because of this, "smart composite materials" come extremely close to meeting all of the aforementioned requirements. Their benefits and flexibility in meeting the aforementioned design criteria have resulted in an abundance of innovative products. Basically, there are two kinds:

- A composite material manufactured by humans that is uniquely designed. This substance is meant to strengthen or stiffen anything. The examples that follow will give you some background on the subject. One product is created by mixing a solid with tiny glass, ceramic, or polymer spheres, another by combining a solid with boron or silicon, and a third product is created by turning polymer, glass, and certain metals into durable foams. In syntactic foams, a resin is mechanically mixed with bubbles to create a composite material. To form laminated composite or sandwich construction, these foams can be joined with thin panels or outer skins. Another illustration of assessing the data, coming to the correct conclusion, and deciding what has to be done.
- Data Instructions (motor nerves): This component's role is to return the structure's decisions and related instructions to its constituent parts.
- Action Devices (muscles): This component's function is to initiate an action by activating the controlling units/devices.

7.9 RESULTS AND DISCUSSION

Smart or intelligent materials have the potential of designing intelligent materials and structures. the variety of potential products, including those with fresh designs, quality assurance, multifunctionality, security features, and field values that are applied externally, such as pressure, heat, and electromagnetic fields [37]. Smart composite materials ingrained with **micro electro mechanical systems (MEMS),** fiber optics, actuators, sensors, sound, vibration, and shape control, product health or lifetime monitoring, intelligent processing, cure monitoring, active and passive controls, artificial organs, self-repair (healing), novel indicating devices, designed magnets, damping flexibility, and pressure distribution are all part of this process [38].

The structural integrity of smart composites with embedded sensors and actuators must not be considerably compromised by the inclusions, which are now much bigger in diameter than the carbon, aramid, or glass-reinforcing fibers, which are typically about 8–10 micrometers in diameter. The reinforcing fibers near optical fiber sensors, which typically have diameters of 100–300 micrometers, are inevitably damaged when they are embedded in composite laminates [33]. The behavior of this disordering depends on the fiber optic's position in relation to neighboring reinforcing plies and the embedded sensor's diameter. If the diameter of sensor is half the thickness of the ply, sensing fibers that are parallel to the local reinforcement, for instance, create the least amount of disruption. A resin-rich area forms locally surrounding the sensor by the deformation of reinforcing fibers that are orthogonal to the sensors. The fiber sensor needs to meet the following requirements in order to be considered:

In order for the sensing technique to be used, the laser light must:

i. cause the least amount of disruption in the placement of reinforcing fibers;
ii. not considerably weaken the composite's mechanical characteristics;
iii. not experience undue attenuation or harm as a result of the embedding procedure; and
iv. incorporate a correct way to insert and eject the laser light inside the system via clampers and wires. Such solutions need to be reliable and work along the manufacturing route.

Before using sensing technology in actual constructions, a number of aspects still need to be thoroughly researched (e.g., space, air, or naval structures).

i. Material aspects cover the impact of the fiber on the properties of the composite material's strength as well as the suitability of the sensors for the composite setting.
ii. Fabrication considerations include determining the best sensor deposition in the composite, maintaining exact placement during production, monitoring, and automating the fabrication process, handling connection difficulties, and constructing and repairing structures affordably.
iii. The type of sensor to use, how to distinguish significant parameters, how to perform assigned measurements to the necessary resolutions, and how to build trustworthy connection schemes and small systems are all examples

of sensing/multiplexing approaches. In comparison to alternative fiber-optic sensing methods, in-line FBG sensors have definite benefits in several areas.

iv. System features include the relationship between strain sensing and planned maintenance, the advancements in artificial intelligence methods necessary to analyze the data, and the potential for system redundancy. Many of the aforementioned elements are crucial when it comes to actuation. In comparison to sensor development, actuator development is less progressed. Despite this, there is a lot of success being had.

7.10 CONCLUSION

Today, the application of intelligent materials and structures is one of the technologies with the greatest potential to boost lifetime effectiveness and dependability. The ultimate goals of research in this field are to comprehend and control any new materials' composition and microstructure, which is crucial for the creation of high-quality smart materials. Superconducting wires, nanostructured materials, and structural ceramics are a few examples of the sophisticated materials that will shape nanotechnology. Smart structures and systems will unquestionably improve our quality of life thanks to new or upgraded materials that can take more heat, lower weight, absorb more sound, reflect more light, and handle vibration and noise better [39].

Although still in their infancy, smart materials systems have a lot to offer, but there are a lot of technological hurdles to be solved. However, if current predictions are accurate, "smart materials" will be used underwater, above ground, in the air, and even in space! The timing of these breakthroughs is still unknown [40].

REFERENCES

[1] Vladimir Pokropivny, Rynno Lohmus, Irina Hussainova, Alex Pokropivny, and Sergey Vlassov. *Introduction to nanomaterials and nanotechnology, volume 1.* Tartu University Press Ukraine, 2007.

[2] Ibrahim Nazem Qader, Kök Mediha, Fethi Dagdelen, and Yıldırım Ay-Dogdu. A review of smart materials: Researches and applications. *El-Cezeri*, 6(3):755–788, 2019.

[3] Gianni Ciofani. *Smart nanoparticles for biomedicine*. Elsevier, 2018.

[4] Igor Galaev and Bo Mattiasson. *Smart polymers: Applications in biotechnology and biomedicine.* CRC Press, 2007.

[5] Jim Davis, Thomas Edgar, Robert Graybill, Prakashan Korambath, Brian Schott, Denise Swink, Jianwu Wang, and Jim Wetzel. Smart manufacturing. *Annual Review of Chemical and Biomolecular Engineering*, 6:141–160, 2015.

[6] Andrew Kusiak. Smart manufacturing. *International Journal of Production Research*, 56(1–2):508–517, 2018.

[7] T Tadaki, K Otsuka, and K Shimizu. Shape memory alloys. *Annual Review of Materials Science*, 18(1):25–45, 1988.

[8] L McDonald Schetky. Shape-memory alloys. *Scientific American*, 241(5):74–83, 1979.

[9] Andreas Lendlein and Steffen Kelch. Shape-memory polymers. *Angewandte Chemie International Edition*, 41(12):2034–2057, 2002.

[10] Marc Behl and Andreas Lendlein. Shape-memory polymers. *Materials Today*, 10(4):20–28, 2007.

[11] Yasubumi Furuya. Design and material evaluation of shape memory composites. *Journal of Intelligent Material Systems and Structures*, 7(3):321–330, 1996.

[12] CC Wang, WM Huang, Z Ding, Y Zhao, and H Purnawali. Cooling- /water-responsive shape memory hybrids. *Composites Science and Technology*, 72(10):1178–1182, 2012.

[13] Li Sun, Wei Min Huang, Zhi Ding, Y Zhao, Chang Chun Wang, Hendra Pur-nawali, and Cheng Tang. Stimulus-responsive shape memory materials: a review. *Materials & Design*, 33:577–640, 2012.

[14] D Damjanovic and RE Newnham. Electrostrictive and piezoelectric materials for actuator applications. *Journal of Intelligent Material Systems and Structures*, 3(2):190–208, 1992.

[15] Kui Chen, Jian Ma, Juan Wu, Xiaoyi Wang, Feng Miao, Yi Huang, Caiyun Shi, Wenjuan Wu, and Bo Wu. Improve piezoelectricity in batio3-based ceramics with large electrostriction coefficient. *Journal of Materials Science: Materials in Electronics*, 31(15):12292–12300, 2020.

[16] Steffen Gloeckner, Rolf Goering, Bernt Goetz, and Andreas Rose. Piezoelectrically driven micro-optical fiber switches. *Optical Engineering*, 37(4):1229–1234, 1998.

[17] P Vinay, SS Venkata, M Hemanth, and A Saiteja. Design and simulation of mems based accelerometer for crash detection and air bags deployment in automobiles. *International Journal of Mechanical Engineering and Technology*, 8(4):424–434, 2017.

[18] Youliang Cheng, Xiaoqiang Zhang, Changqing Fang, Jing Chen, and Zhen Wang. Discoloration mechanism, structures and recent applications of thermochromic materials via different methods: A review. *Journal of Materials Science & Technology*, 34(12):2225–2234, 2018.

[19] Xiaojie Wang and Faramarz Gordaninejad. Magnetorheological materials and their applications. *Intelligent materials*, pages 339–385, 2008.

[20] Mark R Jolly, J David Carlson, and Beth C Munoz. A model of the behaviour of magnetorheological materials. *Smart Materials and Structures*, 5(5):607, 1996.

[21] Martin D Hager, Peter Greil, Christoph Leyens, Sybrand van der Zwaag, and Ulrich S Schubert. Self-healing materials. *Advanced Materials*, 22(47):5424–5430, 2010.

[22] Mitsuhiro Ebara, Yohei Kotsuchibashi, Koichiro Uto, Takao Aoyagi, Young-Jin Kim, Ravin Narain, Naokazu Idota, and John M Hoffman. Smart hydrogels. In *Smart biomaterials*, pages 9–65. Springer, 2014.

[23] Mutsumi Yoshida and Joerg Lahann. Smart nanomaterials. *Acs Nano*, 2(6):1101–1107, 2008.

[24] Narender Ranga, Ekta Poonia, Shivani Jakhar, Ashok K Sharma, Atul Kumar, Sunita Devi, and Surender Duhan. Enhanced antimicrobial properties of bioactive glass using strontium and silver oxide nanocomposites. *Journal of Asian Ceramic Societies*, 7(1):75–81, 2019.

[25] Allan S Hoffman, Patrick S Stayton, Volga Bulmus, Guohua Chen, Jingping Chen, Chuck Cheung, Ashutosh Chilkoti, Zhongli Ding, Liangchang Dong, Robin Fong, et al. Really smart bioconjugates of smart polymers and receptor proteins. *Journal of Biomedical Materials Research*, 52(4):577–586, 2000.

[26] Shukufe Amukarimi, Seeram Ramakrishna, and Masoud Mozafari. Smart biomaterials—A proposed definition and overview of the field. *Current Opinion in Biomedical Engineering*, 19:100311, 2021.

[27] Rudolf Hufenus, Yurong Yan, Martin Dauner, Donggang Yao, and Takeshi Kikutani. Bicomponent fibers. *Hu, Jinlian, Bipin Kumar, and Jing Lu (Eds), Handbook of fibrous materials, Wiley*, pages 281–313, 2020.

[28] CC Chu and LE Lecaroz. Design and in vitro testing of newly made bicomponent knitted fabrics for vascular surgery. In: *Gebelein, C.G. (eds) Advances in Biomedical Polymers*, pages 185–213. Springer, 1987.

[29] Minoo Naebe and Kamyar Shirvanimoghaddam. Functionally graded materials: A review of fabrication and properties. *Applied Materials Today*, 5:223–245, 2016.

[30] Rasheedat M Mahamood, Esther T Akinlabi, Mukul Shukla, and Sisa L Pityana. Functionally graded material: An overview, In Proceedings of the World Congress on Engineering, vol. 3, London, UK. 2012.

[31] María Rosa Aguilar and Julio San Román. *Smart polymers and their applications*. Woodhead Publishing, 2019.

[32] Atul Kumar, Narender Ranga, Surender Duhan, and Rajesh Thakur. In vitro study of aripiprazole loading and releasing efficiency of sba-16. *Journal of Porous Materials*, 27(5):1431–1437, 2020.

[33] Narender Ranga, Suman Gahlyan, and Surender Duhan. Antibacterial efficiency of zn, mg and sr doped bioactive glass for bone tissue engineering. *Journal of Nanoscience and Nanotechnology*, 20(4):2465–2472, 2020.

[34] Atul Kumar, Sunita Devi, Satish Khasa, and Surender Duhan. Biosurfactant as antibiofilm agent. In:*Inamuddin, Charles Oluwaseun Adetunji, and Abdullah M. Asir (Eds) Green sustainable process for chemical and environmental engineering and science*, pages 515–527. Elsevier, 2022.

[35] RE Newnham, QC Xu, S Kumar, and LE Cross. Smart ceramics. *Ferroelectrics*, 102(1):259–266, 1990.

[36] Ajay Kumar Mishra. *Smart ceramics: Preparation, properties, and applications*. CRC Press, 2018.

[37] J Gopi Krishna and JR Thirumal. Application of smart materials in smart structures. , *International Journal of Innovative Research in Science Engineering and Technology*, 4(7), 2015.

[38] Vijay K Tomer and Surender Duhan. Ordered mesoporous ag-doped tio 2/sno 2 nanocomposite based highly sensitive and selective voc sensors. *Journal of Materials Chemistry A*, 4(3):1033–1043, 2016.

[39] Georges Akhras. Smart materials and smart systems for the future. *Canadian Military Journal*, 1(3):25–31, 2000.

[40] R Davidson. Smart composites: Where are they going? *Materials & Design*, 13(2):87–91, 1992.

8 Smart Biomaterials in Industry and Healthcare

Dharmender Kumar and Nidhi Chaubey
Department of Biotechnology, Deenbandhu Chhotu Ram University of Science and Technology, Murthal, Sonipat, Haryana, India

Dinesh Kumar Atal
Department of Biomedical Engineering, Deenbandhu Chhotu Ram University of Science and Technology, Murthal, Sonipat, Haryana, India

CONTENTS

DOI: 10.1201/9781003333760-8

8.1 INTRODUCTION

The term "biomaterials" are used in biomedical applications and is distinct from biological materials like the bone formed from biological systems [1,2]. They are nonviable materials utilized in medical instruments that are intended to correlate with the biological system [1]. Smart biomaterials have the potential to react to the changes occurring in physiological parameters. They also respond to the exogenous stimuli that have an impact on many characteristics of allopathic medicines (modern medicine) [3]. These biomaterials also can be defined as artificial or natural substances that may be used over an interval of time to strengthen, treat, or replace different tissues or organs and the functioning of the body [1,4] or "a synthetic or artificial material used to substitute the parts of the living system or to function or work in close contact with the living tissues" [1,5]. Smart biomaterials have a large range of potential applications from hydrogels for the storage of water to biodegradable materials for the greenhouses, such as in capturing carbons and also working as an antimicrobial packaging [6]. They have distinctive responsive properties towards external (environmental) changes such as pH, light, temperature, and many more. From the material's point of view, biological tissues can consider as a composite made up of nano biomaterials, like—nano-apatite grain, nano muscle fiber, nano-membrane, and many more [1,7]. Nano biomaterials have been focused on in active research owing to their distinct special characteristics and properties, such as the high-surface areas of the nano-sized (1–100 nm) biomaterials or their particles; similarly, polymer-based "smart" biomaterials show predictable and switchable capabilities which can be utilized for various biomedical applications. These applications include gene and drug delivery, tissue engineering, bioimaging, cancer therapy, biosensing, medical implant, wound healing, and in diagnostic systems such as DNA microarrays and proteins [1]. The properties of the biomaterials at the nanoscale are different from that of the properties of biomaterials at the macroscopic scales, which led to the development of nanoscience in wide applications ranging from medical to the clinical scale [8]. As nanobiomaterials are comprised of nano-particles in their core, they are covered by the monolayer of silicas like inert substances or materials. The rapid expansion of nanotechnology has empowered researchers to construct nanomaterials, nanocoating, nanocomposites, and nanofibers for biomedical applications. Hence, the clinical exercises of biomaterial can be enhanced by constructing a nanostructured surface [1]. Liu et al. (2010) have indicated that nano-functionalized surfaces have a promising biological property, whereas the Chitosan-based encapsulation technique that has the emulsifications layer-by-layer self-assembling, ionotropic gelation, hydrogel formation, and film formation can be studied

TABLE 8.1
Different Smart Biomaterials and Their Applications

Types of Smart Biomaterials	External Stimuli	Applications
Poly(N-isopropylacrylamide)	Temperature	Patterned cell, seedings, and co-culture 1561 [10]
Pluronic's (poly(ethylene oxide)-poly (propylene-oxide)-poly(ethylene oxide))	Temperature	Tissue engineering process (new formation of cartilage) [11]
PNIPAm-Arg-Gly-Asp (RGD)	Temperature	Controlling osteoblast adhesion and proliferation (Stile et al. (2003))
Self-assembling peptide	Temperature and pH	Neural tissue engineering [12]
Poly(2-acrylamido-2-methyl propane sulphonic acid-co-N butyl methacrylate)	Electric field	Controlled drug delivery and cells [13]
Self-assembling peptide	Temperature and pH	Peptide (P_{11}-4) supports the primary human dermal fibroblast growth and proliferation [14]
Spiropyran-containing polymers brushes/ graft copolymer	Light	Cell capture and release [15]
Chitosan/Polyethyleneimine (CS/PET) blend	pH	Scaffolds for cellular functioning and cartilage tissue engineering [16–18]

for increasing the encapsulation efficiency, and stability so that they can be used in the healthcare sectors [9]. Various types of smart biomaterials with their applications in various fields have been explained in Table 8.1.

Chitosan is currently the most attractive biopolymer with sustainable qualities and is being used in various applications because of its availability and intrinsic characteristics. They are easily digestible, have non-inflammatory effects, are easily degradable and biocompatible, and also limit bacterial growth. They originated from the chitin taken out from the shells of crustaceans like prawns, crabs, lobsters, etc. These chitins are later on partially or completely deacetylated for the formation of chitosan. Its formation can be seen in Figure 8.1A. These are present in various distinct forms such as powder, foams, gels, and many more. All of these nanofibers based on it are useful in a wide range of applications because of their specific properties like large surface/area-to-volume ratio and high mechanical strength when combined with other polymers. These are also biocompatible and biodegradable; hence, useful in packaging or biomedicinal applications. Because of their versatility, they are used in electrospinning which is a common technique used for the formation of fiber-sized sub-microns. The main principle of these techniques is the production of an induced liquid jet by the electrical fields, as can be seen in Figure 8.1B. Recently for the fabrication of nanofibers, the disruptive technique used most commonly is solution blow spinning (SBS), as can be seen in Figure 8.1C [19].

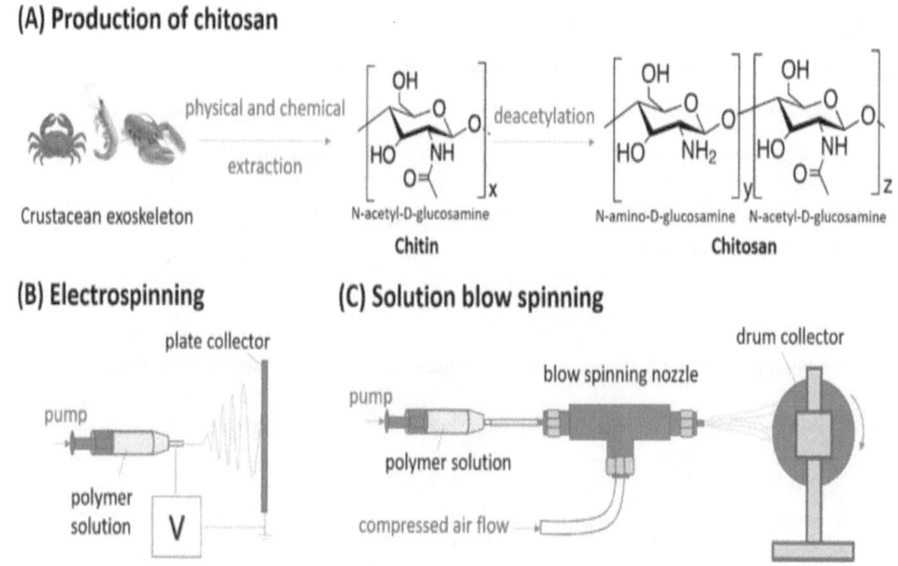

FIGURE 8.1 (A) Production of chitosan as it is made by the deacetylation of chitin extracted from the crustacean's exoskeleton. (B). Electrospinning. (C) Solution blow spinning. (Reproduced from Tian et al. 2021 [19]).

8.2 TYPES OF SMART BIOMATERIALS

The various types of biomaterials with their subtypes are discussed in the following points.

8.2.1 Conventional Biomaterials

They are further subdivided into subtypes such as metals, polymers, composites, and ceramics. These biomaterials have various applications and advantages in different fields such as medical and clinical, etc.

8.2.1.1 Polymers

It is a compound comprised of a very long chain of molecules and every single molecule is formed from the repeating molecule or units that are connected via a bond together. There are hundreds, thousands, or maybe millions of repeating units comprising of a single polymeric molecule or polymer. Many of the polymers are formed from carbon as per the specific characteristic of carbon called catenation and are also considered organic compounds. These are highly preferable materials in life sciences and biological applications. There are some exceptions where they cannot be used, such as in orthopedics as they are having high compatibility and adaptability in biological tissues, molecules, cells, and systems. They are found in a large range of applications, such as in the medical field from the liver and kidney, to the parts of the heart, from dentures to knee joints and hip, from tracheal tubes to facial prostheses, and as sealants and medical adhesive for coating purposes. The polymer-based nanoparticle can be used to prepare an intelligent and smart packaging material with novel characteristics in the packaging industries [20].

8.2.1.2 Metals

Metal is a type of material mainly characterized by its unique properties and features, such as malleability, ductility, thermal conductivity, electrical conductivity, and luster. They consist of metallic elements and their alloys. An alloy is a metal comprised of two or more metallic elements together. While forming an alloy at least one of the elements should be metallic, as it increases and provides higher strength to the compound. Some of the metallics used in forming alloys are cobalt-chromium, stainless steel (316L), pure titanium, etc. Titanium alloys are being used for permanent or long-lasting load-bearing metallic implants because of their fatigue resistance and their large tensile strength. They are involved in a huge range of applications, such as simple wire to the screw to artificial joints such as knees, hips, ankles, and shoulders to the fracture fixation plate, and many more. In addition to that, they are useful in orthopedics as metal implants are used in cardiovascular surgeries, maxillofacial surgeries, and dental implants.

8.2.1.3 Ceramics

These are the materials that are inorganic and comprised of a semimetal/metal with one or more non-metals. They are brittle and hard materials with higher elastic modulus compared to bone. Traditionally, they have very large compressive strengths but have low tensile strengths. They are mostly preferred for cement and dentures, and as a restorative material in dentistry such as crowns, etc. Their applications are less extensive compared to that of metals and polymers. They are sometimes used in joint replacements, augmentation, and in bone-repair surgeries, but to some extent only due to their poor fracture toughness, which limits their use in load-bearing applications.

8.2.1.4 Composites

Composites are a material that is comprised or formed by combining two or more physically distinct or different phases that in turn produce an aggregate property that is distinct from its constituent. These composites have been widely used in the field of dentistry as restorative material or as dental cement for a very long time. They are a mixture of low density by weight and have strength characteristics or properties that make them ideal for making prosthetics such as limbs, etc. The carbon-carbon and the carbon-reinforced composite polymers have attracted researchers, mainly for joint replacement and bone repair, due to their low elastic moduli. However, they do not have shown the biological and mechanical characteristics of the selected biomaterials and the human tissue. These are some uses of conventional biomaterial in the medical sectors with their characteristic properties. Another type of biomaterial is described in the following points [1].

8.2.2 Natural Biomaterials

Biomaterials that originated from animals and plants are termed *natural biomaterials*. They are very similar to the host materials present in their body and they do not cause any toxicity compared to synthetic biomaterials [1]. These natural biomaterials are mainly comprised of polysaccharides that are later on structurally modified by adding the extra functional groups, which in turn provides the additional benefits such as immunogenicity, biodegradability, and bioavailability [21]. There are many different types of natural biomaterials available such as collagen, agarose, alginate, gellan gum, dextran, silk, cellulase, fibrin, and gelatin, etc. [22].

8.2.2.1 Collagen

They are largely used natural biomaterials that have a role in the connective tissues of humans like ligaments, tendon, cartilage, bones, etc. [1]. They are the most common protein found in mammals and mainly consist of approximately 30% of the total mass protein of mammalians. These are important structural proteins found in the extracellular matrix. There are 28 different types formed by different combining quantities of their triple helical domains that are α chains. They do not cause the immunological response and have the domains for the binding of integrins that can increase the cell attachment, growth, and adhesion [22]. Among them, there are the type I, mainly found in skin, tendons, and bones; whereas type II is mainly present in particular cartilages in the joints; and type III is the major part of blood vessels [1]. They remain in the liquid phase at a temperature below 37°C. They can be mixed with other natural biomaterials such as agarose, fibrin, cellulase, etc. to increase their structural integrity and bioactive characteristics [22].

8.2.2.2 Agarose

Agarose is a naturally occurring polysaccharide that is being driven from red seaweeds consisting of repeated disaccharide molecules that are D-galactose. These are a class of the polymer family of carbohydrates and are used in applications of tissue engineering because of their thermoreversible gelling properties and their biocompatibility. Agarose has a similar structure to the extracellular matrix because of its similar macro-molecular property. Its hydrogels are used in tissues like cartilage. Agarose hydrogels, when mixed with 2% of the polycaprolactone, can lead to an increase in their stiffness [23]. Natural agarose does not react with the normal biological human tissues; hence, sometimes it is combined with other natural biomaterials such as fibrin, collagen, cellulase, etc. that in turn increase its supporting ability for the cell survival [22]. Hence, they are widely used in medical and clinical applications for transplants, grafts, etc.

8.2.2.3 Cellulase

Cellulase is most commonly used as plant biomaterial as it is extracted from the plant's cell walls. These are the primary structural unit of the plant cell wall that provides rigidity. They are polysaccharides comprised of (1–4) linked beta-D-glucopyranosyl units joined together. Carboxymethyl cellulose is also written as (CMC); it's a type of cellulose that is soluble in the water and that can be used for modifying or enhancing the viscosities of the other polymers with a low ideal rheologic property. CMC is mixed with the glycol-chitosan hydrogels for creating an ink based on gels that have both cell compatibility and cell stability [24]. They are biocompatible and nanocellulose does not enable the bacterial growth over them, hence making them attractive for applications like dressing wounds and many more. Cellulose nanocrystals form when the high-order chains of cellulose are linked, which can provide high mechanical strengths with shear thinning behavior [22].

8.2.2.4 Fibrin

Fibrinogen is found in the blood as a form of soluble protein and the thrombin enzyme catalyzes the digestion of the fibrinogen into the simpler unit, which is the fibrin monomer. They are insoluble, biodegradable, and biocompatible natural biopolymers or biomaterials with properties or characteristics that can be changed by adjusting the concentration of fibrinogen and the thrombin. They can be mixed with other substances like PCL to modify their properties, which allows the fibrin to have characteristics that imitate soft tissues and hard tissues. They allow communication between the cells because of their nonlinear elasticity also. Hence, they are used in a large range of applications in tissues including skin, heart tissues, neural tissues, and vascularized tissues [1,22].

8.2.3 Nanostructured Biomaterials

Biomaterials like nanofibers, nanocomposites, nanoparticles, and nanosurfaces have various useful applications in tissue repair or regeneration, in gene and drug delivery, in cancer therapeutics, in medical imaging, and some in biomedical engineering. These are the polymer, composite, metal, and ceramics that are made nano-sized or turned into nanostructured biomaterial. There are various methods of making a nanostructure biomaterial. Nano biomaterials formed from the carbon source for e.g., nanotubes of carbon and graphite conserve the internal electrical property and show the excellent biocompatible characteristic requires for biosignaling monitorization [25,26]. Nano-particles mainly form the core of the nano biomaterials. Its shape is mostly spherical but sometimes it can be platelike, cylindrical, and other shapes can also be there. Its core can have multiple layers and can be multifunctional. By using a combination of magnetic and luminescent layers, one can easily detect and manipulate the nanostructured particles. The core of the nanostructured particle is mainly covered by the monolayers. These monolayers are made up of inert materials such as silica. The organic molecule that gets absorbed on the surfaces of the particle may act as biocompatible materials [1], which later on can be used in different medical and clinical applications.

8.3 APPLICATION OF SMART BIOMATERIALS IN DIFFERENT HEALTH SECTORS

The applications of smart biomaterials have been expressed in various forms, as given below.

8.3.1 Clinical Applications

Artificial polymers consist of uniform and less immunogenicity, well-known structures and characteristics, and the dependability of the core material. Polyurethanes foams and polytetra-fluoro-ethylenes (PTFE) are a group of materials with characteristics ranging from very tough, tacky, and soft to hard, brittle ones and viscous, too. The molecular structures of these materials can be changed as per their ability to modify themselves for their desired or required properties. Their processing is easier and cheaper compared to the conventional biomaterials for the advancement of micro- to nano-macro structures for the engineering industries and biomedical industries. These artificial biomaterials are biocompatible; hence, they are used in multiple applications [27]. Some of their major applications include implants and prostheses.

8.3.1.1 Arterial Prostheses

Artificially formed biomaterials with the same elastic and mechanical characteristics as the normal tissues of the body perform the same need and work as the replacement and are cost-effective in comparison to the native amputations of limbs [28]; arterial prosthetics with artificially formed biomaterials offer a better quality of patient's life. Polyethylene terephthalate (PET) and polytetrafluoroethylene (PTFE) are highly hydrophobic polymers and highly crystalline in nature and have been used in various synthetic arterials prosthetics for a very long time [29]. A type of biomaterial called auxetic biomaterial behaves in the same way as the primary arterial materials that not only offer a better replacement for the blood vessel's prosthetics but also overcomes the issue of wall thinning too [30]. Auxetic materials are unique because of their distinct unique deformations in mechanisms to become thicker to respond to blood flow and become thinner as per the favorable environments and decrease the rupturing of the blood vessel prosthesis because of enhanced mechanical characteristics like toughness in fracture. Where else, when the blood flows through the non-auxetic blood vessels results in becoming thinner due to a positive Poisson's ratio [27].

8.3.1.2 Implants

The auxetic structure has potential applications in various implants like in knee, hip, and vertebral disc replacements [31–33] with enhanced mechanical properties in terms of resistance to indentation. They are about three times as resistant compared to synthetic or conventional biomaterials [34,35]. Implants have a higher value of the coefficient of friction or have a cross-section of circular shape observed to have a low distribution of strain and along with that have smaller critical volumes [36]. They are used in the replacements of the femoral components or of a complete hip replacement or complete hip arthroplasty surgery. This material, when enfolded around an aimed material, provides more cushioning to the areas where implants are inserted. Similarly, they improve the fracture toughness along with the indentation resistance, which proved their applications for the implant linings by improving the wear resistances. Implant failures and the implant loosening are the other problems and challenges that are caused because of the stiffness mismatched between the implant materials and the bone [34,35]. So, overcoming the issue of matching the structural implants and the mechanical properties of the architecture and the stiffness with the natural bone is also essential. Nowadays, the potential uses of auxetic structure in the field of biomedical implants have been explained through the 3D modeling assays to evaluate the effective strain and stiffness behavior to the cellular structures such as the periodic, regular, hexagonal and FGA honeycombs with different structures and characteristics and porosities [37]. Some biomaterials involved in implant formation are SiC as it is hemocompatible as well as biocompatible too and are already in use for a long time in the human in-vivo application such as cardiac stents coatings, or as dental implants, and also used for short-term sensors and neural implants [38].

8.3.1.3 Auxetic Stents

For the opening of the blockages, and for providing strength to the arteries wall, and for preventing the elastic recoils following the arterial dilution and a small structural mesh of wires folded in a cylindrical shape termed the cardio-vascular stent, these kinds of structures are inserted into arteria mostly via the balloon angioplasty procedure. Although vascular injuries take place while the stent installment or dispositioning or recognition of stent-like an outer foreign substance triggering neointimal hyperplasia, which causes the reclosing of arteries. Current advances for counteracting the restenosis to employing drug-eluting stents to locally delivering anti-proliferative and immunosuppressants [39]. Over the last few decades, large attention is given to research and development in the improvements of the quality of stents in upgrading the quality of lives of the patients who suffer from arterial blockages. Across many different generations of a stent-like bare metal stent, balloon angioplasty, biosensor able stents, and drug-eluting stents, especially the drug-eluting stent have decreased the early stages and delayed the complication or problems along with the restenosis and stenting rate by 80% [40]. For overcoming the existing risk, many developments toward advancing the long-term efficiency and safety of the stent need to be focused on. Major challenges of recent times come in the development, designing, and manufacturing of such types of stents that are biocompatible and physiologically compatible with strain conditions, mechanical and elastic responses to that of biomaterial formed to be matching with the biological functioning, and mechanical characteristics of the normal human body cells and tissues, therefore minimizing the major threat of stenting for e.g., restenosis, thrombosis, in-stent restenosis, and other early or delayed problems [41–45]. Currently, polyurethane oesophageal auxetic stent hence prevents the tumor from growing [46]. Auxetic material formed from the macrostructures and now extended to the micro- and then to the nanostructured scale have variable biomedical applications. Fabrication and designing of micro-structured auxetic stents along with non-woven nanofibrous drug, and drug delivery systems [47].

8.3.1.4 Auxetic Scaffolds

Regeneration of the bone tissue using the scaffolds is the major focus and receiving the increased interest in the tissue engineering application. The elastic property of the scaffolds is critical as its efficiency in the regenerative tissues and in reducing the inflammatory responses match with the elastic property of the native/local tissues before the implantation. For the development of weight-bearing orthopedics tissue, and for scaffolds for providing the tissues the support so that the force is applied over them, one has to consider the tissues that are undergoing the mechanical strain and stress; their mechanical and elastic properties should be considered too.

In a few applications, an auxetic scaffold is appropriate for the emulation of the local tissues, for the transmitting and in accommodating forces for host tissue sites [48–53]. These scaffolds can be very effective for the proliferation of the chondrocyte with the compressive loads' stimulations [50]. A 3D polyethylene glycol auxetics scaffold has been fabricated using the digital micromirror devices projection printings (DMD-PP) for printing the single-layer construct composed of the cellular pores and structures having different and unique special arrangements, geometries, and deforming mechanisms [51]. While using the biomaterials are based on polyethylene glycol, a multi-layer along with simultaneous, the negative and the positive Poisson's ratio behavior has been fabricated [51] that can be used in various applications in the field of biomedical engineering. Moreover, in the current study, it is found that along with the transitions, the multiple nuclei of the embryonic stem cell (ESC) have auxetic behaviors [54]. It is reported that the phenotype of auxetics in the transition of embryonic stem cell nuclei is derived by the global chromatin decompensations and across the regulation of the molecular turnovers at a differentiating nucleus by the outer external force; hence, authenticity can be a major factor element in the mechanical transduction or mechanotransduction.

8.3.1.5 Dilators

A dilator formed from an auxetic biomaterial used for the opening of the cavity of arteries and other similar blood vessels made it play very different biomedical applications such as heart surgery or coronary angioplasty [55]. The axial expansions put in by the movements of a central-guide wire while using a simple finger–thumb mechanism open the lumens of the arteries; hence, the whole mechanism works similarly to as of a hypodermics syringe. Moreover, the radial expansions can be easily and manually controlled by physicians. These have various advantages over conventional balloon catheters as that in these auxetic expansion membranes there is no need for the inflation of the balloon nor there is any leakage of inflation mediums other than that they have other benefits of auxetic sheaths [56].

8.3.1.6 Auxetic Bandages

In providing protection to the wound from harmful pathogens by using protective fabrics that are having biocompatible and antimicrobial activity should be designed [57]. Designing the auxetic behavior abilities into a structural fabric and foam gives us the potential for using these biomaterials as in the smart-compression bandage and in drug-delivering bandages. The mechanism of this smart bandage formed from auxetic materials is because of shape memory abilities and unique deformation mechanisms that are used under the auxetic effects. The drug-loaded bandages are placed over infected wounds or swellings, which causes auxetic bandages to stretch and deliver the drugs to the wound and, once healing has taken place, the pores will get closed and hence swelling will get reduced; this can show the drug delivery action of the auxetic bandages [58–60]. The increase in porosity and breathability makes wounds heal faster. Recent studies on the smart wound dressings of the auxetic bandages and auxetic foams have been established well [59,60]. Regulating the release of the guest active pharmaceutical ingredients (APIs) form within the microstructure of the host auxetic fibers that are present in the bandages themselves.

These textiles formed from the auxetics are used as a smart compression bandage. Conventionally, the available fibers like knit fabrics etc. are used but now as an alternative, the auxetic fabrics are being prepared by the double helix yarns [3,27,61,62].

8.3.2 Medical Applications

8.3.2.1 In Tissue Engineering

Tissue engineering is comprised of multiple fields where the different fields like medicine, chemistry, biology, material science, and engineering converge to form or develop the solutions that can, promote, restore, or enhance tissue functions [63]. Recent progress in tissue engineering has been focusing on the development of smart biomaterials and their responses toward the external stimuli in terms of the cellular systems that is cells and tissue microenvironments [64]. Smart biomaterials can be designed by focusing on incorporating the specific functional groups in the biomaterials, and by allowing the control over their physical, biological, and chemical properties. This promotes the desired function like enhanced cell adhesion, migration, and growth [65]. Recent approaches of tissue engineering that deal with the dressing of the wounds using the fabrications formed from various smart biomaterial stem cells and the growth factor that has shown increased healing results [66]. Therefore, the synthetic approach employs the fabrication of the smart biomaterials and for that, the two best strategies that are being used are the bioinspired and the natural approach or by a click-based orthogonal method [67]. For tissue engineering, biomaterials should possess cell compatibility, too, rather than chemical and mechanical versatility [68].

8.3.2.2 Hydrogels

Hydrogels are the three-dimensional hydrophilic polymeric network that swells in aqueous environments and is used for various applications in tissue engineering such as scaffolds and as a cellular encapsulations system, and these are the materials that can be modified for mimicking the extracellular matrices or the (EMC) of the native/local tissues [69]. The combination of the distinct specific functional group within hydrogels makes them able to control the chemical, biological, and physical properties and can be introduced often by using the click-based orthogonal method because of their simplicities, efficiency, and ease [67]. The structures of hydrogels can be cross-linked by using reversible or irreversible cross-linking methods [70]. Reversible cross-linking can be done through the physical cross-linking by the thermally induced polymeric chain reaction or entanglement, and by the self-assembling or enabling the hydrogel to go through the structural change in response to the external environment or stimuli [71–73]. A combination of both irreversible and reversible cross-linking's is important. These reversible bondings contribute to self-healing properties, whereas the irreversible bondings contribute or are involved in the maintenance of structural integrity of the smart biomaterials. Moreover, the stability and the degradation of the biomaterials are dependent upon the degree of reversible and irreversible bonding. Hence, biomaterials with only irreversible cross-linking can be limited in respect of the control of the advanced property that is required for its functioning and the biocompatibility. Therefore, this shear thinning system and self-healing property is now being used or applied over the biomaterial that is used for injections or making injectable biomaterials that can be loaded along with the cells, drugs, or biologics, etc. [74]. The injectable smart biomaterials are loaded with cells, bioactive molecules, or drugs that can be used in tissue engineering to promote the healing of damaged tissue.

8.3.2.3 In Medical Devices

Smart biomaterials have the capability to revolutionize the fabrications of medical devices because of their advanced functionalization, design, and development of three-dimensional printing and other processing technology. These smart medical devices are incorporated with the stimuli-responsive biomaterials either on the surfaces of the device or as part of bulk designs of the medical device. The shape-memory polymer is a leading class of smart biomaterials that can have a potential impact on the engineering of medical devices [75]. Polymers with dual, triple, or multiple shapes have been developed and have the potential for enabling multitasking abilities [75–78]. These devices utilize both a chemical and physical approach to grid up the tissue at the body's temperature with their sharp tips that change shape and then release the drug in a localized and controlled sustained manner for the upcoming weeks. Three-dimensional printings enabled the manufacturing of medical devices with different sizes, shapes, and forms [79]. The recent advancement and development in wearable device technologies have been expanded largely to the health sectors [80].

For implantable medical devices like orthopedic implants, accumulators of microbes over the device surfaces and catheters are considered a serious issue and health hazard. Hence, the device surface modifications with biomaterials possess the antimicrobial properties that are capable of the self-cleaning and trigger the release of the antimicrobial agent that can help in minimizing or preventing the device's infections. Therefore, the pH-responsive copolymer coated with the zwitterions is composed of the quaternary amine's monomers and the carboxylic monomers [81]. These copolymer's coatings are found to be a switch between the bacterial adhesives and the bacterial-resistant forms in response to the pH change and allowing the detection of the microorganisms and the removal of the inactivated microorganism [3,82].

8.3.2.4 In Immune Engineering

Immune engineering plays a major role in the maintenance of homeostasis and in the resolution of various diseases in our bodies. The use of biomaterial in the body is recognized as a foreign material, and hence causing a negative effect on the immune system that in turn causes an immune response that can interfere with the transplantation of medical devices or any other drug delivery systems. However, the development of the smart biomaterial solved the understanding between the immune system and the foreign synthetic biomaterials, therefore preventing the immune response in the case of biomaterial-based transplants or implants. On implanted objects, the foreign body responds mainly in two-step processes consisting of firstly the inflammation reaction and after that the wound healing [83], which triggers the deposition of a thick collagen layer covering the implant [83–85], which causes the efficiency of the implant to be compromised in the case of encapsulated living cells. For example, for the treatment of the type-1 diabetes, the pancreatic islet cells are used where the fibrosis surrounds the device and interferes with the nutrient supply and with the exchange of waste products, resulting in cell death [83,86].

The development of immune-interactive smart biomaterials has increased the understanding of the mechanism involved in the immune response, which causes the progress of achieving the designing of smart biomaterials with intrinsic immunogenicity and the variation in shapes, sizes, and the chemistry of the biomaterial that can be used to increase the response towards the vaccines and distinct immunotherapy. Different classes of biomaterials comprise of lipid-modified photoactive or polymer, pH sensitive, and another reactive group, and helped in the delivery of therapeutic to the immune cells [87–90]. Recent research has focused on the activation of the patient's immune system for fighting the diseases like cancer and the therapy is known as cancer immunotherapy. This is done by utilizing the controlled target delivery of the small molecules, nucleic acids, or antibodies. These technologies have emerged as a highly effective treatment for cancer. Smart biomaterials have the potential to promote the immunomodulation for personalized treatment of cancer. The utilization of the antigen-specific T cell in the backpacking of the interleukin proteins and nanogel responses to the change in T cell surface reduction potential result from the antigens' recognition in the tumor microenvironment [91].

Moreover, the smart biomaterials offer a unique opportunity for modernized vaccines. By forming smart biomaterials that can in single injections have the ability to inject the vaccine in the pulsatile fashion maximize efficiency and, hence, solve the issue related to the engineered vaccine that requires repeated injections for the production of the memory in the T cell to recognize the specific antigen [3,92]. Such a strategy for enhanced cancer immunotherapy is using targeted delivery of the chimeric antigen receptor (CAR) T cell.

8.3.2.5 In Drug Delivery

The oligonucleotide-based technology that includes antisense oligonucleotide and mRNA, short interfering RNAs, and the genome-editing CRISPR therapy is the major technique for the development of drugs along with small molecules [93]. Once the drug is developed, their delivery in the targeted cell, tissue, or organ becomes the major issue. To facilitate the effective intracellular deliveries of nucleic acid–based drugs are often engineered to have pH-responsive biomaterials like polymers or lipids containing the ionizable functional groups, zwitterion, or tertiary amines [94,95]. These tertiary amines undergo protonation in weakly acidic pH of the endosomal compartments and in turn helps to facilitate the intracellular releases and limits the toxicity associated along with the cationic carrier [3,96,97]. Polymeric-based nanoparticles are widely used for drug delivery applications because of their physiochemical properties that can be controlled easily, and the chemical functionality of the building blocks forming the polymer can be altered easily. The nanoparticles formed using polymers facilitate controlled drug releases at the point of target sites without having any toxic effect. Some different types of synthetic and natural polymers are used for this purpose such as poly (lactic-co-glycolic acid) (PLGA), poly (N-vinylpyrrolidone) (PVP), poly(e-caprolactone) (PCL), poly (ethylene glycol) (PEG), and polyvinyl alcohol (PVA). These all are biocompatible, biodegradable, and non-toxic. They can be used for controlling the rate of destruction and mainly in the drug release for several factors can be optimized such as molecular weight, nanoparticle size, and functionality of the end group [98]. Other external factors, e.g., temperature and pH, can affect the drug-releasing rates, and thus can be regulated for the development of the stimuli response on the delivery of drug systems. Many studies have shown that copolymers can also be used for both the loading of hydrophilic and hydrophobic drugs together in a single drug delivery system [99] [Figure 8.2].

Nanoparticles formed by using lipids as a building block are divided into solid lipid nanoparticles and liposomes. Liposomes consist of lipid molecules that are hydrophilic and lipophilic and which contain a huge hydrophobic group, for example phospholipids and glycolipids, which spontaneously rearrange once held in watery surroundings for making a vesicle-type structure with an outer hydrophobic lipid bilayer shell and an inner hydrophilic core. Because of their unique organizations, liposomes are used for drug delivery with changing aqueous solubilities degrees, both within the shell and the core. Since they are comprised of the naturally occurring lipid molecules, they are being highly non-immunogenic, biocompatible, and are highly biodegradable in the human body without causing any toxic effects. Therefore, they are largely used in the cancer treatments or therapeutics applications as in the advancement of polymeric chemistry and synthetic chemistry. Researchers have figured out the use of nanoparticles in the delivery of drugs [100,101]. As the nanostructured delivery of drugs (NDDS) is very versatile in function and has the ability to get incorporated into the human body such as having characteristics like water solubility, stability, specific targeting, and also being resistant to reverse drug systems [102,103].

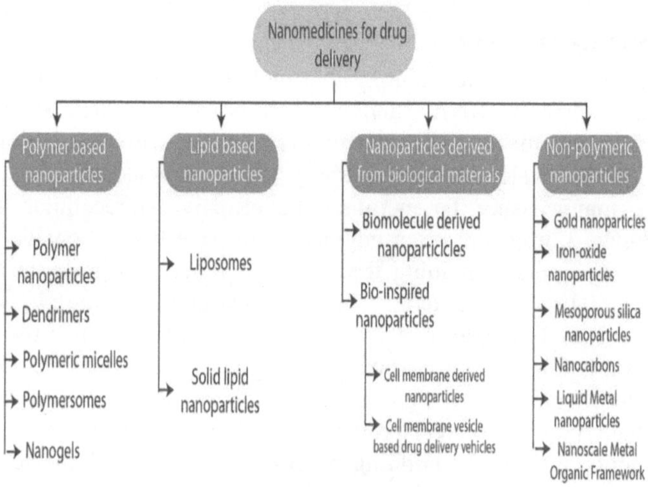

FIGURE 8.2 Use of nanomedicine in drug delivery (Reproduced from Kumar et al. 2021 [100]).

8.4 ADVANTAGES OF BIOMATERIALS

Smart biomaterials have various advantages as they are formed by biocompatible compositions of chemicals so that they can avoid adverse effects on the tissue or in turn do not cause tissue reactions [1] and they can respond to the external signal of the surrounding environment [3]. They are formed in such a way that they become resistant to biological degradations as in biopolymers. Biomaterial metals are made resistant to corrosion and are made highly resistant to wear with acceptable mechanical and physical characteristics [1,104].

The biomaterial nanomeric-based polymers such as nano polymers, nanogels, micelles, and dendrimers have various advantages like high chemical functionality and easily biodegradable in nature, with high drug loading capacities, too. Inorganic nano biomaterials show inherited multifunctional capabilities and robust synthetic chemistry, which makes them useful for a wide range of applications [100]. SERS are using the three-dimensional stacked nanowires made up of silver on the glass-fibered filter sensors that are useful in the diagnosis of various cancers such as prostate or pancreatic [105]. Other than this there are other lipid-based natural biomaterials also available that have various advantages over the conventional biomaterials, such as they are highly degradable with the least toxicity and very useful in drug delivery or easy drug loading; some of the natural biomaterials are turned into nanoscale so that they can become highly biocompatible and do not react with the human tissue, cells, and organs. These are the some of the useful advantages of the smart biomaterial which make them suitable for various medical and clinical applications [100]. In tissue engineering, the smart biomaterials shouldn't only mimic the extracellular matrix but should also know the way to interact with the tissues promoting the adhesions and the cells or be able to be disintegrative at the point of demand, while in the delivery of drugs, the smart biomaterials can provide the stimuli-responsive deliveries and the tissue-selective deliveries over a long period, and after that, these all get cleared out from the body once their work is completed. In immune engineering, smart biomaterials should be provided the delivery to the targeted immune cells and in preventing the negative immune response or some unwanted inflammations [3].

8.5 DISADVANTAGES OF BIOMATERIALS

Smart biomaterials have multiple disadvantages such as in case of biomaterials formed from polymeric nanoparticles that show the issues in some cases because of their difficult scaling processes and also their complicated method of processing and synthesis of the polymers makes the requirement of the very precise or accurate skilled labor for the production of these types of nanomaterials so that this kind of problem can be solved and the formed biomaterial can be used for its applications. Other than this, the biomaterials formed by the lipid-based nanoparticles show issues due to their low stability for the long term; hence, the biomaterials formed using these lipids can't be used for the longer term and it's difficult to scale them up; moreover, there is a chance of infection as it's hard to control the quality of the biomaterials formed by these materials. Hence, for this, proper maintenance has to be provided while forming these kinds of biomaterials so that these issues can be solved. Many of the nanoparticles that are based on biological materials show problems in their scaling up and are also less stable for long-term usage, whereas the nanoparticle base inorganic biomaterials show low biodegradability and also have some of the toxic effects on the human body; other than this, they have a low capacity for drug loading and drug delivery. These are some of the problems, issues, or disadvantages that are involved with the different types of biomaterials that can be resolved by skilled techniques and precautions [100].

8.6 CONCLUSIONS AND FUTURE PERSPECTIVES

One of the most important goals of biomaterials is their role or application in the different medical clinical fields like tissue and immune engineering, medical devices, and the drug delivery system [63]. Specifically, smart biomaterials can be used in achieving specific targets and in reducing the side effects or toxicities and delivering the drugs or agents in a controlled manner, and increasing treatment efficiency [63]. They mimic nature, increasing good understanding of their fundamentals and can encourage the advancement of the materials inspired biologically and having unique and distinct biological characteristics that involve biomedical applications [3,106]. Moreover, the naturally available biomaterial with exploring cutting-edge chemistry and fabrication technology could lead to the development of many novel artificially designed biomaterials. These natural smart biomaterials that are obtained from natural substances or products like polysaccharides xanthan gum, commonly known as gelatin, have various advantages such as they are biodegradable, biocompatible, and have a cell recognition site. The future scope of smart biomaterials is being considered as they will be able to respond to and interpret the complex signaling mechanism in human cells, tissues, organs, or the body. This in turn allows the instances to detect, react, and cure the malignant changes at cellular levels such as in the initial stages of cancer, etc.

Further effort should be focused also on the betterment of the understanding of the instructions of biomaterials used in the body for helping and developing the good immunoprotective biocompatibility of materials, and later on, the utilization of the biocompatible chemistry and biorthogonal for the biosynthesizes of biomaterials and for the biomaterial designing [107]. Another approach is to learn a good understanding of interactions among materials and cells, including a decrease of the complexities among the two distinct components and employing the sample in the in-vitro/in-vivo model for quantifying the desired interaction. Some examples are a few of the materials containing the functional group can act for protection and can be useful for the modifications that allow the recognition of major entities involved in the modulation of cellular behavior and contact, which can later on be estimated by the reporter assay's coupling with the activities of the specific biological pathways, e.g., inflammation and proliferation.

Furthermore, the incorporation of the computational model over surface chemical structures of materials can encourage a good comprehension of this interaction. To make sure a good translation of the pre-clinical study requires smart biomaterials in the clinic, in that various factors are being considered, which involve smart biomaterials regulation and manufacturing constrain; for example, batch-batch reproducibility, the complexity of materials, and the ease of scalability. The consistency is in the material performance and its biocompatibility. Smart biomaterials have a combination of biological, drug, and tissue entities, every one of it is having their own distinct regulatory pathways and can face manufacturing and regulatory challenges, which may cause the struggle to demonstrate high-level consistencies in the patient. In tissue engineering, these smart biomaterials can be used in the three-dimensional printings of the organ, encouraging facile and stable printings along with the increased integration among the host tissues for avoiding rejections and inflammations. The use of computational learning and machine learning technology may also have a potential impact on the designing of biomaterials and in improving the prediction of their interaction and their physiological component, which includes the serum protein, cellular receptors, and cell membranes [3].

Recently, as per the advancement seen in the field of semiconductor technologies, the implantation of electronic chips or devices that act or mimic normal tissue or perform cellular functions can be used for various treatments, e.g., treatment of retinopathies. Many in-vitro studies have shown the non-toxic effects of smart nano biomaterials like nanowires, nanotubes, etc. Their long-term study in in-vivo gave a critical understanding of the importance of nano biomaterials in regenerative medicines. Hybrid nano biomaterials are very useful as they carry the merit while overpowering the demerit of nano biomaterials. Furthermore, these hybrid nano biomaterials penetrate deeper into the ocular segment, hence resulting in the localization of the delivery of drugs at the higher dosage, enriching the local concentration of drugs [108].

REFERENCES

[1] Singh, N., & Sharma, U. C. (2014). Introduction to nano-biomaterials In: Navani, N. K., & Singh S. (eds.), *Nanotechnology*, Vol. 11, pp. 1–12 Biomaterials, Texas, Studium Press.

[2] Wang, M. (2004). Bioactive materials and processing, In: Shi, D. (ed.), *Biomaterials and Tissue Engineering*, pp. 1–82, Springer-Verlag, Berlin Heidelberg, Germany.

[3] Kowalski, P., & Bhattacharya, C. (2018). Smart biomaterials: Recent advances and future directions, Langer R. *ACS Biomaterials Science and Engineering*, 4(11), 3809–3817.

[4] Hulbert, S. F. et al. (1987). Ceramics in clinical applications: Past, present and future, In: Vincezini, P. (ed.), *High Tech Ceramics*, pp. 189–213, Elsevier, Amsterdam, Netherlands.

[5] Ratner, B. D. et al. (2004). *Biomaterial Science: An Introduction to Materials in Medicine*, Academic Press, San Diego, US.

[6] Kondoyanni, M., Loukatos, D., Maraveas, C., Drosos, C., & Arvanitis, K. G. (2022). Bio-Inspired robots and structures toward fostering the modernization of agriculture. *Biomimetics (Basel, Switzerland)*, 7(2), 69. 10.3390/biomimetics7020069

[7] Zhang, X. (2013). Biological effects and risks of nano-bioceramics. (www.cityu.edu.hk). Retrieved on April 12, 2013.

[8] Alshabanah, L., & Hagar, M. (2021). Hassanin, Hybrid nanofibrous membranes as a promising functional layer for personal protection equipment: Manufacturing and antiviral/antibacterial assessments. *A Polymers*, 13(11), pp. 1776.

[9] Zhang, H., Feng, M., Fang, Y., Wu, Y., Liu, Y., Zhao, Y., & Xu, J. (2022). Recent advancements in encapsulation of chitosan-based enzymes and their applications in food industry. *Critical Reviews in Food Science and Nutrition*, 1–19. Advance online publication. 10.1080/10408398.2022.2086851

[10] Yamato, M., Konno, C., Utsumi, M., Kikuchi, A., & Okano, T. (2002). Thermally responsive polymer-grafted surfaces facilitate patterned cell seeding and co-culture. *Biomaterials*, 23, 561–567.

[11] Cao, Y., Ibarra, C., & Vacanti, C. (2006). Preparation and use of thermoresponsive polymers. In: Morgan, J. R., Yarmush, M. L., (eds.), *Tissue Engineering: Methods and Protocols*, pp. 75–84, Humana Press, New York. ISBN 978-1-59259-602-7.

[12] Koss, K. M., & Unsworth, L. D. (2016). Neural tissue engineering: Bioresponsive nanoscaffolds using engineered self-assembling peptides. *Acta Biomater.*, 44, 2–15.

[13] Zhao, X., Kim, J., Cezar, C. A., Huebsch, N., Lee, K., & Bouhadir, K. (2011). Active scaffolds for on-demand drug and cell delivery. *Proc. Natl. Acad. Sci. USA*, 108, 67–72.

[14] Kyle, S., Felton, S., McPherson, M. J., Aggeli, A., & Ingham, E. (2012). Rational molecular design of complementary self-assembling peptide hydrogels. *Adv. Healthc. Mater.*, 1, 640–645.

[15] Cao, Z., Bian, Q., Chen, Y., Liang, F., & Wang, G. (2017). Light-responsive Janus-particle-based coatings for cell capture and release. *ACS Macro Lett.*, 6, 1124–1128.

[16] Khan, F., Tare, R. S., Oreffo, R. O. C., & Bradley, M. (2009). Versatile biocompatible polymer hydrogels: Scaffolds for cell growth. *Angew. Chem. Int. Ed.*, 48(5), 978–982.

[17] Williams, D. (1990). An introduction to medical and dental materials. In: Williams, D. (ed.), *Concise Encyclopedia of Medical and Dental Materials*, pp. Xvii–xx, Pergamon Press and MIT Press, England.

[18] Saenz, A., Brostow, W., & Rivera-Muñoz, E. (1999). Ceramic biomaterials: An introductory overview. *Journal of Materials Education*, 21, 297–306.

[19] Tien, N. D., Lyngstadaas, S. P., Mano, J. F., Blaker, J. J., & Haugen, H. J. (2021). Recent developments in chitosan-based micro/nanofibers for sustainable food packaging, smart textiles, cosmeceuticals, and biomedical applications. *Molecules (Basel, Switzerland)*, 26(9), 2683. 10.3390/molecules26092683.

[20] Akhila, V., & Badwaik, L. S. (2022 April 1). Recent advancement in improvement of properties of polysaccharides and proteins based packaging film with added nanoparticles: A review. *Int J Biol. Macromol.*, 203, 515–525. 10.1016/j.ijbiomac.2022.01.181. Epub 2022 Feb 2. PMID: 35122798.

[21] Luo, M., Zhang, X., Wu, J., & Zhao J. (2021 August 15). Modifications of polysaccharide-based biomaterials under structure-property relationship for biomedical applications. *Carbohydrate Polym.*, 266, 118097. 10.1016/j.carbpol.2021.118097. Epub 2021 Apr 24. PMID: 34044964.

[22] Benwood, C., & Chrenek, J. (2021). Natural biomaterials and their use as bioinks for printing tissues, Willerth. *Source-Bioengineering*, 8(2), p. 27.

[23] Daly, A. C., Critchley, S. E., Rencsok, E. M., & Kelly, D. J. (2016). A comparison of different bioinks for 3D bioprinting of fibrocartilage and hyaline cartilage. *Biofabrication*, 8, 045002.

[24] Janarthanan, G., Tran, H. N., Cha, E., Lee, C., Das, D., & Noh, I. (2020). 3D printable and injectable lactoferrin-loaded carboxymethyl cellulose-glycol chitosan hydrogels for tissue engineering applications. *Mater. Sci. Eng. C*, 113, 111008.

[25] Manickam, P., Mariappan, S. A., Murugesan, S. M., Hansda, S., Kaushik, A., Shinde, R., & Thipperudraswamy, S. P.. (2022). Artificial Intelligence (AI) and Internet of Medical Things (IoMT) assisted biomedical systems for intelligent healthcare. *Biosensors*, 12(8), 562. 10.3390/bios12080562

[26] Salata, O. (2004). Applications of nanoparticles in biology and medicine. *J Nanobiotechnol.*, 2, 3. 10.1186/1477-3155-2-3
[27] Bhullar, S., Bhullar, S., & Ramkrishna, S. (2015). *Smart Biomaterials – A Review, Rev. Adv. Mater. Sci*, 40(3), 303–314.
[28] Park, J., & Lakes, R. S. (2007). Hard tissue replacement – II: Joints and teeth, In: *Biomaterials: An Introduction*, 395–458, Springer, New York.
[29] Cheshire, N. J. W., Wolfe, M. S., Noone, M. A., Davies, L., & Drummond, M. (1992). *The economics of femorocrural reconstruction for critical leg ischemia with and without autologous vein, Jounral Vasc. Surg.*, 15, 167–175.
[30] Chlupáč, J., Filová, E., & Bačáková, L. (2009). Blood vessel replacement: 50 years of development and tissue engineering paradigms in vascular surgery. *Physiol. Res.*, 58 (Suppl 2), S119–S140. 10.33549/physiolres.931918. PMID: 20131930.
[31] Ali, M. N., Busfield, J. J. C., & Rehman, I. U. (2014). Auxetic oesophageal stents: Structure and mechanical properties. *J Mater. Sci.: Mater. Med.*, 25, 527–553. 10.1007/s10856-013-5067-2
[32] Caddock, B. D., & Evans, K. E. (1995 Sep). Negative Poisson ratios and strain-dependent mechanical properties in arterial prostheses. *Biomaterials*, 16(14), 1109–1115. 10.1016/0142-9612(95)98908-w. PMID: 8519933.
[33] Caddock, B. D., & Evans, K. E. (1989). Microporous materials with negative Poisson's ratios. I. Microstructure and mechanical properties. *Journal of Physics D: Applied Physics*, 22, p. 1877. 10.1088/0022-3727/22/12/012
[34] Chan, N., & Evans, K. E. (1998). *Indentation resilience of conventional and auxetic foams, J Cell. Plast.*, 34, 231–260.
[35] Swanson, S. A. V. (1977). The scientific basis of joint replacement. In: S. A. V. Swanson & M. A. R. Freeman, (eds.), Pitman Medical Publishers, Turnbridge Wells, p. 130.
[36] Kim, K., & Sung, C. (2021). Computational and histological analyses for investigating mechanical interaction of thermally drawn fiber implants with brain tissue. *Park Micromachines*, 12(4), p. 394.
[37] Abdelaal, A. M., & Darwish, M. H. (2012). Analysis, fabrication and a biomedical application of auxetic cellular structure. *International Journal of Engineering and Innovative Technology (IJEIT)*, 2, 218–223.
[38] Saddow S. E. (2022). Silicon carbide technology for advanced human healthcare applications. *Micromachines*, 13(3), 346. 10.3390/mi13030346.
[39] Stretching the Boundaries, that is, ideas and information on Physics, Science, Technology, Archaeology, Arts and Literatures. Physics at http://physics-sparavigna.blogspot.com
[40] Zhang, X. & Phil, J. *J. Drug-Eluting Bioresorbable Stents, Med-Tech Innovation* (http://www.med-techinnovation.com).
[41] Ali, M. N., & Rehman, I. U. (2011). An Auxetic structure configured as oesophageal stent with potential to be used for palliative treatment of oesophageal cancer; development and in vitro mechanical analysis. *J Mater. Sci.: Mater. Med.*, 22, 2573–2581. 10.1007/s10856-011-4436-y.
[42] Scarpa, F., Smith, C. W., Ruzzene, M., & Wadee, M. K. (2008). Mechanical properties of auxetic tubular truss-like structures. *Physica Status Solidi (b)*, 245(3), 584–590.
[43] Raamachandran, J., & Jayavenkateshwaran, K. (2007). Modeling of stents exhibiting negative Poisson's ratio effect. *Computer Methods in Biomechanics and Biomedical Engineering*, 10(4), 245–255. 10.1080/10255840701198004
[44] Baker, C. E. (2011). *Auxetic Spinal Implants: Consideration of Negative Poisson's Ratio in the Design of an Artificial Intervertebral Disc* (Thesis, Master of Science, University of Toledo).

[45] Hengelmolen, R. (2004). AU2003249001.
[46] Bhullar, S. K., & Jun, M. B. G. (2013). In: Proc. International Workshop on Smart Materials & Structures. *SHM and NDT for the Energy Industry*, p. 1.
[47] Jun, P. & Koo, K. (2013). The effect of negative poisson's ratio polyurethane scaffolds for articular cartilage tissue engineering applications. *J. Advances in Materials Science and Engineering, ID853289*. 10.1155/2013/853289
[48] Fozdar, D. Y., Soman, P., Lee, J. W., Han, L. H., & Chen, S. (2011). Three-dimensional polymer constructs exhibiting a tunable negative Poisson's ratio. *Advance Functional Material*, 21, 2712. 10.1002/adfm.201002022
[49] Soman, P., Lee, J. W., Phadke, A., Varghese, S., & Chen, S. (2012). Spatial tuning of negative and positive Poisson's ratio in a multi-layer scaffold. *Acta Biomaterialia*, 8(7), 2587–2594.
[50] Yong, P., Shiwu, D., Yong, H., Tongwei, C., Changqing, L., Zhengfeng, Z., & Yue, Z. (2010). Demineralized bone matrix gelatin as scaffold for tissue engineering. *African Journal of Microbiology Research*, 4(9), 865–870.
[51] Pagliara, S., Franze, K., McClain, C. R., et al. (2014 Jun). Auxetic nuclei in embryonic stem cells exiting pluripotency. *Nature Materials*, 13(6), 638–644. 10.1038/nmat3943. PMID: 24747782; PMCID: PMC4283157.
[52] Alderson, K. L., & Evans, K. E. (1992). The fabrication of microporous polyethylene having a negative Poisson's ratio. *Polymer*, 33(20), 4435–4438.
[53] Chludzinski, M. & Hammill, E. (2005). US Patent 6837890.
[54] Choi, J. B., & Lakes, R. S. (1992). Non-linear properties of metallic cellular materials with a negative Poisson's ratio. *Journal of Materials Science*, 27(19), 5375–5381.
[55] Simkins, V. R., Alderson, A., Davies, P. J., & Alderson, K. L. (2005). Single fibre pullout tests on auxetic polymeric fibres. *Journal of Materials Science*, 40(16), 4355–4364.
[56] Evans, K. E., Alderson, A., & Christian, F. R. (1995). Auxetic two-dimensional polymer networks. An example of tailoring geometry for specific mechanical properties. *Journal of the Chemical Society, Faraday Transactions*, 91(16), 2671–2680.
[57] Bhattacharjee, S., Joshi, R., Yasir, M., Adhikari, A., Chughtai, A. A., Heslop, D., Bull, R., Mark Willcox, M., & Macintyre, C. R. (2021). Graphene- and nanoparticle-embedded antimicrobial and biocompatible cotton/silk fabrics for protective clothing. *ACS Applied Bio Materials*, 4(8), 6175–6185. 10.1021/acsabm.1c00508
[58] Liu, Y., Hu, H., Lam, J. K., & Liu, S. (2010). Negative Poisson's ratio weft-knitted fabrics. *Textile Research Journal*, 80(9), 856–863.
[59] Miller, W., Hook, P. B., Smith, C. W., Wang, X., & Evans, K. E. (2009). The manufacture and characterisation of a novel, low modulus, negative Poisson's ratio composite. *Composites Science and Technology*, 69(5), 651–655.
[60] Alderson, A., Rasburn, J., Ameer-Beg, S. G., Mullarkey, P., Perrie, W., & Evans, K. E. (2000). An auxetic filter: A tuneable filter displaying enhanced size selectivity or defouling properties. *Industrial & Engineering Chemistry Research*, 39(3), 654–665. 10.1021/ie990572w
[61] Alderson, A., Rasburn, J., & Evans, K. E. (2007). Mass transport properties of auxetic (negative Poisson's ratio) foams. *Physica Status Solidi (b)*, 244(3), 817–827.
[62] Alderson, A., Davies, P. J., Evans, K. E., Alderson, K. L., & Grima, J. N. (2005). Modelling of the mechanical and mass transport properties of auxetic molecular sieves: An idealised inorganic (zeolitic) host–guest system. *Molecular Simulation*, 31(13), 889–896.
[63] Langer, R., & Vacanti, J. P. (1993). Tissue Eengineering. *Science*, 5110, 920–926. 10.1126/science.8493529
[64] Lee, S. J., Yoo, J. J., & Atala, A. (2018). Biomaterials and tissue engineering. In:*Kim, B. (eds), Clinical Regenerative Medicine in Urology*, pp. 17–51, Springer, Singapore.

[65] Ruskowitz, E. R., & DeForest, C. A. (2018). Photoresponsive biomaterials for targeted drug delivery and 4D cell culture. *Nat. Rev. Mater.*, 3(2), 17087.
[66] Bhar, B., Chouhan, D., Pai, N., & Mandal, B. B. (2021). Harnessing multifaceted next-generation technologies for improved skin wound healing. *ACS Applied Bio Materials*, 4(11), 7738–7763. 10.1021/acsabm.1c00880
[67] Truong, V. X., Ablett, M. P., Richardson, S. M., Hoyland, J. A., & Dove, A. P. (2015). Simultaneous orthogonal dual-click approach to tough, in-situ-forming hydrogels for cell encapsulation. *J. Am. Chem. Soc.*, 137(4), 1618–1622.
[68] Sadtler, K., Singh, A., Wolf, M. T., Wang, X., Pardoll, D. M., & Elisseeff, J. H. (2016). Design, clinical translation and immunological response of biomaterials in regenerative medicine. *Nat. Rev. Mater.*, 1(7), 16040.
[69] Zhang, Y. S., & Khademhosseini, A. (2017). Advances in engineering hydrogels. *Science (Washington, DC, U. S.)*, 356(6337), eaaf3627.
[70] Eslahi, N., Abdorahim, M., & Simchi, A. (2016). Smart polymeric hydrogels for cartilage tissue engineering: A review on the chemistry and biological functions. *Biomacromolecules*, 17(11), 3441–3463.
[71] Zhao, X. (2014). Multi-scale multi-mechanism design of tough hydrogels: Building dissipation into stretchy networks. *Soft Matter*, 10(5), 672–687.
[72] Akhtar, M. F., Hanif, M., & Ranjha, N. M. (2016). Methods of synthesis of hydrogels – A review. *Saudi Pharm. J.*, 24(5), 554–559.
[73] Parhi, R. (2017). Cross-linked hydrogel for pharmaceutical applications: A review. *Adv. Pharm. Bull.*, 7(4), 515–530.
[74] Liu, M., Zeng, X., Ma, C., Yi, H., Ali, Z., Mou, X., Li, S., Deng, Y., & He, N. (2017). Injectable hydrogels for cartilage and bone tissue engineering. *Bone Res.*, 5, 17014.
[75] Razzaq, M. Y., Behl, M., & Lendlein, A. (2012). Memory-effects of magnetic nanocomposites. *Nanoscale*, 4(20), 6181–6195.
[76] Balk, M., Behl, M., Wischke, C., Zotzmann, J., & Lendlein, A. (2016). Recent advances in degradable lactide-based shape-memory polymers. *Adv. Drug Delivery Rev.*, 107, 136–152.
[77] Behl, M., Razzaq, M. Y., & Lendlein, A. (2010). Multifunctional shape-memory polymers. *Adv. Mater.*, 22(31), 3388–3410.
[78] Neffe, A. T., Hanh, B. D., Steuer, S., & Lendlein, A. (2009). Polymer networks combining controlled drug release, biodegradation, and shape memory capability. *Adv Mater*, 21 (32–33), 3394–3398.
[79] Kuang, X., Chen, K., Dunn, C. K., Wu, J., Li, V. C. F., & Qi, H. J. (2018). 3D printing of highly stretchable, shape-memory, and self-healing elastomer toward novel 4D printing. *ACS Appl. Mater. Interfaces*, 10(8), 7381–7388.
[80] Kim, H., Kim, S., Lim, D., & Jeong, W. (2022). Development and characterization of embroidery-based textile electrodes for surface EMG detection. *Sensors (Basel, Switzerland)*, 22(13), 4746. 10.3390/s22134746
[81] Mi, L., Bernards, M. T., Cheng, G., Yu, Q., & Jiang, S. (2010). PH responsive properties of non-fouling mixed-charge polymer brushes based on quaternary amine and carboxylic acid monomers. *Biomaterials*, 31(10), 2919–2925.
[82] Yu, Q., Wu, Z., & Chen, H. (2015). Dual-function antibacterial surfaces for biomedical applications. *Acta Biomater.*, 16, 1–13.
[83] Anderson, J. M., Rodriguez, A., & Chang, D. T. (2008). Foreign body reaction to biomaterials. *Semin. Immunol.*, 20(2), 86–100.
[84] Wick, G., Grundtman, C., Mayerl, C., Wimpissinger, T.-F., Feichtinger, J., Zelger, B., Sgonc, R., & Wolfram, D. (2013). The immunology of fibrosis. *Annual Reviews*, 31, 104–135.
[85] Wynn, T. A., & Ramalingam, T. R. (2012). Mechanisms of fibrosis: Therapeutic translation for fibrotic disease. *Nat. Med.*, 18 (7), 1028–1040.

[86] Ward, W. K. (2008). A review of the foreign-body response to subcutaneously-implanted devices: The role of macrophages and cytokines in biofouling and fibrosis. *J. Diabetes Sci. Technol.*, 2(5), 768–777.

[87] Sahdev, P., Ochyl, L. J., & Moon, J. J. (2014). Biomaterials for nanoparticle vaccine delivery systems. *Pharm. Res.*, 31(10), 2563–2582.

[88] Shao, K., Singha, S., Clemente-Casares, X., Tsai, S., Yang, Y., & Santamaria, P. (2015). Nanoparticle-based immunotherapy for cancer. *ACS Nano*, 9(1), 16–30.

[89] Andorko, J. I., Hess, K. L., & Jewell, C. M. (2015). Harnessing biomaterials to engineer the lymph node microenvironment for immunity or tolerance. *AAPS J.*, 17(2), 323–338.

[90] Sharp, F. A., Ruane, D., Claass, B., Creagh, E., Harris, J., Malyala, P., Singh, M., O'Hagan, D. T., Petrilli, V., Tschopp, J., et al. (2009). Uptake of particulate vaccine adjuvants by dendritic cells activates the NALP3 inflammasome. *Proc. Natl. Acad. Sci. U. S. A.*, 106(3), 870–875.

[91] Tang, L., Zheng, Y., Melo, M. B., Mabardi, L., Castaño, A. P., Xie, Y.-Q., Li, N., Kudchodkar, S. B., Wong, H. C., Jeng, E. K., et al. (2018). Enhancing T cell therapy through TCR-signaling-responsive nanoparticle drug delivery. *Nat. Biotechnol. 36(8), 707–716.* 10.1038/nbt.4181.

[92] Mchugh, K. J., Nguyen, T. D., Yang, D., Behrens, A. M., Rose, S., Tochka, Z. L., Tzeng, S. Y., Norman, J. J., Tomasic, S., Taylor, M. A., et al. (2017). Fabrication of fillable microparticles and other complex 3D microstructures. *Science*, 357, 1138–1142.

[93] Morrison, C. (2018). Alnylam prepares to land first RNAi drug approval. *Nat. Rev. Drug Discovery*, 17(3), 156–157.

[94] Yin, H., Kauffman, K. J., & Anderson, D. G. (2017). Delivery technologies for genome editing. *Nat. Rev. Drug Discovery*, 16(6), 387–399.

[95] Kowalski, P. S., Capasso Palmiero, U., Huang, Y., Rudra, A., Langer, R., Anderson, D. G. (2018). Ionizable amino-polyesters synthesized via ring opening polymerization of tertiary amino-alcohols for tissue selective MRNA delivery. *Adv. Mater.*, 30(34), 1801151.

[96] ur Rehman, Z., Hoekstra, D., & Zuhorn, I. S. (2013). Mechanism of polyplex- and lipoplex-mediated delivery of nucleic acids: Real-time visualization of transient membrane destabilization without endosomal lysis. *ACS Nano*, 7(5), 3767–3777.

[97] Rehman, Z., Zuhorn, I. S., & Hoekstra, D. (2013). How cationic lipids transfer nucleic acids into cells and across cellular membranes: Recent advances. *J. Controlled Release*, 166(1), 46–56.

[98] Jaraswekin, S., Prakongpan, S., & Bodmeier, R. (2007). Effect of poly(lactide-co-glycolide) molecular weight on the release of dexamethasone sodium phosphate from microparticles, *Journal of Microencapsulation*, 24(2), 117. 10.1080/0265204 0701233655

[99] Winkler, J. S., Barai, M., & Tomassone M. S. (2019). Dual drug-loaded biodegradable Janus particles for simultaneous co-delivery of hydrophobic and hydrophilic compounds. *Experimental Biology and Medicine*, 244(14), 1162–1177. 10.1177/1535370219876554.

[100] Kumar, N., Fazal, S., Miyako, E., Matsumura, K., & Rajan, R. (2021). Avengers against cancer: A new era of nano-biomaterial-based therapeutics. *Materials Today*, 51, 317–349, ISSN 1369-7021, 10.1016/j.mattod.2021.09.020. (https://www.sciencedirect.com/science/article/pii/S1369702121003321).

[101] Yechezkel (Chezy) Barenholz (2012). Doxil® — The first FDA-approved nano-drug: Lessons learned. *Journal of Controlled Release*, 160(2), 117–134, ISSN 0168-3659, 10.1016/j.jconrel.2012.03.020

[102] Khan, M. I., Hossain, M. I., Hossain, M. K., Rubel, M., Hossain, K. M., Mahfuz, A., & Anik, M. I. (2022). Recent progress in nanostructured smart drug delivery systems for cancer therapy: A review. *ACS Applied Bio Materials*, 5(3), 971–1012. 10.1021/acsabm.2c00002

[103] Naidu, N., Wadher, K., & Umekar, M. (2021). An overview on biomaterials: Pharmaceutical and biomedical applications. *Journal of Drug Delivery and Therapeutics*, 11, 154–161. 10.22270/jddt.v11i1-s.4723.

[104] Davis, J. R. (ed.) (2003). Overview of biomaterials and their use in medical devices, In: *Handbook of Materials for Medical Devices*, 1–11, ASM International, United States.

[105] Phyo, J. B., Woo, A., Yu, H. J., Lim, K., Cho, B. H., Jung, H. S., & Lee, M.-Y. (2021). *Analytical Chemistry*, 93(8), 3778–3785. 10.1021/acs.analchem.0c04200

[106] Zhao, W., Cui, C. H., Bose, S., Guo, D., Shen, C., Wong, W. P., Halvorsen, K., Farrokhzad, O. C., Sock, G., & Teo, L. (2012). Bioinspired multivalent DNA network for capture and release of cells. *Proc. Natl. Acad. Sci. U. S. A.*, 109, 19626.

[107] Capasso Palmiero, U., Sponchioni, M., Manfredini, N., Maraldi, M., & Moscatelli, D. (2018). Strategies to combine ROP with ATRP or RAFT polymerization for the synthesis of biodegradable polymeric nanoparticles for biomedical applications. *Polym. Chem.*, 9, 4084.

[108] Sharma, R., Sharma, D., Hazlett, L. D., & Singh, N. K. (2021). Nano-biomaterials for retinal regeneration. *Nanomaterials*, 11, 1880. 10.3390/nano11081880

9 Ferroelectric Polymer Composites and Evaluation of Their Properties

Sergey M. Lebedev and Olga S. Gefle
National Research Tomsk Polytechnic University, Tomsk, Russia

CONTENTS

9.1 INTRODUCTION

Ferroelectric and piezoelectric polymer composites are widely used in electric field control systems in electrical insulation [1,2], energy storage devices [3,4], transducers and sensors in smart systems [5,6], absorbing materials in microwave technique [7,8], tissue engineering as scaffolds for bone tissue regeneration [9–13], etc. These composite polymer materials consist of a polymer matrix filled with high permittivity ferroelectric ceramic filler. However, the availability of a large number of interfaces between ceramic particles and the polymer matrix leads to forming double charge layers at these boundaries [14,15] and the appearance of different kinds of relaxation polarization. Furthermore, electroactive ferroelectric fillers themselves possess spontaneous polarization, which causes additional dielectric losses due to the formation of domain polarization under the electric field in a certain frequency range.

To significantly increase the permittivity of such multiphase composites (reaching the "dielectric percolation threshold"), there are four possible ways:

- the content of ceramic filler should be more than 50 vol%;

DOI: 10.1201/9781003333760-9

- the filler particles should be subjected to additional chemical surface treatment;
- the matrix should be polar dielectric of a high permittivity;
- development of multiphase composites filled with two filler types.

However, the first of them leads to deterioration in the rheological properties of composites and the impossibility of processing them by extrusion. The second one leads to deterioration of the filler properties. Dispersion of fillers may be improved by heat treatment, purification, graphitization, surface functionalization, and the oxidation of their surface. However, these technologies lead not only to a reduction of disadvantages, but also to a degradation of composite properties. For example, dispersion of filler particles in the polymer matrix is closely related to functionalization of their surface, but surface functionalization strongly affects the properties of fillers. Moreover, it has been earlier reported [16,17] that composites filled with functionalized fillers have either improved and worsened properties as compared to untreated ones. The third way is difficult to realize in practice since the range of polymer dielectrics with a high permittivity is limited. Grunlan et al. [18,19] reported that the electrical properties of composites may be changed by the application of three-phase composites filled with two different fillers. The authors explain the enhancement in properties of filled composites by the formation of the so-called segregated network. In order to modify the basic properties of polymer composites, we have proposed a new approach to design polymer composites with improved electrical and thermal properties due to filling with two different types of fillers [20,21].

The first part of this chapter is devoted to the ferroelectric polymer composites based on the conventional petroleum-based polymers such as low density polyethylene (LDPE), elastomers, and polyvinylidene fluoride (PVDF). Conventional polymer materials are widely applied in various fields of industry and science owing to low price, good processability, chemical and corrosion resistance, and etc. However, these polymeric materials are not biodegradable [22].

The problem of environmental pollution with synthetic polymer waste is becoming increasingly threatening, so the development of new biodegradable polymer composites is a very urgent task. In this regard, the second part of this work is related with the development of polymer composites based on biodegradable poly(lactic acid) (PLA) matrix.

Among biodegradable polymers, PLA has been extensively researched for the previously mentioned potential applications. PLA is a biodegradable semi-crystalline polymer with high mechanical properties, produced from reproducible resources (corn, sugarcane, sugar beets, etc.). Degradation time of PLA in environmental conditions varies from 6 to 24 months, while for petroleum-based polymers such as polyethylene or polypropylene, this time varies from 500 to 1,000 years [23].

PLA is widely used in biomedicine for the manufacture of biodegradable implants and scaffolds [24–26]. Regardless of PLA possessing excellent properties, it is an insulating polymer material. Often this does not allow PLA to be applied in other fields [27–32]. The addition of various fillers with specific properties into the neat PLA allows its thermal, mechanical, and electrical properties to be significantly changed.

This chapter shows recent advances in the development, processing, and evaluation of the basic properties of ferroelectric polymer composites. These polymer composites may be widely applied in different high-technology fields such as additive manufacturing (AM), biomedicine, electronics, and electrical engineering.

Future investigations can be focused on improving compatibility of polymer matrix and fillers, uniform distribution of fillers, and transfer of new laboratory technologies to various fields of industries. The use of new biodegradable polymers will expand the range of ferroelectric polymer composites with desired properties. Finally, newly developed biodegradable ferroelectric composites filled with barium titanate are "environmentally friendly" compared to conventional polymer composites filled with lead zirconate titanate ceramic powder (PZT).

PLA is widely applied in various fields of industries and science because of its properties. This biodegradable polymer has become one of the alternatives to conventional petroleum-based polymers. According to many authors, PLA can become one of the main polymeric materials produced on an industrial scale, taking into account the constant increase in the cost of oil and oil products.

Until recently, lead zirconate titanate (PZT) was used in most cases as filler in ferroelectric polymer composites. This is due to the fact that PZT possesses very high permittivity and excellent ferroelectric and piezoelectric properties. However, lead and its compounds are very toxic. Therefore strict environmental protection requirements make it necessary to develop lead-free ferroelectric and piezoelectric polymer materials. Strict requirements are also imposed on polymer matrices on which these composites are based since environmental pollution from petroleum-based polymer waste is also a very important ecological issue.

In 1945, Wul and Goldman [33,34] discovered ferroelectricity in barium titanate (BT, $BaTiO_3$); unlike other ferroelectrics, its properties at the Curie temperature do not change smoothly, but abruptly. $BaTiO_3$ possesses high permittivity and relatively high piezoelectricity. Moreover, it is also non-toxic, unlike PZT. A large number of ferroelectric and piezoelectric polymer composites filled with lead-free fillers have been developed recently.

Smart ferroelectric and piezoelectric scaffolds are used to transform a mechanical load into an electric field (the so-called direct piezoelectric effect) and vice versa (the so-called inverse piezoelectric effect). The latter effect is the basis of the electric stimulation method of tissue growth by the external electric field. Yasuda and Fukada [35,36] were the first to investigate piezoelectricity in bone tissue in the 1950s. Bassett [37] later showed that a mechanical load applied to a bone can generate an electrical field that accelerates the growth of new bone tissue until the load is removed.

Biodegradable scaffolds prepared of biopolymers alone (usually it is PLA, polycaprolactone (PCL), etc.) are not appropriate as materials for electric stimulation of bone tissue growth, since these polymers are insulating materials. To increase the conductivity of polymer scaffolds, many authors have investigated binary biopolymer/CNT composites, which allow the electrical conductivity to be improved for the electric stimulation of bone tissue growth [38,39].

Ferroelectric $BaTiO_3$-based polymer composites are widely used in regenerative medicine [4,9–13]. Barium titanate can increase the polarity of PLA-based composites due to its higher permittivity, resulting in increased biocompatibility of composite and bone tissue. Fan et al. [4] showed that the permittivity of PLA/$BaTiO_3$ nanocomposites increased up to 7–8 at 20% $BaTiO_3$ content compared to that for the neat PLA. Li et al. [11] found that the permittivity of PLA/$BaTiO_3$ porous fibrous scaffold was about 1.19, i.e. very close to that of the natural bone. Moreover, it is well known that the greater the polarity of the polymer matrix and filler, the stronger an interaction between components of the composite, and the better the biocompatibility between the composite scaffold and bone tissue. However, the greatest interest for tissue engineering is in multiphase composites biopolymer/CNT/$BaTiO_3$ [8,40,41]. These multifunctional three-phase composites combine different properties of polymer matrix and fillers, such as biodegradability (biopolymer matrix), ferroelectricity and piezoelectricity (barium titanate), and high conductivity (CNT).

It would be very interesting to combine the advantages of PLA, CNTs, and $BaTiO_3$ in multiphase biodegradable composites not only for biomedicine but also for other fields of industries. However, information on the development of three-phase PLA/$BaTiO_3$/CNT composites and research of their properties is practically absent in the literature. This work is the first to develop and characterize multiphase PLA-based ferroelectric composites with high permittivity filled with barium titanate and a small amount of CNT.

9.2 EXPERIMENTAL TECHNIQUE AND SAMPLES

Polymeric and elastomeric dielectrics such as low density polyethylene (LDPE), non-polar and polar rubbers (natural, butadiene-styrene, and butadiene-nitrile rubbers), and polyvinylidene fluoride (PVDF) were applied as polymer matrices in this study.

Lead zirconate titanate (PZT) ceramic powder has been used as ferroelectric fillers. The filler content in all composites was varied from 0 to 55 vol%.

Composites based on LDPE and PVDF were prepared by Plastograph Brabender EC Plus. The stock temperature and processing time were 190°C–210°C and 10 min, respectively. The speed of the mixer blades was varied from 30 rpm to 90 rpm.

Composites based on elastomers were prepared by a hot rolling method at a temperature of 150°C.

To prepare samples, the resulting polymeric composites were prepared by a hot pressing method by means of a hydraulic press under a pressure of 12.5 MPa.

The real part of admittance $\gamma_a = \omega \cdot \varepsilon_0 \cdot \varepsilon' \cdot \tan\delta$ ($\omega = 2\pi f$ is the frequency; $\varepsilon_0 = 8.854 \cdot 10^{-12}$ F/m is the electric constant; ε' is the real part of the complex permittivity; $\tan\delta$ is the loss factor), ε', $\tan\delta$, and phase angle φ were measured under 3 V AC voltage in the frequency range from 0.01 Hz to 10^6 Hz by the dielectric spectroscopy method. The dielectric spectra for all composites were measured at room temperature.

The hysteresis loops (dependencies $D = f(E)$, where D and E are the dielectric displacement and electric field strength) for composites studied have been measured by the modified method of Sawyer-Tower [42].

DSC and TGA methods have been performed for the investigation of temperature phase transitions for studied composites by means of the combined analyzer

Q600. The sample weight loss and heat flow were controlled at temperature varied from 20°C to 500°C at a temperature change rate of 3°C/min.

Polylactic acid (PLA, Ingeo 4043D) was used as a biopolymer matrix to prepare of composites. $BaTiO_3$ powder was supplied by Aril LLC (Russia). The powder composition and particle size of $BaTiO_3$ are shown in Table 9.1. Pristine single-walled carbon nanotubes (SWCNT Tuball^{TM}, Figure 9.1) were supplied by OCSiAl LLC (Russia).

All composites were produced by Plastograph Brabender EC Plus. The stock temperature and processing time were 190°C–210°C and 10 min, respectively. Materials used in this work are shown in Table 9.2.

TABLE 9.1
$BaTiO_3$ Powder Composition

Basic substance content, %	not less than 99.8
BaO/TiO_2 molar ratio	0.996 ± 0.002
SrO content, %,	not more than 0.003
Al_2O_3 content, %	not more than 0.003
Na_2O content, %	not more than 0.01
SiO_2 content, %	not more than 0.03
Nb_2O_5 content, %	not more than 0.01
Free BaO content, %	not more than 1.0
Humidity, %	not more than 0.2
Losses on ignition, %	not more than 0.2
Specific surface (BET method), m^2/g	2.4
Particle size: D50 (Sedigraph), μm	1.0
Particle size: SEM, μm	0.5
Specific gravity, g/cm^3	6.02
The structure	tetragonal

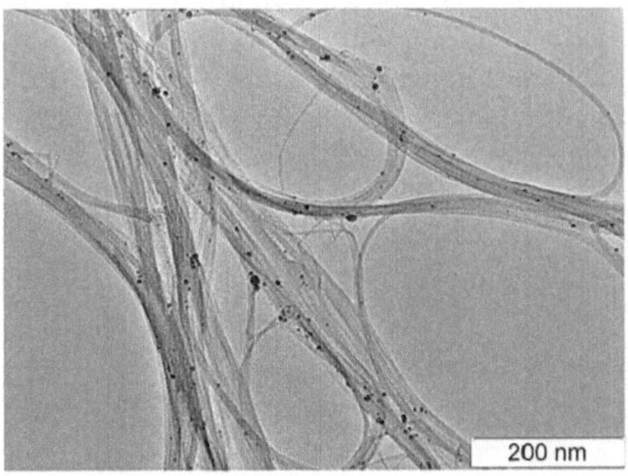

FIGURE 9.1 Micrographs of pristine CNTs.

TABLE 9.2
Composition of Materials Applied

No.	Composition
1	Neat PLA
2	PLA+0.05 wt% CNT
3	PLA+5 wt% $BaTiO_3$
4	PLA+10 wt% $BaTiO_3$
5	PLA+20 wt% $BaTiO_3$
6	PLA+40 wt% $BaTiO_3$
7	PLA+5 wt% $BaTiO_3$+0.05 wt% CNT
8	PLA+10 wt% $BaTiO_3$+0.05 wt% CNT
9	PLA+20 wt% $BaTiO_3$+0.05 wt% CNT
10	PLA+40 wt% $BaTiO_3$+0.05 wt% CNT

A small content of CNTs in studied composites (see Table 9.2) is caused by two reasons. First, a higher content of CNTs in composites results in a sharp increase in conductivity by several orders of magnitude [21], which leads to significant dielectric losses (losses due to through conduction). Second, the cost of CNTs is very high.

The real part of admittance γ_a, ε', $\tan\delta$, and phase angle φ were measured under 3 V AC voltage in the frequency range from 0.1 Hz to 10^6 Hz by the dielectric spectroscopy method. The dielectric spectra for all composites were measured at room temperature.

To study a sub-molecular structure of neat PLA and PLA-based composites, all samples were controlled by a scanning electron microscopy method (SEM).

The mechanical properties of composites were controlled via tensile tests as stress-strain curves. Measurements were carried out by means of an Instron 3345 universal testing machine equipped with a 5-kN load cell. The measurements were performed at a constant crosshead rate of 1.0 mm/min at room temperature. No less than ten samples were tested for each composite.

The melt flow index (MFI) for the PLA matrix and all composites were studied by a tester Instron MF20 (ISO 1133 and ASTM D1238). The MFI was measured for all composites (die diameter of 2.09 mm at 190°C and a load of 2.16 kg).

The dielectric hysteresis loops for all composites were monitored with aix ACCT TF Analyzer 2000 with varying voltage from 0 to 1,200 V. The sample thickness was 0.85 mm.

9.3 RESULTS AND DISCUSSION

9.3.1 Ferroelectric Composites Based on LDPE, Elastomeric, and PVDF Matrices

To evaluate the main electrical characteristics of composites, the experimental and calculated values of the effective permittivity ε_{eff} in a low electric field were estimated.

TABLE 9.3
Calculated and Experimental Values of the Effective Permittivity for Composites Studied

		ε_{eff}	
Number of Composite	**Composition**	**Calculation**	**Experiment**
No. 1	LDPE+10 vol%PZT	3.1	3.2
No. 2	LDPE+30 vol%PZT	6.7	7.1
No. 3	LDPE+40 vol%PZT	10.9	12.2
No. 4	Polar rubber+30 vol%PZT	25.1	29.2
No. 5	Polar rubber+45 vol%PZT	47.6	51.7
No. 6	Polar rubber+55 vol%PZT	81.8	71.5
No. 7	PVDF+30 vol%PZT	28.7	28.0

It has been established that the most adequate prediction of the ε_{eff} of composites with different filler content and accuracy acceptable for engineering assessment is provided by the Bruggeman model [43]:

$$\frac{\varepsilon_f - \varepsilon_{eff}}{\varepsilon_{eff}^{1/3}} = \frac{(1-\phi)\cdot(\varepsilon_f - \varepsilon_m)}{\varepsilon_m^{1/3}}, \tag{9.1}$$

where ϕ is the volume filler content; and ε_f, ε_{eff}, and ε_m are permittivities of a filler, composite, and matrix, respectively.

Calculated and experimental values of ε_{eff} for composites studied are presented in Table 9.3. It is obvious that the difference between the experimental and calculated values of ε_{eff} does not exceed 15%.

The significant difference in the permittivity values for elastomeric and polymer composites is due to the fact that when highly polar filler (PZT) is introduced into a non-polar LDPE matrix, the cohesive interaction between the components is very weak, in contrast to composites based on elastomers.

Since the change in the permittivity was achieved owing to a higher volume content of the filler, and the adding a dispersed filler significantly modifies the structure and properties of matrix due to interfacial interactions and the formation of a boundary layer near the filler particles [14,15], the improvement in the properties of composites is possible only with a certain degree of heterogeneity of composites. However, even at the optimal volume filler content the electrical characteristics of composite in a low and high electric field can differ significantly. This is especially true for composites filled with ferroelectric ceramics. In particular, the value of the critical field, at which the nonlinear growth of tanδ and real part of admittance γ_a begins, can decrease with an increase in the filler content, owing to an increase in the local field at the boundary between the matrix and ceramic phase. In this regard, the dependences of ε' and tanδ of LDPE-based and elastomer composites with different filler content on the external electric field strength were studied.

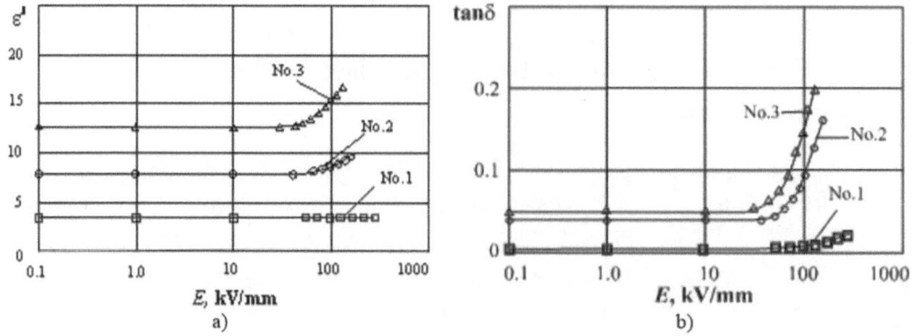

FIGURE 9.2 Relationships ε′ (a) and tanδ (b) on the field strength E for composites based on LDPE.

The results of studies of ε′ and tanδ = $f(E)$ for composites based on LDPE are presented in Figure 9.2. It can be seen that the value of ε′ for composites at filler content (C) of 10 vol% (No. 1) is practically independent of the field, and the average values of ε′ within the measurement deviation coincide with ε′ measured in a low electric field. When the filler content is changed up to 30 vol% (No. 2) and E ≥ 10 kV/mm, the average values of ε′ exceed the permittivity of the composite in a low field by 15%, and at C = 40 vol % (No. 3), an increase in ε′ at E ≥ 10 kV/mm is about 45%. When the electric field changes from 4 to 14 kV/mm, tanδ of composites with different filler content increases by more than 100%. Similar results were obtained for composites based on elastomers (Figure 9.3).

An increase in tanδ may be also associated with an increase in the conductivity of the composites.

The value of the local field E_L (Lorentz field) at the phase boundaries of the filler particles of a spherical shape and polymer matrix can be calculated using Eqs. 9.2 and 9.3 [44]:

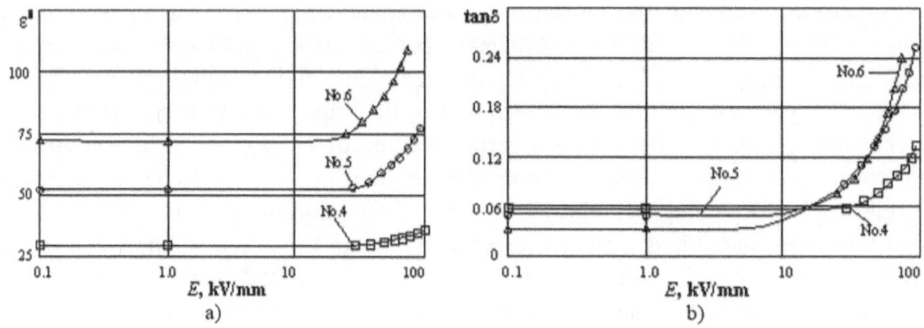

FIGURE 9.3 Relationships ε′ (a) and tanδ (b) on the field strength E for elastomeric composites.

$$E_L = \frac{3\varepsilon_f}{2\varepsilon_c + \varepsilon_f} E_0, \tag{9.2}$$

where ε_f and ε_c are permittivity of the filler and polymer composite; E_0 is the field strength. In this case, the field acting on the filler particles is given by:

$$E_f = \frac{3\varepsilon_c}{2\varepsilon_c + \varepsilon_f} E_0 \tag{9.3}$$

The results of the E_L evaluation showed that with an increase in the filler content from 10 to 40 vol%, the value of the local field is about three times higher than the value of the external field. At the same time, it is possible to achieve a decrease in E_L without increasing the filler content either by increasing the matrix permittivity or by decreasing the filler permittivity. However, a decrease in ε' of the filler will result in a decrease in the effective value of the composite permittivity. Therefore, obtaining a relatively homogeneous material filled with ferroelectric ceramics is a rather difficult challenge.

Another feature of ferroelectric composites is the presence of electrical hysteresis in the relations of ε' and $\tan\delta$ on the electric field strength E represented in Figure 9.4. It can be seen that the values of ε' and $\tan\delta$, measured with changing the AC voltage upwards and downwards, differ significantly.

Availability of hysteresis may be caused by the fact that when measuring ε' и $\tan\delta$ in the upwards direction, the vectors of electric moments of domains of the filler are oriented in the field direction (domain polarization). With a decrease in voltage (downwards direction), an increase in ε' and $\tan\delta$ is due to the delay of polarization, an increase in the energy required to changing the direction of the polarization vector of ferroelectric domains.

Figure 9.5 shows typical oscillograms of electrical hysteresis loops for composite No. 4, obtained at various values of the external field. It can be seen that in the region of weak fields, linear relation of the electric displacement on the magnitude of the external field is observed. However, with a further increase in the field ($E > 6$ kV/mm), an electrical hysteresis is observed, that is, a delay in polarization relative to the applied external field.

Moreover, the form of hysteresis loops for polymer composites even with filler the content 55 vol% significantly differs from the typical dependences $D = f(E)$ for ceramic ferroelectrics in that for all composites saturation of hysteresis loops $D = f(E)$ is not observed even in the region of a high electric field.

The magnitude of the electrical displacement D is given by Eq. 9.4 [1]:

$$D = \varepsilon'\varepsilon_0 E = D_o + P = \varepsilon_o E + P, \tag{9.4}$$

where ε_o is electric constant, ε' is the real part of the complex permittivity, P is total polarization, and E is instantaneous values of the electric field, in particular, for AC voltage $E = E_m \cdot \sin(\omega t + \varphi)$.

Figure 9.5 shows that when $E = 0$ (at t = 0; 0.5T; T; etc., where T is the period of external AC voltage), the electric displacement (the polarization) has the value P_{rem},

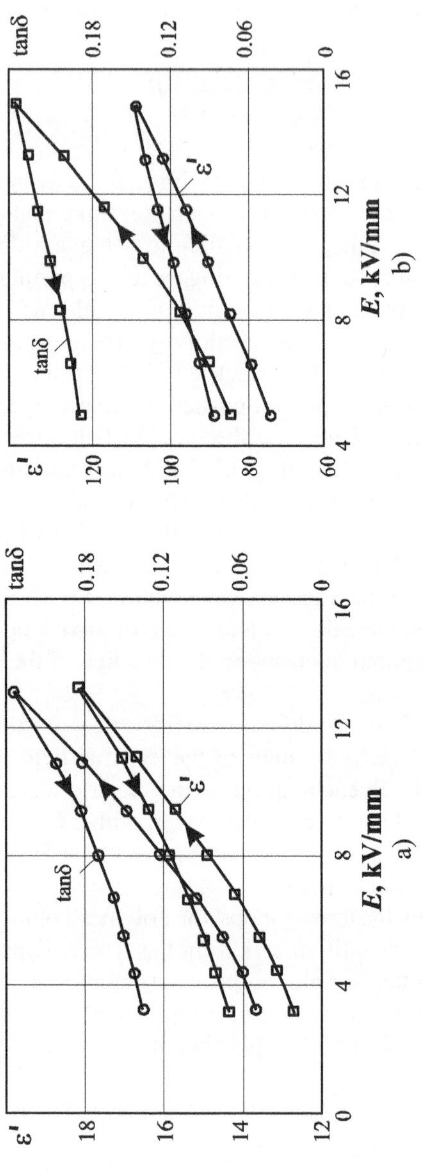

FIGURE 9.4 Dependencies of ε′ and tanδ on the field strength for composites No. 3 (a) and No. 6 (b).

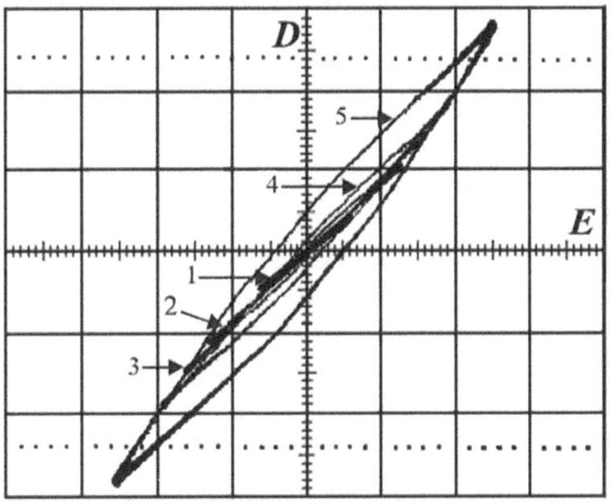

FIGURE 9.5 Typical shapes of dielectric hysteresis loops for the composite No. 4: 1, 2.5 kV/mm; 2, 5.0 kV/mm; 3, 6.0 kV/mm; 4, 7.5 kV/mm; 5, 10.0 kV/mm.

the so-called remnant polarization. In order to reduce the composite polarization to zero, it is necessary to apply electric field the opposite direction E_c (the coercive field, by analogy with ferromagnetics).

Figure 9.6 shows the field dependences of the total polarization, calculated according to eq. 9.4, for elastomeric composites. It can be seen that the field dependences of the total polarization are nonlinear in character. Moreover, the total polarization values increase with the filler content. The nonlinear growth of the total polarization is caused by the fact that at the threshold value of the external field, the orientation of the polarization vectors of the ferroelectric domains of the filler is observed resulting in sharp increase in the total polarization of the composites.

The third group of composites developed in this part of work, are composites based on polyvinylidene fluoride (PVDF) filled with PZT. PVDF is a highly polar polymer with a permittivity of about 10, so it is often applied as a matrix for the preparation of PZT-filled composites.

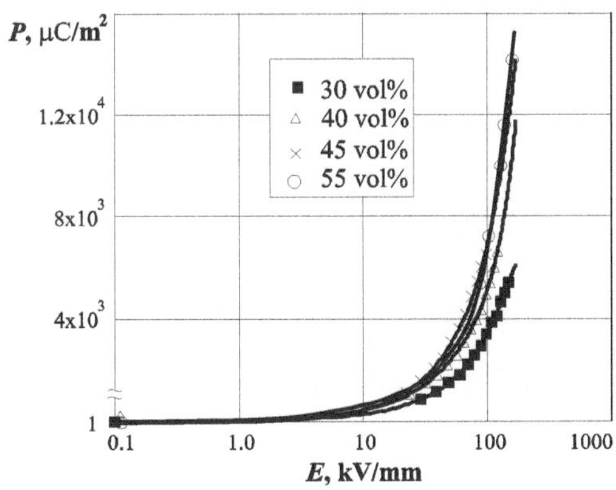

FIGURE 9.6 Relationships between the total polarization and the field strength for elactomeric composites.

To compare properties of composites based on LDPE, elastomers, and PVDF comparative tests were carried out by the method of dielectric spectroscopy, the results of which are shown in Figures 9.7–9.9.

It can be seen that for composites based on LDPE (No. 1–3), filling the polymer matrix with PZT powder leads in the formation of a maximum of dielectric losses in the high-frequency range (Figure 9.7). The values of ε′ and tanδ increase and the maximum itself shifts to lower frequencies at the PZT content 10–40 vol%. The shift of the position of the tanδ peak to lower frequencies with an increase in the filler content is a known phenomenon from the viewpoint of the behavior of relaxation types of polarization. The addition of filler (PZT) to the matrix can lead to the formation of weakly bound polar side groups and radicals, the mobility of which decreases with increasing the filler content. In turn, this can result in a gross in the activation energy W of the polarization process. The relaxation time of polarization τ (or the characteristic frequency $F = 1/\tau$ corresponding to the maximum of tanδ) will change as [45]:

$$\tau = \tau_0 \cdot e^{W/kT}, \tag{9.5}$$

where τ_0 is a constant, W is the activation energy of polarization, k is the Boltzmann constant, and T is temperature. Thus, the relaxation time will decrease with increasing

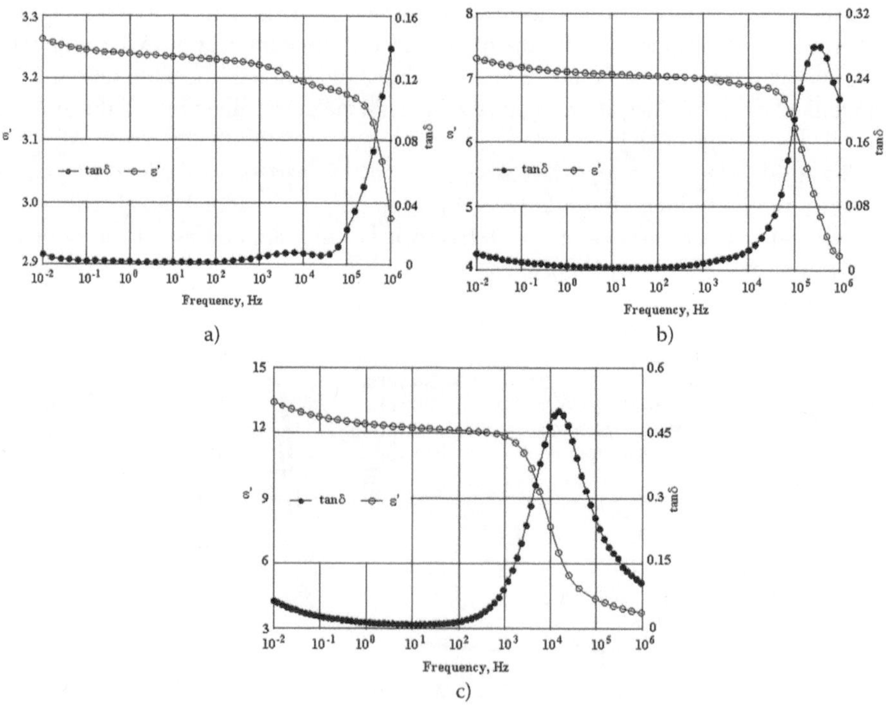

FIGURE 9.7 Dependencies of ε′ and tanδ on the frequency for composites based on LDPE with different filler content: a, 10 vol% PZT (No.1); b, 30 vol% PZT (No. 2); c, 40 vol% PZT (No. 3).

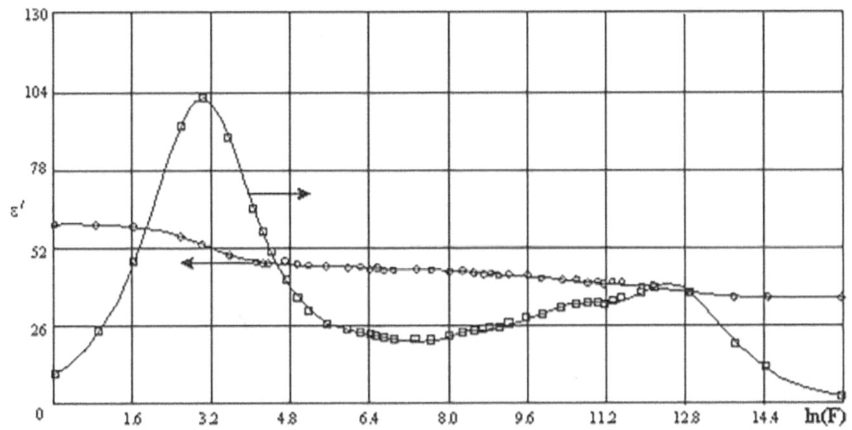

FIGURE 9.8 Typical dependencies of ε′ and ε″ on ln(F) for elastomeric composite No. 5.

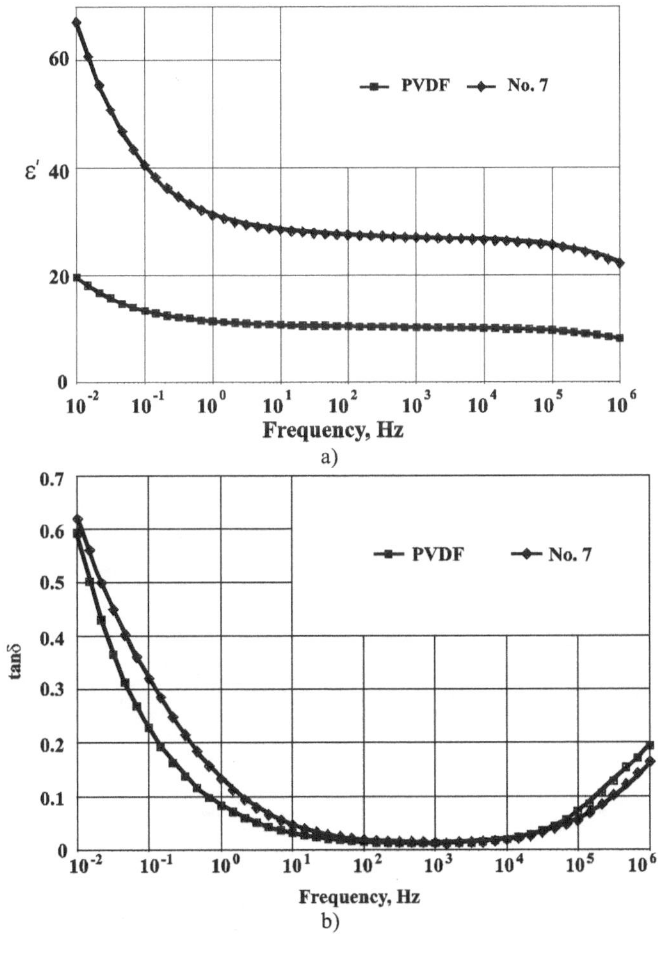

FIGURE 9.9 Dependencies of ε′ and tanδ on the frequency for PVDF-based composites.

temperature (while the characteristic relaxation frequency F will increase) and increase with increasing filler content in the composite.

It should be noted that the increase in ε′ for composites based on LDPE at the frequency 10^3–10^{-2} Hz does not exceed 12.0–13.5 even with a filler content of 40 vol%, despite the fact that the permittivity of PZT is approximately 1,700. This indicates that the interfacial interaction of the non-polar matrix and the polar filler is very weak. However, the value of ε′ increases compared to the polymer matrix. That is, composites No. 1–3 can be used in the low-frequency range (10^{-2} to 10^3 Hz).

Elastomeric composites No. 4–6 (see, for example, Figure 9.8) possess a high permittivity, but high dielectric losses limit their application in the studied frequency range.

Composites based on PVDF are characterized by the presence of a dispersion of the complex permittivity both in the range of low and high frequencies. Typical dependences ε′ and tanδ for PVDF and composite No. 7 are shown in Figure 9.9. It is obvious that the addition of a filler with high permittivity into the PVDF matrix results in an increase in ε′ by a factor of 3 for composite No. 7 compared to PVDF in the frequency range from 10° to 10^5 Hz. In this case, dielectric losses for PVDF and composite No. 7 practically do not change in frequency range studied. That is, PVDF-based composites can be used at the frequency 10°–10^5 Hz, since at other frequency ranges their dielectric losses increase sharply.

To study the effect of the filler on the temperature of phase transitions, all composites were controlled by DSC-TGA. The experimental results of the DSC-TGA analysis are shown in Figures 9.10 and 9.11.

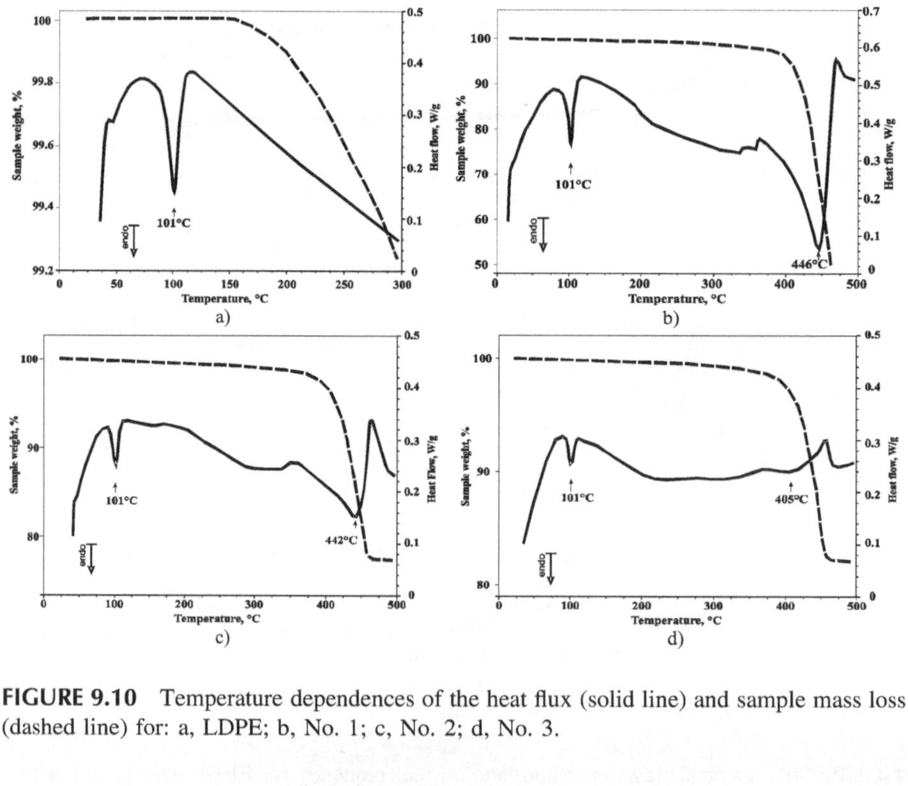

FIGURE 9.10 Temperature dependences of the heat flux (solid line) and sample mass loss (dashed line) for: a, LDPE; b, No. 1; c, No. 2; d, No. 3.

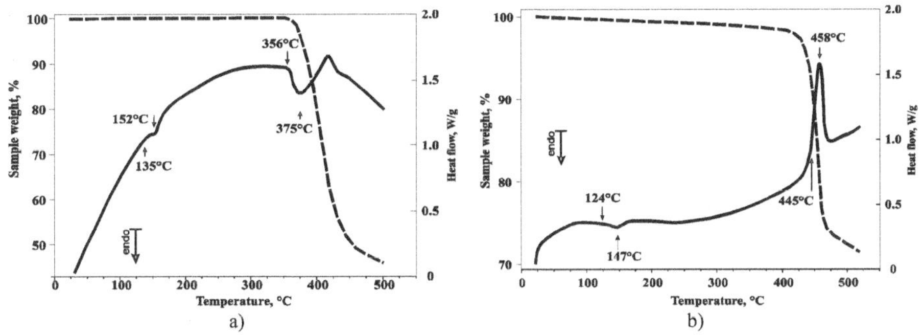

FIGURE 9.11 Temperature dependences of the heat flux (solid line) and sample mass loss (dashed line) for: a, PVDF, and b, composite No. 7.

As can be seen from Figure 9.10, the temperature of the first phase transition, corresponding to the melting process of the studied composites, practically does not change and is approximately 101°C. This indicates that the adding PZT powder into LDPE practically does not result in a change of the melting point of polymer composites compared to that for the LDPE matrix. The second endothermic minimum, corresponding to the decomposition process, is observed at temperatures of 405°C–446°C. In this case, the decomposition temperature decreases from 446°C to 405°C with a change in the content of PZT with LDPE from 10 to 40 vol%.

Similar results of DSC-TGA analysis of PVDF and composite No. 7 are shown in Figure 9.11. The thermogram for PVDF also shows two phase transitions. The first one, at a temperature of about 152°C, corresponds to the melting point, and the second one, at 375°C, corresponds to the decomposition temperature of PVDF.

The melting point for composite No. 7 is reduced by 4°C–5°C compared to PVDF. In addition, an exothermic maximum associated with the thermal oxidation reaction is observed at 375°C in contrast to the endothermic process in PVDF. This is caused by the fact that PZT contains oxygen. The interaction between polar filler and polar polymer matrix is much stronger compared to LDPE-based composites because PZT is an active filler. This result in both a slowdown in the thermal oxidative degradation of the matrix and an increase in the temperature at which the decomposition process of composite No. 7 begins (~445°C) compared to PVDF.

9.3.2 Ferroelectric Composites Based on Biodegradable Poly(Lactic Acid)

Rheological studies (MFI measurements) are very important for ferroelectric composite materials for tissue engineering, since the majority of scaffolds for regenerative medicine are prepared by the extrusion method using 3D-printing technology [13,46–48]. These results show the suitability of a particular composite material for extrusion processing. The results of MFI measurements are shown in Figure 9.12.

As expected, the MFI values for both PLA/$BaTiO_3$ and PLA/$BaTiO_3$/CNT composites decrease with increasing barium titanate content. Notably, however, the MFI values both for PLA/$BaTiO_3$ composites with 5 wt% and 10 wt% of filler

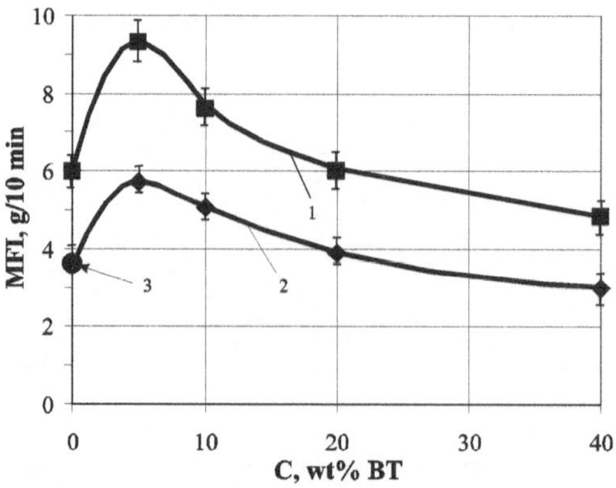

FIGURE 9.12 Melt flow index as a function of the BT content for: 1, PLA/BT composites; 2, PLA/BT/CNT composites; and 3, PLA/CNT composite.

content are significantly higher compared to neat PLA and PLA/$BaTiO_3$/CNT composites. This may be due to the fact that the $BaTiO_3$ particles in composites may act as a dry lubricant (similar to graphite particles in friction pairs), resulting in a decrease in the friction coefficient, an increase in the mobility of polymer chains of the matrix, and an increase in the MFI compared to the neat PLA. An indirect confirmation of this assumption is that the MFI for the PLA/0.05 wt% CNT composite (pointed with an arrow) is only 3.67 g/10 min, which is about 60% less compared to neat PLA.

Besides, it can be seen that the MFI values of PLA/$BaTiO_3$ composites are higher compared to those for PLA/$BaTiO_3$/CNT composites, despite the same $BaTiO_3$ content. It can be due to the fact that CNT particles act as reinforcing elements, hindering the movement of macromolecular chains of the matrix, which results in a decrease of MFI for PLA/$BaTiO_3$/CNT composites. As shown in Figure 9.12, the MFI values for all developed composites are more than 3 g/10 min even at 40 wt% $BaTiO_3$, which implies that these composites may be processed by extrusion, in particular, by 3D-printing technology.

The dielectric spectroscopy results for PLA/$BaTiO_3$ composites are shown in Figure 9.13. It can be seen that the real part of the complex permittivity ε' for PLA/$BaTiO_3$ composites at 40 wt% of filler increases by about two times compared to the neat PLA both at the low- and high-frequency ranges, despite the high permittivity of $BaTiO_3$.

All composites show a weak peak of $\tan\delta$ near 10^5 Hz. This weak peak for polymer composites is associated with the presence of –OH groups, since the measurements are carried out in air. It is obvious that all PLA/$BaTiO_3$ composites are non-conductive materials as the phase angle φ is close to 90° while the angle $\delta = (90° - \varphi)$ is less than 2°, and $\tan\delta$ does not exceed $4 \cdot 10^{-2}$.

This can be explained by the fact that particles of the filler are surrounded by the thin layer of the matrix and interact weakly with each other. In this case, even at the filler content of 40 wt%, these composites have an "islands-in-ocean" structure, when the filler particles are separated from each other by insulating layers of a

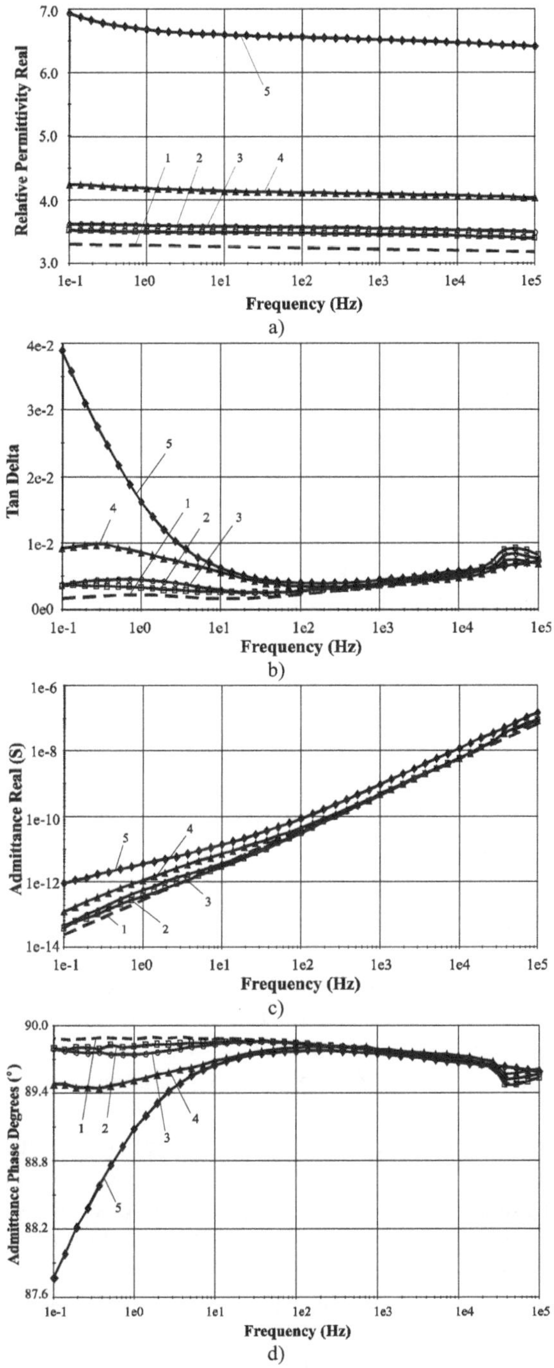

FIGURE 9.13 Frequency dependencies of ε′ (a), tanδ (b), γ_a (c), and φ (d) for PLA/BT composites: 1, PLA; 2, 5% BT; 3, 10% BT; 4, 20% BT; and 5, 40% BT.

polymer matrix and do not form a volume network. This situation is similar to that which occurs near the percolation threshold in the percolation theory of electrical conductivity [49,50], or in the case of interfacial thermal resistance, which is known as the Kapitza interfacial resistance in the theory of thermal conductivity [51,52].

For comparison, Figure 9.14 demonstrates similar frequency dependences for the PLA/CNT composite.

It is obvious that the permittivity of this composite sharply increases compared to that both for the neat PLA (by about 6.3 times at the low frequency range) and for PLA/40% $BaTiO_3$ composite (by about three times at the low-frequency range). A similar situation is characteristic for the admittance (the increase in γ_a at the low frequency reaches about 700 times in comparison with the PLA) and for loss factor tanδ (it increases from 0.002 for the PLA and from 0.039 for PLA/40% $BaTiO_3$ composite to 0.26 for the PLA/CNT composite).

The dielectric spectra of ε′, tanδ, γ_a, and φ for PLA/$BaTiO_3$/CNT composites with a fixed CNT content of 0.05 wt% and various content of $BaTiO_3$ are shown in Figure 9.15. The value of ε′ for three-phase composites is significantly changed with increasing the $BaTiO_3$ content. Indeed, the value of ε′ at 0.1 Hz increases from 3.25 for the neat PLA to 134 for PLA/$BaTiO_3$/CNT composites with 40wt% $BaTiO_3$, while at 10^6 Hz it changes from 3.15 to 10, respectively (Figure 9.15a). The loss factor tanδ for all three-phase composites does not exceed 0.62 (Figure 9.15b). High dielectric losses for all three-phase composites can be caused by the domain polarization of the filler. Besides, the increase in dielectric losses may be due to dielectric losses caused by through conductivity (Figures 9.14c and 9.15c).

The admittance of three-phase composites is increased both at the low-frequency range (by about four orders, from $2.3\cdot10^{-14}$ S for neat PLA to $2.7\cdot10^{-10}$ S for the three-phase composite at 40 wt% of $BaTiO_3$) and at the high-frequency one (by more than one order of magnitude) with increasing the filler content. This phenomenon can be caused by the high conductivity of CNTs, despite their low content in three-phase composites.

In this regard, a small amount of filler ($BaTiO_3$) in the matrix does not result in an increase in the permittivity of two-phase composites, despite the very high permittivity of the filler (Figure 9.13a). The higher the filler content, the more particles/clusters can be formed, which tends to overlap and form a continuous network throughout the composite volume in accordance with the percolation model.

Figure 9.16 demonstrates the SEM microscopy results for the cryo-fractured surface of samples, while Figure 9.17 demonstrates the schematic model of the structure of composites studied. It can be seen that separate particles/clusters of $BaTiO_3$ at a low filler content (< 20 wt%) are separated by thin layers of the PLA matrix. As the filler content increases, the distance between the filler particles/clusters decreases and the probability of direct contact between them increases (Figure 9.16d). This fact explains the significant increase in the permittivity of PLA/$BaTiO_3$ composites at the filler content of 40 wt%.

Figures 9.16e and 9.16(f–i) demonstrate the surface structure of the PLA/CNT and PLA/$BaTiO_3$/CNT composites. It is obvious that the addition of CNTs into PLA/$BaTiO_3$ composites results in the formation of local conductive bridges between

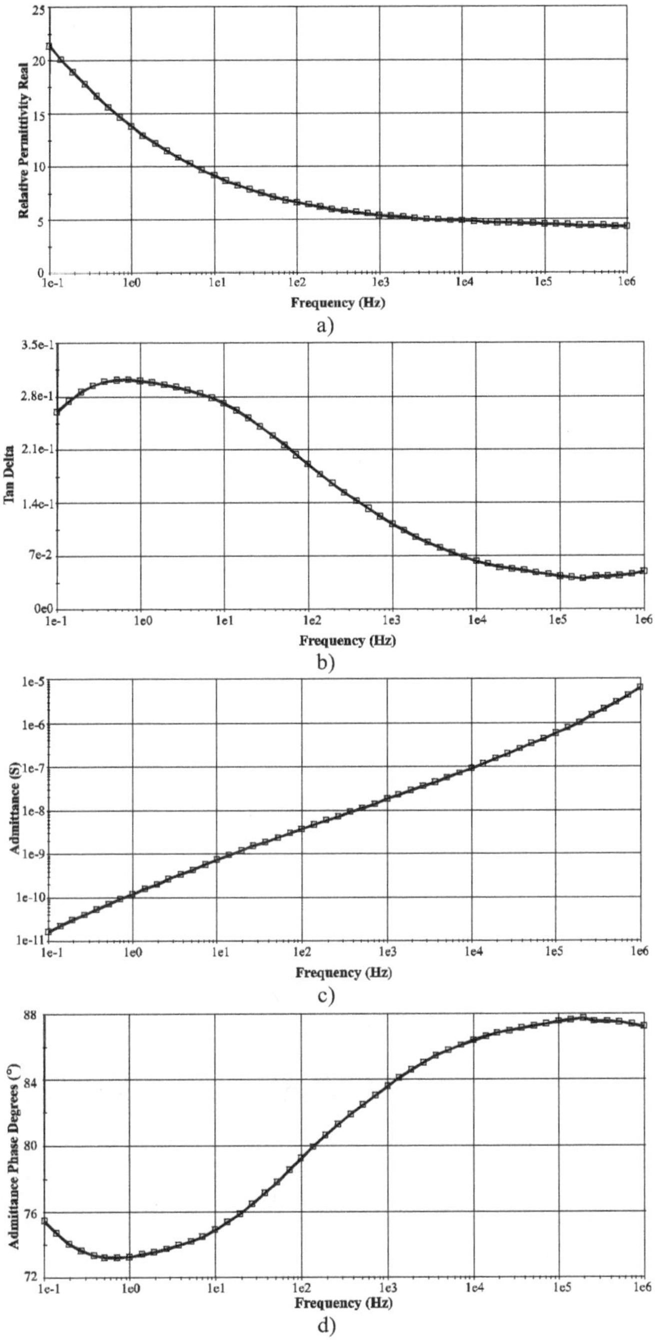

FIGURE 9.14 Frequency dependencies of ε′ (a), tanδ (b), γ_a (c), and φ (d) for PLA/CNT composite.

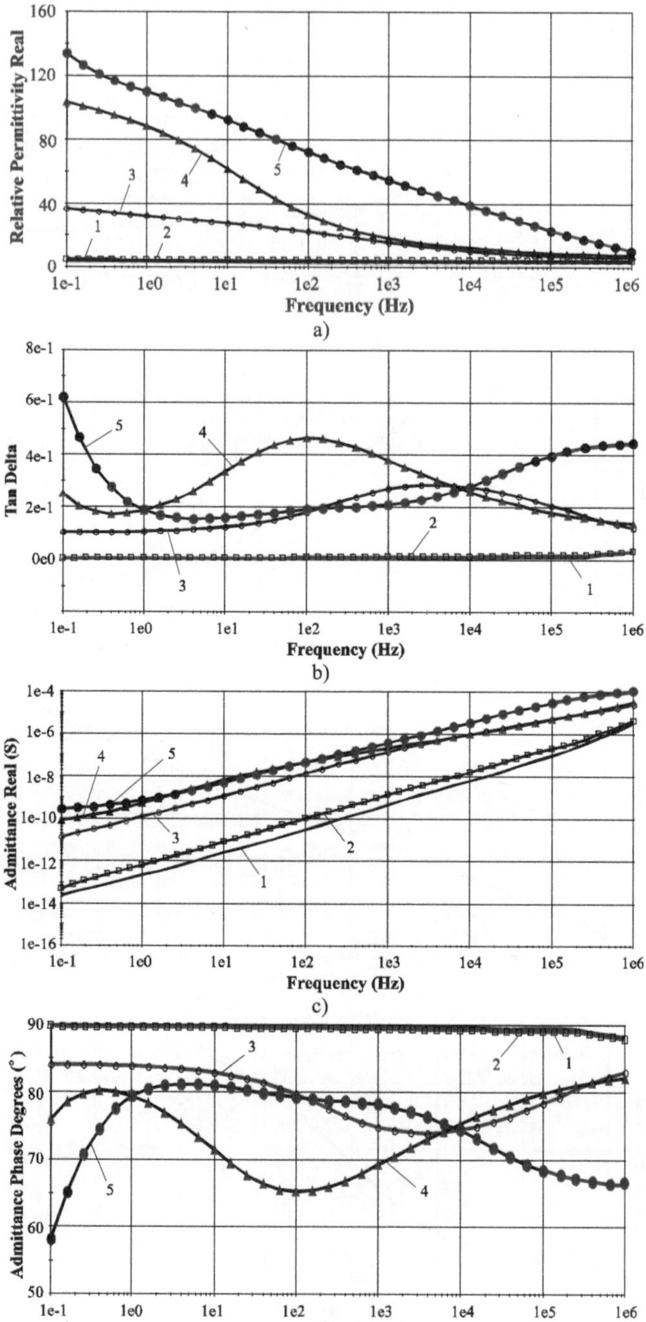

FIGURE 9.15 Frequency dependencies of ε′ (a), tanδ (b), γ_a (c), and φ (d) for PLA/BT/CNT composites with different BT content: 1, PLA; 2, 5% BT; 3, 10% BT; 4, 20% BT; and 5, 40% BT.

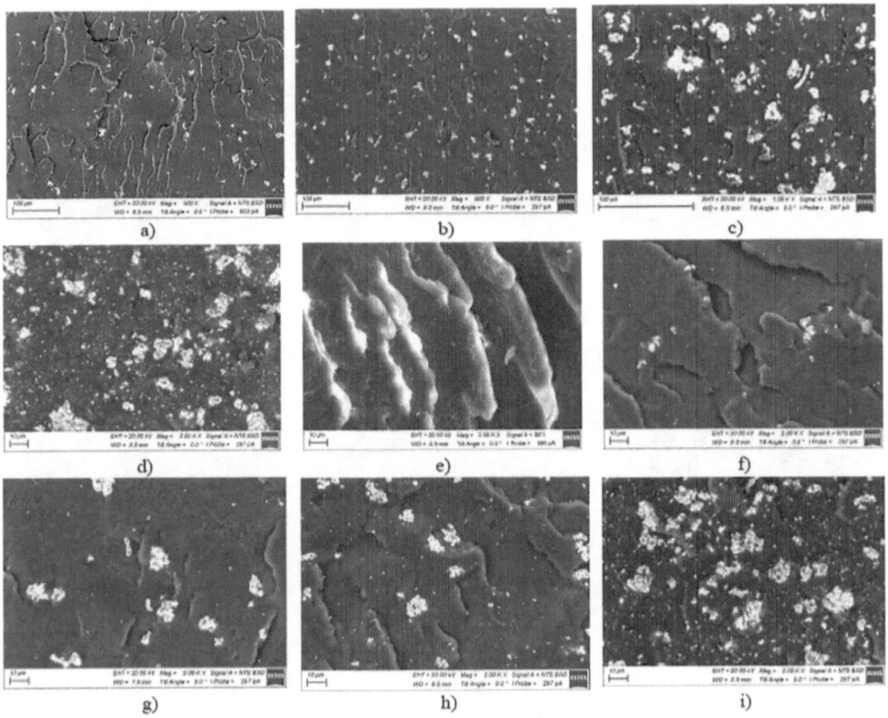

FIGURE 9.16 SEM images of composite surface structures: (a–d) PLA/BT; (e) PLA/CNT; (f–i) PLA/BT/CNT.

$BaTiO_3$ particles/clusters. These conductive bridges connect $BaTiO_3$ particles/clusters to each other, forming continuous pathways or networks due to the high aspect ratio (the ratio of CNTs length to their diameter is more than 3,000), which leads to an increase in the permittivity.

On the other hand, an increase in the $BaTiO_3$ content in three-phase PLA/ $BaTiO_3$/CNT composites leads to an improvement in the distribution of CNTs and to the partial destruction of CNT bundles due to high shear deformation during compounding. In turn, a more uniform distribution of CNTs in the volume of the three-phase composites leads in an increase of both the permittivity and admittance as compared to two-phase ones. In contrast to two-phase composites, an increase in the value of ε' for three-phase composites is visible even at the $BaTiO_3$ content of 10 wt%. This situation is similar to lowering the percolation threshold for electrical conductivity in the percolation model.

It is well known that ferroelectrics have a domain structure. Inside each ferroelectric domain, there is a spontaneous polarization, the field strength vector of which is randomly oriented in the ferroelectric volume. The polarization vectors of each domain under the electric field are oriented in the electric field direction, resulting in a sharp increase in the permittivity of the ferroelectric. A consequence of the domain structure of ferroelectrics is the nonlinear dependence of their

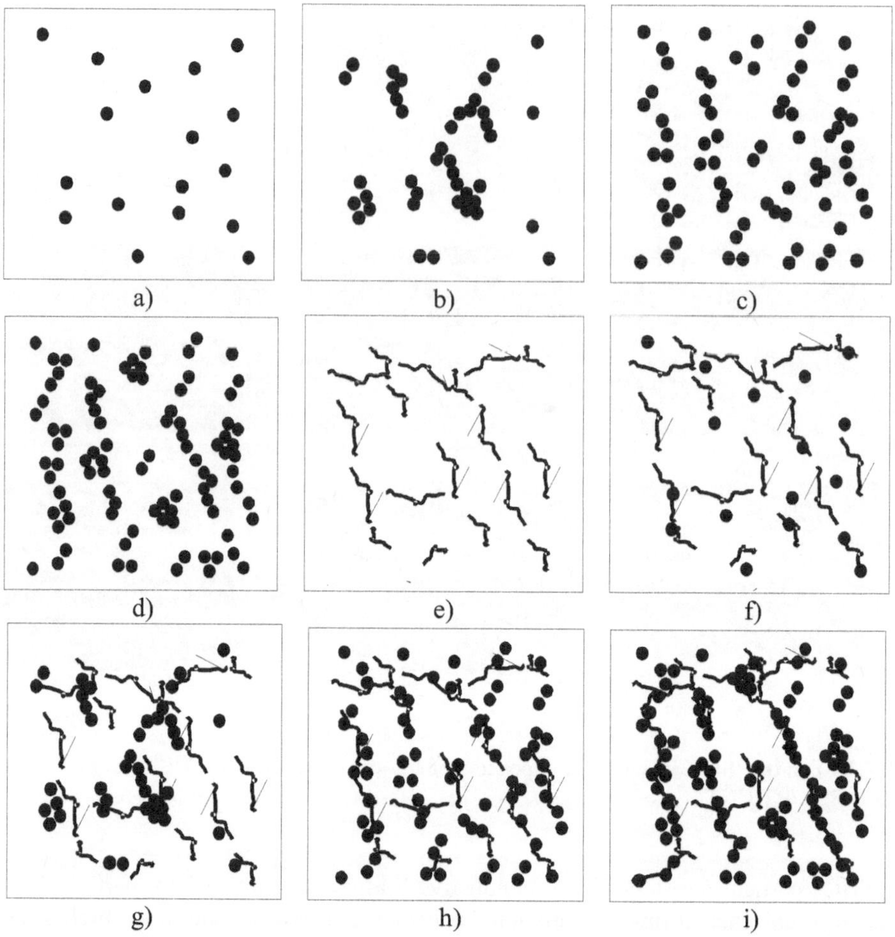

FIGURE 9.17 Schematic diagrams of composite structures: (a–d) PLA/BT; (e) PLA/CNT; (f–i) PLA/BT/CNT; black circles are BT particles; black lines are CNTs; white is PLA matrix.

polarization on the field strength and the polarization delay when the direction of the external field is varied, the so-called dielectric hysteresis loop. Thus, the permittivity value of ferroelectric composites depends on the electric field strength.

Figure 9.18 clearly demonstrates the relation of properties for ferroelectric composites on the field strength, despite the fact that the electric field strength during the measurements was very low. It can be seen that the values of ε' and γ_a for PLA/40%$BaTiO_3$/CNT composites at the low-frequency range (<100 Hz) change by about two and three times, respectively, when the electric field strength varies by a factor of 100. This is explained by the orientation of the dipole moments of $BaTiO_3$, which manage to orient themselves in the field direction at low frequencies of the external electric field, close to the relaxation frequency of domain polarization. The values of ε' and γ_a at the high-frequency range do not practically change for all composites when the electric field strength is varied, which once

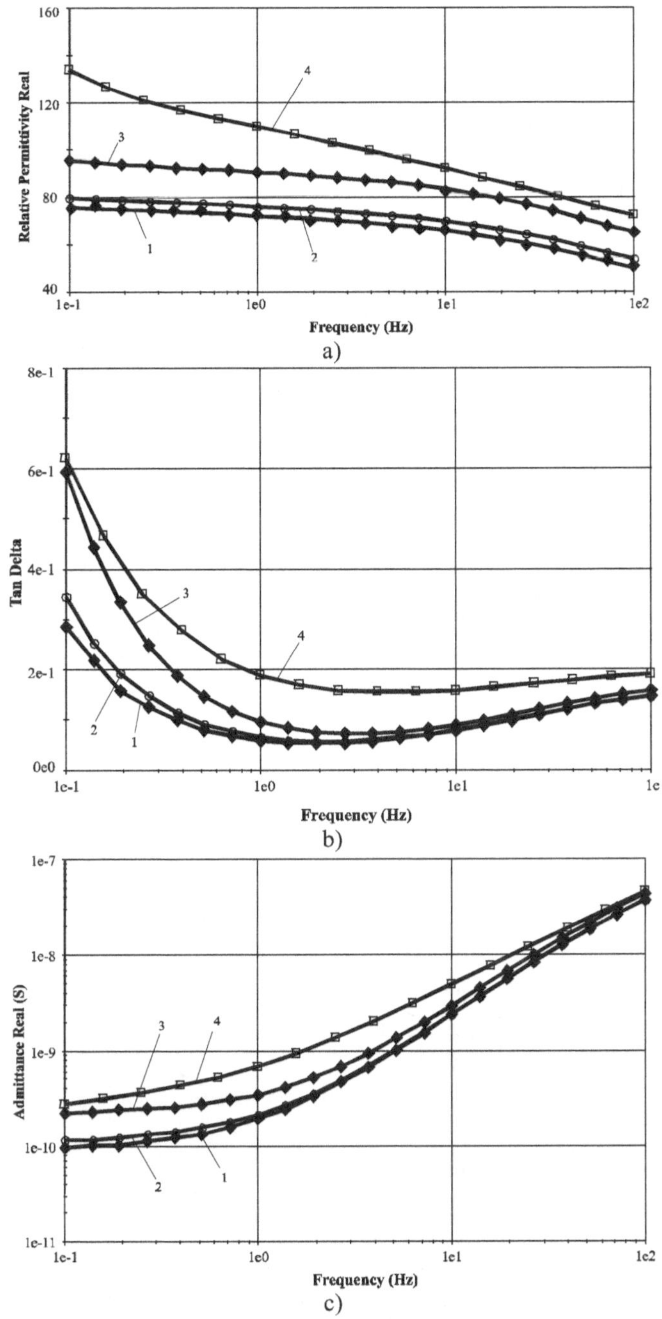

FIGURE 9.18 Frequency dielectric spectra of ε' (a), $\tan\delta$ (b), and γ_a for PLA/40%BT/CNT composites at different field strength: 1, 34.5 V/m; 2, 345 V/m; 3, 1150 V/m; and 4, 3450 V/m.

again confirms the fact that these changes are caused by domain polarization and through conductivity.

Even more convincing proof that the developed composites are ferroelectric materials can be the results of studying the dielectric hysteresis. The hysteresis loops (the relations between the electric displacement and electric field strength $D = \varepsilon_0 E + P$, where $\varepsilon_0 = 8.85 \cdot 10^{-12}$ F/m is the electric constant; $E = E_0 \cdot \sin(\omega t + \varphi)$) are the instantaneous electric field strengths and P is the polarization) for two-phase PLA/$BaTiO_3$ composites; these are presented in Figure 9.19.

The availability of dielectric hysteresis loops (delay of polarization with a change in the magnitude of the field) is a characteristic feature of ferroelectrics.

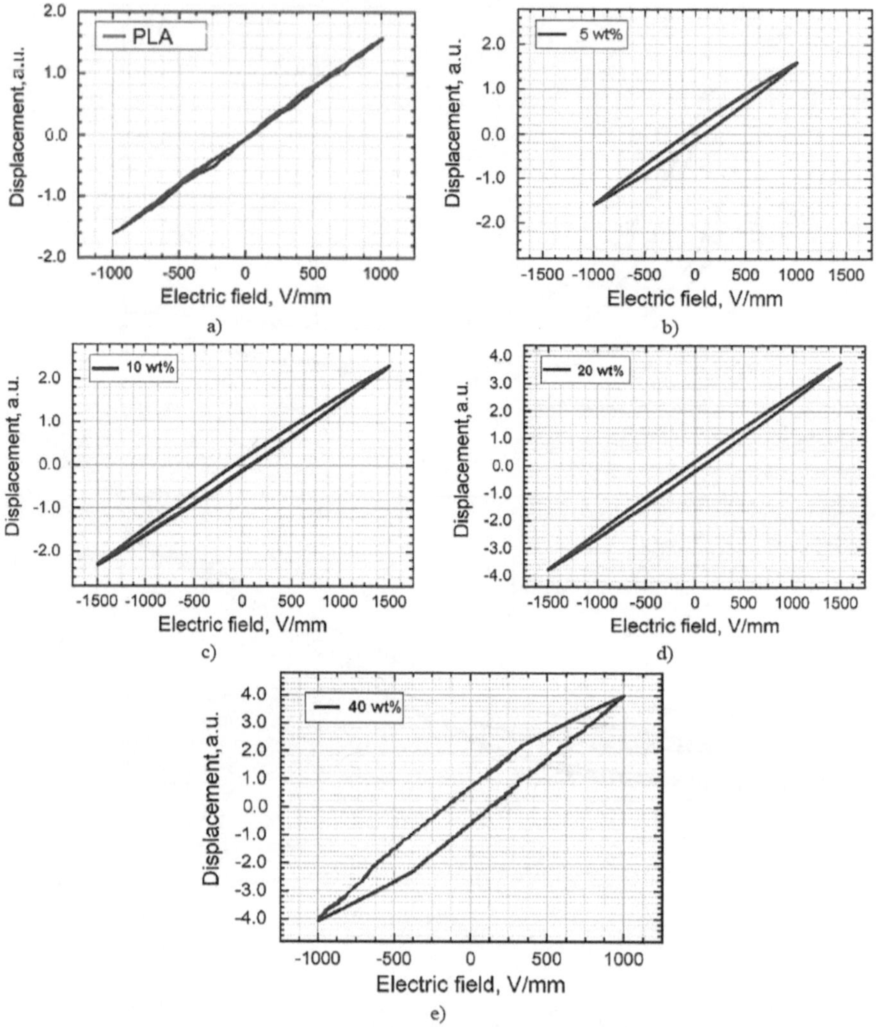

FIGURE 9.19 Dielectric hysteresis loops for: a, PLA; b, PLA+5 wt%$BaTiO_3$; c, PLA+10 wt%$BaTiO_3$; d, PLA+20 wt%$BaTiO_3$; and e, PLA+40 wt%$BaTiO_3$.

It can be seen that for the PLA polymer matrix, the relation of the dielectric displacement on the strength of the external field is practically a straight line. The maximum value of the electric displacement for each composite is increased with increasing the filler content. In addition, for all composites, the presence of a remnant electric displacement (remnant polarization) is observed when the field direction changes. Unfortunately, we were unable to measure dielectric hysteresis loops for three-phase composites because of overlap of samples during the measurement. This is conditioned by the high conductivity of CNTs.

The results of tensile tests in the form of dependences of the relative elongation at break ($\Delta l/l$, where Δl is the elongation at break and l is the initial distance between grips) and Young's modulus (E) on the $BaTiO_3$ content are shown in Figure 9.20. The values of elongation and Young's modulus were derived from the stress-strain curves.

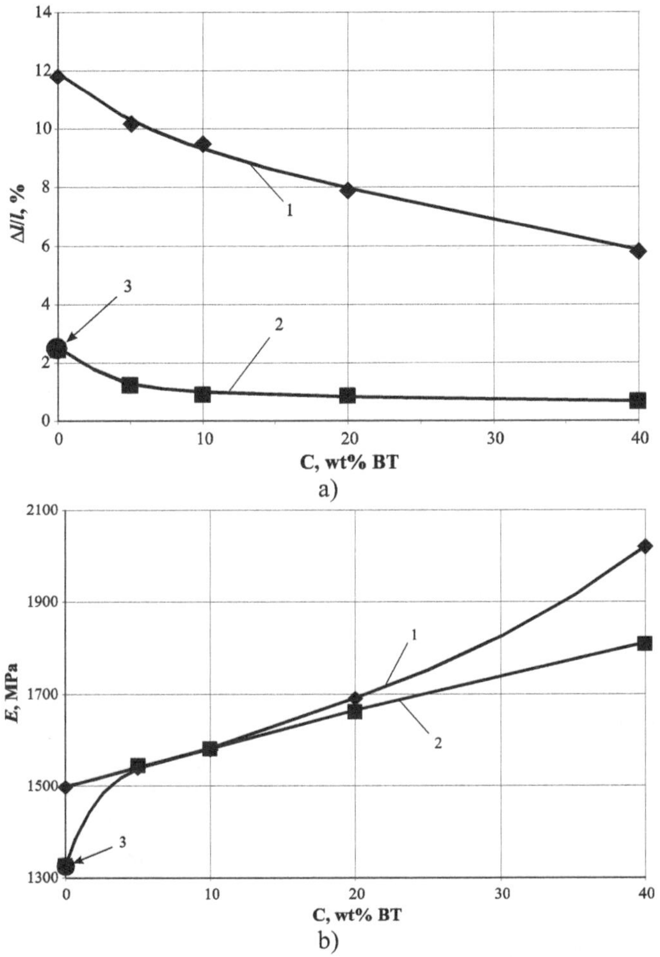

FIGURE 9.20 The dependencies of elongation at break $\Delta l/l$ (a) and Young's modulus (b) on the filler content: 1, for PLA/BT composites; 2, for PLA/BT/CNT composites; and 3, for PLA/CNT composite.

As expected, there are no significant differences in the elongation at break, except for those associated with the sharp decrease in the elongation for PLA/$BaTiO_3$/CNT composites even at the low $BaTiO_3$ content compared to PLA/$BaTiO_3$ composites. However, the change in the rate of decrease in elongation can be conditioned by the presence of a small amount of CNTs, rather than an increase in the $BaTiO_3$ content. For example, an increase in the $BaTiO_3$ content to 40 wt% for PLA/$BaTiO_3$ composites results in a twofold decrease in the elongation compared to the neat PLA, while for three-phase composites with the same $BaTiO_3$ content it is more than 18 times. For comparison, the elongation value for the PLA/0.05 wt% CNT composite is pointed with an arrow in Figure 9.20a. Obviously, the elongation value for this composite is more than five times lower than that for the neat PLA.

At the same time, Young's modulus increases from 1,500 MPa for the PLA matrix to 2,020 MPa and 1,810 MPa for PLA/$BaTiO_3$ and PLA/$BaTiO_3$/CNT composites at 40 wt% $BaTiO_3$, which are 35% and 21% higher than that for the neat PLA. Young's modulus for the three-phase PLA/$BaTiO_3$/CNT composite at 40 wt% $BaTiO_3$ is about 37% higher than that for the PLA/0.05 wt%CNT composite, which is shown in Figure 9.20b with an arrow. As earlier mentioned, this is caused by the fact that the CNT particles act as reinforcing elements, restricting the movement of macromolecular chains of the matrix, which leads to a decrease of their mobility making composites filled with CNTs stiffer and more brittle.

A distinctive feature of composite polymer materials is the appearance of a significant dispersion of the permittivity over a wide frequency range due to the presence of both numerous matrix/filler interfaces and various kinds of polarization, such as Maxwell-Wagner, dipole-relaxation, ion-relaxation, etc. Furthermore, the electroactive ferroelectric fillers themselves possess spontaneous polarization, which causes additional dielectric losses due to the formation of domain polarization under the electric field in a certain frequency range.

Relatively few works are devoted to the study of these problems for ferroelectric composites based on PLA in the literature, although it is the frequency dependences of the dielectric properties that underlie the development of such multiphase ferroelectric composite polymer materials.

Attempts to relate the composition, structure, and properties of multiphase polymer composites filled with barium titanate ($BaTiO_3$) with their dielectric properties were made by Dang et al. [41], Lee at al. [53], and Hammami et al. [54]; however, these attempts were not entirely correct in terms of explaining the obtained experimental data. Thus, when interpreting Cole-Cole plots for polyvinylidene fluoride/multi-walled carbon nanotubes/$BaTiO_3$ composites (PVDF/ $BaTiO_3$/MWCNT), the authors [41] did not take into account the frequency dispersion of the complex dielectric permittivity of composites, assuming that an increase in the value of ε' at the loss factor $\tan\delta$ of 0.6 and 3.0 is owing only to the Maxwell-Wagner-Sillars polarization without regard to conductivity. However, at such values of $\tan\delta$, polymer composites PVDF/MWCNT/$BaTiO_3$ behave like an either as semi-conductive materials (at $\tan\delta = 0.6 \Rightarrow \delta = 31°$, and the phase angle $\varphi = 59°$), or as quasi-conductive ones ($\tan\delta = 3.0 \Rightarrow \delta = 72°$, and $\varphi = 18°$), and the increase in their permittivity is determined mainly by the contribution of through conductivity. On the other hand, Lee at al. [53]

have demonstrated that Nyquist plots for polymer nanocomposites filled with $BaTiO_3$ "... show straight vertical lines ... indicating that the nanocomposite capacitors behave like an ideal capacitor ... ." However, this can be related to the so-called scale effect, and changing the scale along the Z'-axis will cause these plots to transform into completed, uncompleted, or skewed semicircles. Besides the increase in the values of the loss factor tanδ in the high-frequency range, the authors of [53] associate not to the presence of a relaxation maximum of dielectric losses, but "... to the lead resistance of the assembly." Moreover, none of the listed works take into account the domain polarization of electroactive ferroelectric fillers that occurs under an electric field [41,53,54].

The aim of this part of work is establishing the relationship between the composition and dielectric properties of ferroelectric composites filled with barium titanate by the dielectric spectroscopy method.

As can be seen in Figures 9.13 and 9.15, the value of ε′ for the two-phase PLA/$BaTiO_3$ composite at 40 wt% of filler increases about twice, while for the three-phase PLA/$BaTiO_3$/CNT composite at the same filler content it increases by a factor of 40 in comparison with the PLA.

It should be noted that the change in the permittivity has a threshold character with increasing temperature (Figure 9.21). Figures 9.21a and 9.21b are derived from Figure 9.13a and Figure 9.15a, respectively, at a frequency of 0.1 Hz. As shown in Figure 9.21a, the permittivity increases slightly up to 10 wt% of filler content. A further increase in the filler content leads in an increase in the growth rate of the permittivity. The character of the change of ε′ for two-phase and three-phase composites is the same; the only difference being that the rate of change in ε′ for three-phase composites at the filler content more than 5 wt% is significantly higher (Figure 9.21b). This can be conditioned by the fact that the filler particles in two-phase composites weakly interact with each other, since they are surrounded by a thin insulating layer of the matrix. On the other hand, CNTs in three-phase composites form conductive bridges between the barium titanate particles, resulting in the sharp increase of permittivity (see Figure 9.17).

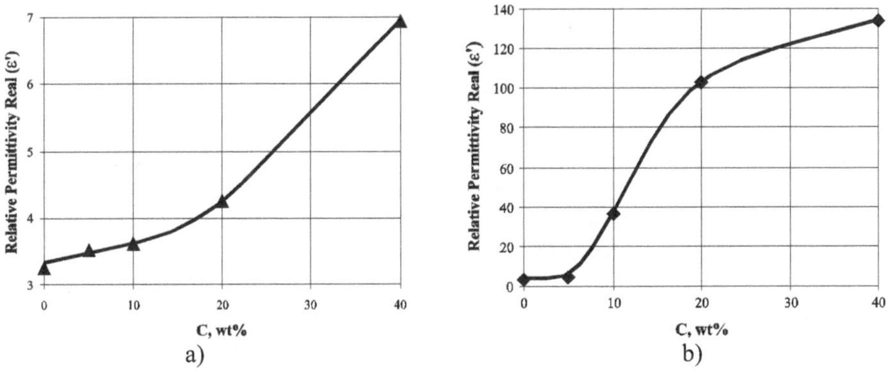

FIGURE 9.21 Dependencies of the permittivity on the BT content for two-phase PLA/BT(a) and three-phase PLA/BT/CNT (b) composites at frequency of 0.1 Hz.

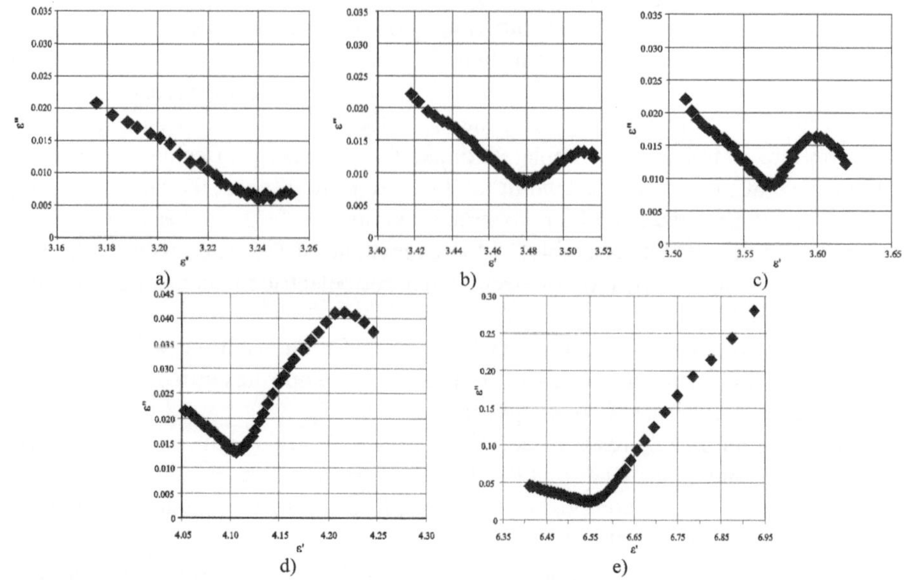

FIGURE 9.22 Frequency relations $\varepsilon'' = f(\varepsilon')$ for the polymer matrix (a) and two-phase PLA/$BaTiO_3$ composites with various filler content: 5 wt% BT (b); 10 wt% BT (c); 20 wt% BT (d); and 40 wt% BT (e).

To establish the relationship between the composition of two-phase and three-phase composites and their dielectric properties, further analysis of the dielectric spectroscopy data was carried out using frequency relationships $\varepsilon'' = f(\varepsilon')$.

Figure 9.22 shows relationships $\varepsilon'' = f(\varepsilon')$ for two-phase PLA/$BaTiO_3$ composites with different filler content. It is obvious that diagrams $\varepsilon'' = f(\varepsilon')$ have an unconventional form of semicircles, but are the result of a superposition of relaxators both at the low-frequency and high-frequency ranges.

As can be seen in Figure 9.22a, the high-frequency component of the permittivity dominates in the dielectric relaxation spectrum of the polymer matrix (the left part of the relationship $\varepsilon'' = f(\varepsilon')$). Adding the filler into the polymer matrix leads to the appearance of a relaxation peak at medium-frequency and low-frequency ranges (right part of relationship $\varepsilon'' = f(\varepsilon')$), which begin to dominate in the dielectric relaxation spectra in two-phase composites with increasing the filler content. As a result, the main contribution into the dielectric relaxation spectrum is conditioned by the low-frequency component of the spectrum (Figure 9.22e). This is stipulated by the maximum of relaxation losses of the neat filler is observed at the low- and medium-frequency range from 10 Hz to 10 kHz [55–57].

Figure 9.23 demonstrates frequency relationships $\varepsilon'' = f(\varepsilon')$ for three-phase PLA/$BaTiO_3$/CNT composites. Similar to the diagrams presented in Figure 9.22, relationships $\varepsilon'' = f(\varepsilon')$ in Figure 9.23 are the result of a superposition of several relaxators with different relaxation times or distribution of relaxation times that are associated with the polymer matrix, the filler ($BaTiO_3$), and CNTs. At the same time, dielectric losses (values of ε'') of three-phase composites increase by several

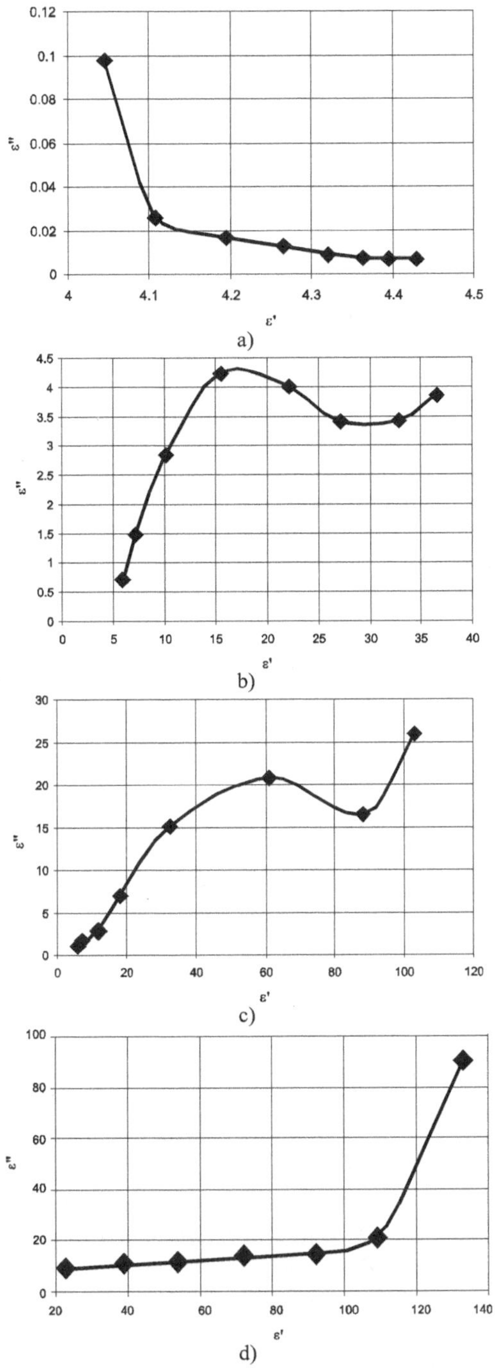

FIGURE 9.23 Frequency relations ε'' = *f*(ε') for three-phase PLA/$BaTiO_3$/CNT composites with various filler content: 5 wt% BT (a); 10 wt% BT (b); 20 wt% BT (c); and 40 wt% BT (d).

orders in comparison with those for two-phase ones. The high-frequency component in the dielectric relaxation spectra for three-phase composites observes only for the composite at 5 wt% $BaTiO_3$. A further increase in the $BaTiO_3$ content leads in an increase in the contribution of the low- and medium-frequency components to the overall dielectric relaxation spectrum, and at a filler content of 40 wt% the main contribution to the dielectric relaxation spectrum is due to conduction losses at the low frequency of the spectrum.

In order to evaluate the effect of temperature on properties of developed ferroelectric composites, the three-phase composite at filler content of 40 wt% was tested in the temperature range from 12°C to 75°C. These tests are very important because it is well known that with elevating temperature, the permittivity of $BaTiO_3$ ceramics increases significantly, reaching a maximum at 120°C (the Curie point). A further increase in temperature results in the same sharp decrease in the permittivity due to the first-order phase transition from ferroelectric to paraelectric condition [33]. The results of these tests are presented in Figures 9.24–9.26.

Figure 9.24 shows temperature dependencies of ε', $\tan\delta$, γ_a, and the phase angle φ. It is obvious that an elevation of temperature leads in the increase in ε' by a factor of 6 (Figures 9.24a and 9.25). The frequency dependence of the loss factor $\tan\delta$ (Figure 9.24b) clearly demonstrates the availability of a frequency dispersion of the permittivity in this composite. It can be clearly seen that there are three regions of dispersion, such as low frequency (10^{-1}–10^{1} Hz), medium frequency (10^{1}–10^{4} Hz), and high frequency (10^{4}–10^{6} Hz). As mentioned earlier, the first of them is conditioned by dielectric losses due to through conductivity (CNTs), the second is caused by dielectric losses owing to slow types of relaxation polarization, and the third is conditioned by a polymer matrix (PLA).

With an increase in temperature, the values of admittance γ_a at the low frequency are increased by a factor of 40. Another confirmation that the frequency dispersion in the low-frequency range of the spectrum is attributed to dielectric losses due to the through conductivity associated with CNTs, is the presence of quasi-steady state DC conductivity, which is observed at a frequency of less than 10^{1} Hz (Figure 9.24c) [58]. A characteristic feature of this kind of conductivity is that the magnitude of the conductivity does not depend on the frequency.

Figure 9.26 provides additional confirmation of the assumption about the relationship the frequency dispersion of the complex permittivity with the composition and properties of multiphase composites. It shows Nyquist plots for PLA/$BaTiO_3$/CNT composites with 40 wt% of filler with elevating temperature. Nyquist plots are the frequency dependencies of the imaginary part Z'' as a function of the real part Z' of the complex impedance $Z'' = f(Z')$ in the form of uncompleted semicircles at the complex plane. Dependencies $Z'' = f(Z')$ for multiphase polymer composites usually carries a negative sign, since these systems are capacitive. It can be seen that both the maximum values of Z'' and the radii of semicircles $Z'' = f(Z')$ decrease with elevating temperature. An increase in temperature leads in an increase in the mobility of carriers of both bound and free charges. The increase in the mobility of charge carriers must be attributed to the rise in the total current, consisting of both the through current (the transfer of free charge carriers) and displacement current (small displacement of bound charges under the field). In turn, an increase in the

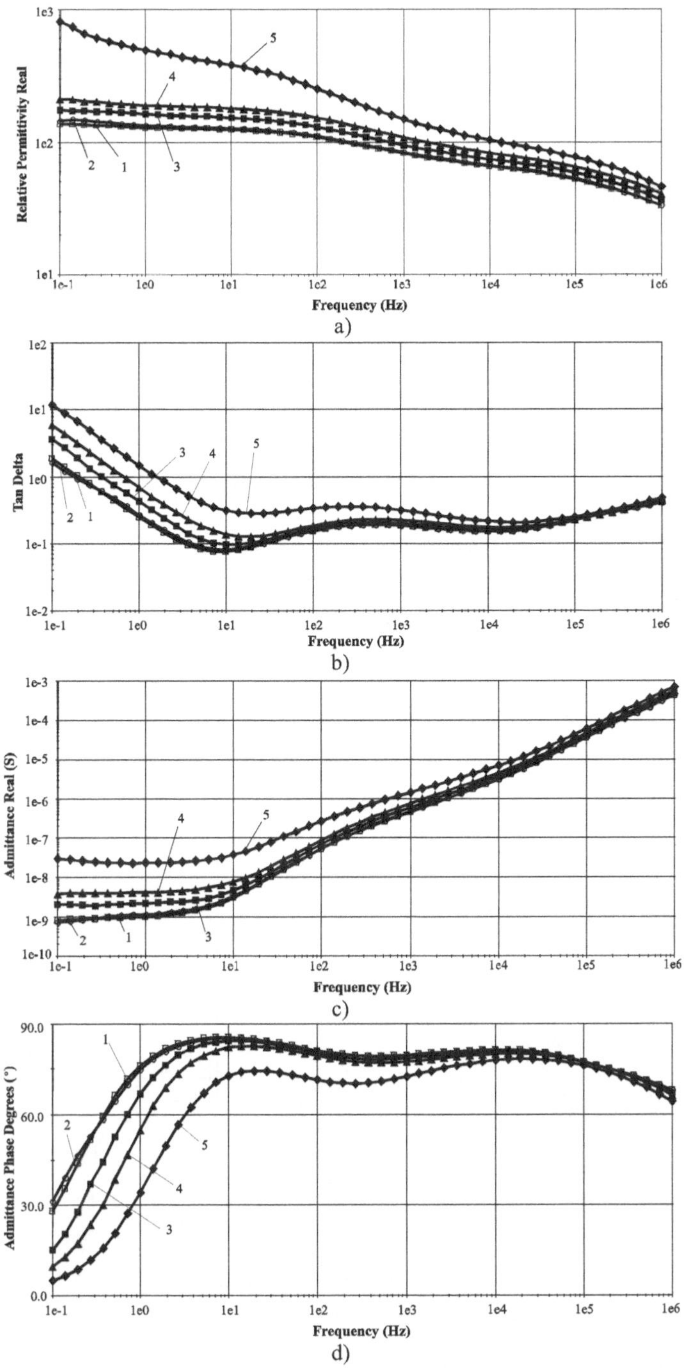

FIGURE 9.24 Frequency dependencies of ε′ (a), tanδ (b), γ_a (c), and φ (d) for PLA/BT/CNT composites at different temperatures: 1, 12°C; 2, 25°C; 3, 40°C; 4, 60°C; 5, 75°C.

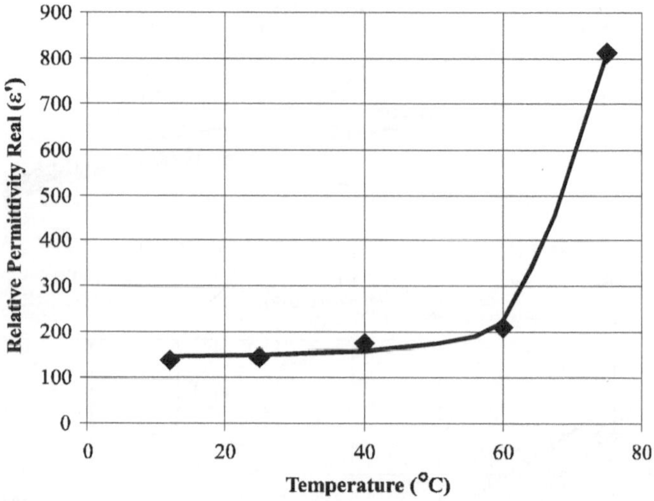

FIGURE 9.25 Temperature dependence of the permittivity for three-phase PLA/BT/CNT composites derived from Figure 9.24a at a frequency of 0.1 Hz.

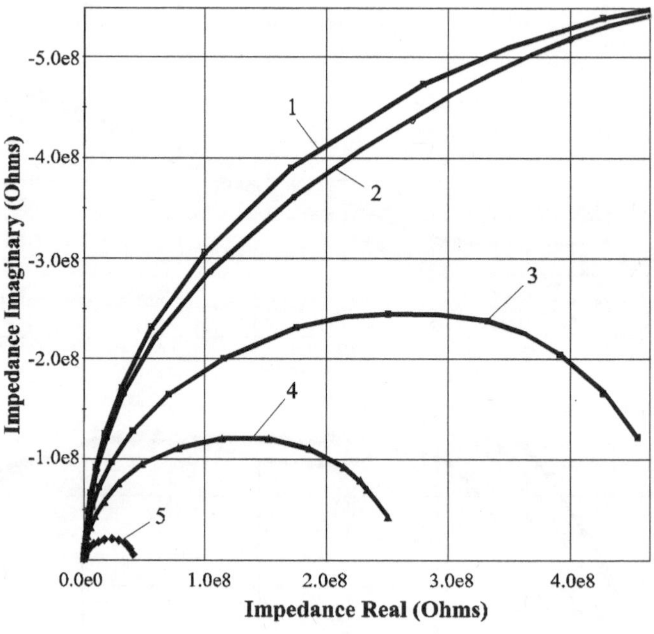

FIGURE 9.26 Nyquist plots for PLA/40wt%BT/CNT composite at different temperature: 1, 12°C; 2, 25°C; 3, 40°C; 4, 60°C; and 5, 75°C.

total current at a fixed measuring voltage should lead to a decrease in the total complex impedance, which is clearly seen in Figure 9.26. At the same time, an increase in temperature results in a narrowing of the width of the dielectric relaxation spectrum and shifts the dielectric relaxation process to lower frequencies.

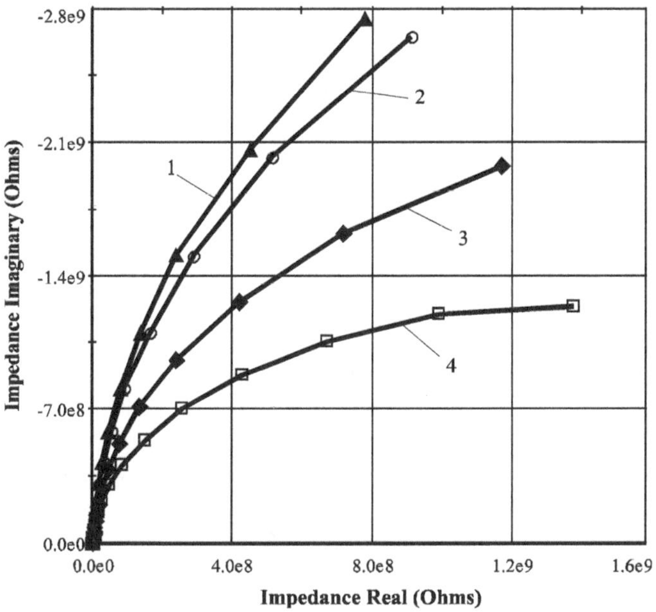

FIGURE 9.27 Nyquist plots for PLA/40wt%BT/CNT composite at different electric field strength: 1, 34.5 V/m; 2, 345 V/m; 3, 1,150 V/m; and 4, 3,450 V/m.

It is well known that $BaTiO_3$ possesses a domain structure. In the initial state, spontaneous polarization exists within each ferroelectric domain. The polarization vector of each domain is randomly oriented in space. The polarization vectors of all ferroelectric domains under the field will be oriented in the electric field direction, resulting in a sharp increase in the permittivity. Another characteristic feature of ferroelectrics is the nonlinear dependence of their polarization on the field strength and the polarization delay when the direction of the external field is changed, the so-called dielectric hysteresis loop. The energy consumption of the field for the dielectric hysteresis contributes to dielectric losses.

Experimental results of studying the effect of the field on the dielectric relaxation spectra in three-phase composites are shown in Figure 9.27.

Figure 9.27 shows Nyquist plots for PLA/$BaTiO_3$/CNT composites with 40 wt% of filler with changing the external electric field strength. It can be clearly seen that an external field affects the dielectric spectra of three-phase PLA/BT/0.05%CNT composites in almost the same way as temperature (see Figure 9.26), activating polarization processes and results in the decrease of the width of dielectric relaxation spectra (decrease in radii of uncompleted semicircles $Z'' = f(Z')$) due to the energy of the external electric field, despite small changes in the external electric field strength.

9.4 CONCLUSIONS

The main conclusions of this work can be summarized as follows:

1. For all studied composites filled with PZT based on LDPE, elastomeric, and PVDF matrices, the phenomenon of dielectric hysteresis is observed, which is caused by the formation of the domain polarization in them under an external field.
2. Composites based on LDPE filled with PZT ferroelectric ceramic powder can be used in the frequency range from 10^{-2} to 10^3 Hz, since relaxation maxima of dielectric losses for these composites are observed at frequency less than 10^{-2} Hz or more than 10^4 Hz.
3. Ferroelectric composites based on PVDF are suitable for use in the frequency range from 10^0 to 10^5 Hz, since in this range they do not have a noticeable frequency dispersion of the complex permittivity.
4. Elastomeric ferroelectric composites possess a high permittivity, but high dielectric losses limit their application in the studied frequency range.
5. This work is the first to develop and characterize multiphase PLA-based ferroelectric composites with high permittivity filled with barium titanate and a small amount of CNT.
6. The addition of CNTs (0.05 wt%) into two-phase PLA/$BaTiO_3$ composites allows both dielectric and mechanical properties of biocomposites to be enhanced. Moreover, such a modification of the structure of ferroelectric composites can significantly reduce the $BaTiO_3$ content in them at the required value of permittivity, which, in turn, can lead to the improvement of the rheological and mechanical properties of biocomposites.
7. The melt flow index of two-phase composites increases by about 55% and 27%, respectively, compared to the neat PLA, when the $BaTiO_3$ content is 5% and 10 wt%.
8. The permittivity of the two-phase PLA/$BaTiO_3$ composite at a filler content of 40 wt% in the low-frequency range increases by about two times in comparison with the PLA, whereas for the three-phase PLA/$BaTiO_3$/CNT composite it increases by more than 40 times at the same filler content. The value of the loss factor tangent delta for these composites is about 0.04 and 0.6, respectively.
9. The experimental data of dielectric spectroscopy in the frequency domain correlate well with the SEM microscopy analysis from the viewpoint of the percolation theory.
10. The Nyquist plots analysis of ferroelectric composites allowed to us to relate the dielectric properties of the developed multiphase composites to their composition.

REFERENCES

[1] Gefle, O. S., Lebedev, S. M., and Pokholkov, Y. P. 2007. *Barrier effect in dielectrics*. Tomsk: TML-Press (in Russian).
[2] Gefle, O. S., Lebedev, S. M., and Uschakov, V. Y. 1997. The mechanism of the barrier effect in solid dielectrics. *J Phys D: Appl Phys* 30:3267–3273.
[3] Chen, Q., Shen, Y., Zhang, S., et al. 2015. Polymer-based dielectrics with high storage energy density. *Ann Rev Mater Res* 45:433–458.

[4] Fan, Y., Huang, X., Wang, G., et al. 2015. Core-shell structured biopolymer@BaTiO3 nanoparticles for biopolymer nanocomposites with significantly enhanced dielectric properties and energy storage capability. *J Phys Chem C* 119:27330. 10.1021/acs.jpcc.5b09619

[5] Yamashita, T., Okada, H., Itoh, T., et al. 2015. Manufacturing process for piezoelectric strain sensor sheet involving transfer printing methods. *Jpn J Appl Phys* 54:10ND08. 10.7567/JJAP.54.10ND08

[6] Ruschau, G. R., Newham, R. E., Runt, J., et al. 1989. 0–3 ceramic-polymer composite chemical sensors. *Sens Actuators* 20:269–275.

[7] Post, J. E. 2005. Microwave performance of MCM-D embedded capacitors with interconnects. *Microwave and Optical Technol Lett* 46:487–492.

[8] Ni, Q.-Q., Zhu, Y.-F., Yu, L.-J., et al. 2015. One-dimensional carbon nanotube@ barium titanate@polyaniline multiheterostructures for microwave absorbing application. *Nanoscale Res Lett* 10:174–181. 10.1186/s11671-015-0875-6

[9] Kapat, K., Shubhra, Q. T. H., Zhou, M., et al. 2020. Piezoelectric nano-biomaterials for biomedicine and tissue regeneration. *Adv Funct Mater* 30:1909045. 10.1002/adfm.201909045

[10] Jacob, J., More, N., Kalia, K., and Kapusetti, G. 2018. Piezoelectric smart biomaterials for bone and cartilage tissue engineering. *Inflammation and Regeneration* 38:2–10. 10.1186/s41232-018-0059-8

[11] Li, Y., Dai, X., Bai, Y., et al. 2017. Electroactive $BaTiO_3$ nanoparticle-functionalized fibrous scaffolds enhance osteogenic differentiation of mesenchymal stem cells. *Int J Nanomedicine* 12:4007–4018.

[12] Rajabi, A. H., Jaffe, M., and Arinzeh, T. L. 2015. Piezoelectric materials for tissue regeneration: A review. *Acta Biomaterialia* 24:12–23. 10.1016/j.actbio.2015.07.010

[13] Kemppi, H., Finnilä, M. A., Lorite, G. S., et al. 2021. Design and development of poly-L/D-lactide copolymer and barium titanate nanoparticle 3D composite scaffolds using breath figure method for tissue engineering applications. *Colloids and Surfaces B: Biointerfaces* 199:111530. 10.1016/j.colsurfb.2020.111530

[14] Lewis, T. J. 2004. In Interfaces and nanodielectrics are synonymous. Proc. Int. Conf. Solid Diel., Toulouse, July 5–9, 2004, pp. 792–795.

[15] Tanaka, T. 2005. Dielectric nanocomposites with insulating properties. *IEEE Trans Dielect Electr Insul* 12:914–928.

[16] Gojny, F. H., Wichmann, M. H. G., Fiedler, B., et al. 2006. Evaluation and identification of electrical and thermal conduction mechanisms in carbon nanotube/epoxy composites. *Polymer* 47:2036–2045.

[17] Wischke, C., and Lendlein, A. 2010. Shape-memory polymers as drug carriers – A multifunctional system. *Pharmaceutical Research* 27:527–529.

[18] Liu, L., and Grunlan, J. C. 2007. Clay assisted dispersion of carbon nanotubes in conductive epoxy nanocomposites. *Adv Funct Mater* 17:2343–2348. 10.1002/adfm.200600785

[19] Miriyala, S. M., Kim, Y. S., Liu, L., et al. 2008. Segregated network of carbon black in poly(vinyl acetate) latex: Influence of clay on the electrical and mechanical behavior. *Marcomol Chem Phys* 209:2399–2409. 10.1002/macp.200800384

[20] Lebedev, S. M., Gefle, O. S., Amitov, E. T., et al. 2016. Conductive carbon nanotube-reinforced polymer composites and their characterization. *IEEE Trans Diel Elect Ins* 23:1723–1731.

[21] Lebedev, S. M., Gefle, O. S., Amitov, E. T., et al. 2017. Poly(lactic acid)-based polymer composites with high electric and thermal conductivity and their characterization. *Polym Test* 58:241–248. 10.1016/j.polymertesting.2016.12.033

[22] Vert, M., Santos, I. D., Ponsart, S., et al. 2002. Degradable polymers in a living environment: Where do you end up? *Polym Int* 51:840–844.

[23] Sinclair, R. G. 1996. The case for polylactic acid as a commodity packaging plastic. *J Macromol Sci Part A: Pure Appl Chem* 33:585–597.
[24] Vainionpaa, S., Rokkanen, P., and Tormala, P. 1989. Surgical applications of biodegradable polymers in human tissues. *Prog Polym Sci* 14:679–716.
[25] Jong, W. H. D., Bergsma, J. E., Robinson, J. E., et al. 2005. Tissue response to partially in vitro predegraded poly-L-lactide implants. *Biomaterials* 26:1781–1791.
[26] Langer, R., and Vacanti, J. P. 1993. Tissue engineering. *Science* 260:920–926.
[27] Gupta, A. P., and Kumar, V. 2007. New emerging trends in synthetic biodegradable polymers – Polylactide: A critique. *Europ Polym J* 43:4053–4074.
[28] Gupta, B., Revagade, N., and Hilborn, J. 2007. Poly(lactic acid) fiber: An overview. *Prog Polym Sci* 32:455–482.
[29] Luckachan, G. E., and Pillai, C. K. S. 2011. Biodegradable polymers - A review on recent trends and emerging perspectives. *J Polym Environ* 19:637–676.
[30] Sullivan, E. M., Gerhardt, R. A., Wang, B., et al. 2016 Effect of compounding method and processing conditions on the electrical response of exfoliated graphite nanoplatelet/polylactic acid nanocomposite films. *J Mater Sci* 51:2980–2990.
[31] Hayashi, S., Kondo, S., Kapadia, P., et al. 1995. Room-temperature-functional shape memory polymers. *Plastics Engineering* 51:29–31.
[32] Ratna, D., and Karger-Kocsis, J. 2008. Recent advances in shape memory polymers and composites: A review. *J Mater Sci* 43:254–269.
[33] Wul, B. M. 1946. Substances with high and ultra-high dielectric constant. *Electricity* 3:12–20 (in Russian).
[34] Wul, B. M., and Goldman, I. M. 1945. Dielectric constant of barium titanate depending on the intensity under AC field. *Dokl Akad Nauk USSR* 49:179–182 (in Russian).
[35] Yasuda, I. 1954. On the piezoelectric property of bone. *J Jpn Orthop Surg Soc* 28:267–279.
[36] Fukada, E., and Yasuda, I. 1957. On the piezoelectric effect of bone. *J Phys Soc Jpn* 12:1158–1162.
[37] Bassett, C. A. L. 1967. Biologic significance of piezoelectricity. *Calcified Tissue Res* 1:252–272.
[38] Supronowicz, P. R., Ajayan, P. M., Ullmann, K. R., et al. 2002. Novel current-conducting composite substrates for exposing osteoblasts to alternating current stimulation. *J Biomed Mater Res* 59:499–506. 10.1002/jbm.10015
[39] Sitharaman, B., Shi, X., Walboomers, H. F., et al. 2008. In vivo biocompatibility of ultra-short single-walled carbon nanotube/biodegradable polymer nanocomposites for bone tissue engineering. *Bone* 43:362–370. 10.1016/j.bone.2008.04.013
[40] Nawanil, C., Panprom, P., Khaosa-ard, K., et al. 2010. Effect of surface treatment on electrical properties of barium titanate/carbon nanotube/polydimethylsiloxane nanocomposites. Int Conf Sci Technol Emerging Mater, AIP Conf Proc, pp. 020029-1–020029-7. 10.1063/1.5053205
[41] Dang, Z.-M., Yao, S.-H., Yuan, J.-K., et al. 2010. Tailored dielectric properties based on microstructure change in BaTiO3-carbon nanotube/polyvinylidene fluoride three-phase nanocomposites. *J Phys Chem C* 114:13204–13209. 10.1021/jp103411c
[42] Sawyer, C. B., and C. H. Tower. 1930. Rochele salt as a dielectric. *Phys Rev* 35:269–273.
[43] Bruggeman, D. A. G. 1935. Berechnung verschiedener physikalischer Konstanten von heterogenen Substanzen. *Ann Phys Lpz* 24:636–679.
[44] Gefle, O. S., Lebedev, S. M., Pokholkov, Y. P., Tkachenko, S. N., Volokhin, V. A., Cherkashina, E. I. 2005. In Study of dielectric relaxation of composite materials by the dielectric spectroscopy method. Proc. ISEIM'05, Kitakyushu, Japan, 5–9 June, 2005, Vol. 3, pp. 85–87.

[45] Bogoroditskiy, N. P., Volokobinskiy, Y. M., Vorob'ev, A. A., et al. 1965. Theory of dielectrics. *Moscow-Energy (in Russian).*
[46] Mancuso, E., Shah, L., Jindal, S., et al. 2021. Additively manufactured $BaTiO_3$ composite scaffolds: A novel strategy for load bearing bone tissue engineering applications. *Mater Sci Eng C* 126:112192. 10.1016/j.msec.2021.112192
[47] Ribeiro, C., Sencadas, V., Correia, D. M., et al. 2015. Piezoelectric polymers as biomaterials for tissue engineering applications. *Colloids and Surfaces B: Biointerfaces* 136:46–55. 10.1016/j.colsurfb.2015.08.043
[48] Tandon, B., Blaker, J. J., and Cartmell, S. H. 2018. Piezoelectric materials as stimulatory biomedical materials and scaffolds for bone repair. *Acta Biomaterialia* 73:1–20. 10.1016/j.actbio.2018.04.026
[49] Kharitonov, E. V. 1983. *Dielectric materials with inhomogeneous structure.* Moscow: Radio and Communication (in Russian).
[50] Stauffer, D., and Aharony, A. 1992. *Introduction to percolation theory.* London: Taylor and Francis, 2nd Ed.
[51] Kapitza, P. L. 1965. *Collected papers of* P. L. Kapitza. D. ter Haar (Ed.), Vol. 2, Oxford: Pergamon Press.
[52] Shenogin, S., Xue, L., Ozisik, R., et al. 2004. Role of thermal boundary resistance on the heat flow in carbon-nanotube composites. *J Appl Phys* 95:8136–8144.
[53] Lee, K. H., Kao, J., Parizi, S. S., et al. 2014. Dielectric properties of barium titanate supramolecular nanocomposites. *Nanoscale.* 6:3526–3531.
[54] Hammami, H., Arous, M., Lagache, M., et al. 2006. Experimental study of relaxations in unidirectional piezoelectric composites. *Composites: Part A* 37:1–8.
[55] Upadhyay, R. H., Argekar, A. P., and Deshmukh, R. R. 2014. Characterization, dielectric and electrical behaviour of $BaTiO_3$ nanoparticles prepared via titanium(IV) triethanolaminato isopropoxide and hydrated barium hydroxide. *Bull Mater Sci* 37:481–489.
[56] Curecheriu, L., Balmus, S.-B., Buscaglia, M. T., et al. 2012. Grain size-dependent properties of dense nanocrystalline barium titanate ceramics. *J Am Ceram Soc* 95:3912–3921. 10.1111/j.1551-2916.2012.05409.x
[57] Walker, E., Akishige, Y., Cai, T., et al. 2019. Maxwell-Wagner-Sillars dynamics and enhanced radio-frequency elastomechanical susceptibility in PNIPAm hydrogel-KF-doped barium titanate nanoparticle composites. *Nanoscale Res Lett* 14:385–396. 10.1186/s11671-019-3171-z
[58] Jonscher, A. K. 1996. *Universal relaxation law.* London: Chelsea Dielectric Press.

10 4D Print Today and Envisaging the Trend with Patent Landscape for Versatile Applications

B. Arulmurugan, Devarajan Balaji, V. Bhuvaneswari, and S. Dharanikumar
Department of Mechanical Engineering, KPR Institute of Engineering and Technology, Coimbatore, Tamil Nadu, India

S. Rajkumar
School of Mechanical and Electrochemical Engineering, Institute of Technology, Hawassa University, Hawassa, Ethiopia

CONTENTS

10.1 INTRODUCTION

3D printing, otherwise called added substance producing, was imagined in 1986 by Charles Hull [1]. This strategy appears differently in relation to the subtractive assembling methods that actually rule the business today; rather than emptying mass raw material into molds or removing profiles from it, added substance fabricating

DOI: 10.1201/9781003333760-10

persistently adds a material layer over layer to deliver 3D articles [2]. With the guide of PCs, added substance assembling can be utilized to create very exact and complex calculations. Since Hull's development, added substance production has added to incalculable fields, including aviation [3], medical industry [4], civil [5], hardware [6], and individual utilization [7]. Objects made through 3D printing have intended to be static; they show no progressions over the long haul and on the off chance that they do, it is viewed as a breakdown.

Only after the TED conference in the year 2013, therein Skylar Tibbits demonstrated that the 3D-printed object that might change its shape over a time period only by the world opens up for 4D printing and was denoted by the name 4D [8]. In the time from that point forward, specialists have shown that the potential applications for 4D printing are boundless. The primary consideration of this is the many savvy materials that can be 3D printed. These brilliant materials can be partitioned into three classifications: polymers, earthenware production, along with metals.

> The thought behind 4D printing is that you take multi-material 3D printing and add another ability, which is change, [so] that right off the bed, the parts can change starting with one shape then onto the next shape, straightforwardly all alone.

4D printing is undoubtedly an augmentation of 3D printing; most papers distributed for this subject include utilizing notable added substance-producing strategies; for example, intertwined statements demonstrating to produce objects enabling their abilities to transform their shape.

Be that as it may, this definition is tested by later headways, especially in polymer 4D printing. Ample instances of polymeric items are 4D printed that might not alter their shape but instead might adjust some parametric element like tone. These articles get along with time, and such must likewise be incorporated beneath the gamp of 4D printing. Time-subordinate impacts are accomplished by implementing a controlled improvement. In the long, early stretches of polymer 4D printing, intensity and water were the main two improvements being used. Nowadays, polymeric constituents have revealed the capacity to respond in astonishingly novel ways in light of non-warm upgrades [9]. Pseudothermal improvements, for example, light along with electric flow are elective approaches to prompting a temperature change in the sample. Promotions, for example, pH and ionic strength, can work close by H_2O to cooperate with the polymers through totally non-thermal components.

According to the Oxford Dictionary, 4D printing is defined as "The activity or cycle of utilizing 3D printing strategies to make an item that can change its shape or properties in an anticipated manner after some time in response to conditions, for example, openness to water, air, heat, or an electric momentum" [10]. This definition is right concerning current 4D printing; it features a requirement for an upgrade to be available and accepts objects that change in that frame of mind rather than shape. It is defined likewise and presents the need for the reaction to be unsurprising; it is non-sufficient to just a 3D print of savvy material. After printing the brilliant substances, specialists examine their response when an upgrade is applied. Numerical models evaluate the reaction as a promotion component [11–14]. The specific answers of a 4D written word must be known before it can be utilized beyond the research center.

With this data, the substance can be transformed to a deep level of accuracy when a painstakingly estimated improvement is implemented. A 4D-printed object deciding the state of limitation of their reaction to modifying their shape is achieved by programming. Versatile programming methods are created for excellent 4D-printable materials.

Despite the Oxford dictionary definition's apparent conclusiveness, research findings may challenge it. According to Mathews and colleagues, an electrochemically printed cell along with bio-nano ink created hydrogen when irradiated with visible light [15]. The object's properties were not altered so much as chemical spawn was generated. Using novel monomers inserted into the conventional polymer network, our group and Boyer's group developed another cutting-edge branch of 4D printing, allowing 3D-printed components to be reactivated and expanded [16–18]. It unlocks the door to many unique features that can be added after printing. As a result, we offer the following concept of 4D printing: "The 3D printing of an artifact from smart material which is programmed to modify its characteristics in response to stimulation, expand in size or mass, or create chemical spawn."

Digital light processing (DLP) [19], stereo-lithography apparatus (SLA) [20], multi-photon lithography (MPL) [21], direct ink writing (DIW) [22], as well as fused deposition modeling (FDM) [23], are the five main 3D printing methods currently used for polymer 4D printing. Extrusion printing, such as FDM and DIW, uses ink excreted out of a nozzle to print objects in three dimensions as they move across the printer's bed. Additionally, extrusion printers can be equipped with multiple printheads so that they can print various materials at the same time. When it comes to polymer 4D printing, this process is commonly referred to as "multi-material printing." Photopolymerizing species include SLA, MPL, and DLP; they require light for a polymerization reaction of the monomers, which also solidifies the process. In recent years, advances in additive manufacturing have made it possible to create complex objects that would have been impossible to make in any other way prior to this time. The goal of 4D printing would be to produce things with highly detailed physical properties that can be reconfigured.

Polymer 4D printing has advanced from simple shape transformations, simple stimuli such as water and heat, and primitive printing methods to more advanced techniques. A wide range of topics is covered, including popular ones of 4D-printable polymers, their activation mechanisms, and novel programming methods and stimuli. Among the topics we cover are novel polymeric intelligent materials and 4D-printing techniques, which we assume investigators must concentrate on to advance the field. 4D is geared toward chemists, so we avoid discussing founded additive manufacturing techniques. There's no "applications" section because of this. The area is still in its infancy, and we leave the reader's imagination to fill in the blanks. Similar papers have recently been published, including reviews on the current events in 4D printing [24–33]. It is important to note that this study does not focus solely on polymers but rather on as many examples as we could find of intelligent polymer 4D printing as we could find. Others [34–42] look at polymer 4D printing for specific applications or just one type or class of smart polymer and report similar findings. We don't limit our discussion of polymer 4D printing to just a few specific applications in this study. We also plan to review less commonly used intelligent polymers, which are expected to see a lot of development in the future. This review aims to accurately depict how it has progressed and what we can expect in the near term.

10.2 4D TODAY

10.2.1 Shape-Modifying Polymers

Shape-modifying polymers (SMPs) are one of the supreme common types of 4D-printed components because they can be deformed into diverse shapes and then return to their standard condition. Polymers exhibit a glassy-to-rubbery transition, which results in this memory effect [43]. Even though SMPs can be found in a variety of polymers, not all are SMPs due to factors such as the material's structure, morphology, and additional conditions [44].

SMPs are set to their primary shape after being 3D printed. When polymer chains take on this shape, they are in an entropically favored state of randomly coiled coils. Ten to 50 chain atoms of polymer in the object become mobile when heated above their glass transition temperature. Things can change shape when they are subjected to force. After cooling the object back to the Tg, polymer chains will no longer be able to move freely, so that this new configuration will remain. A more aligned arrangement of polymer chains makes this secondary shape less stable thermodynamically (entropy has decreased). Because of this, chains are used to store stress in the second configuration. Deformed chain segments in an SMP must not recoil in the second arrangement, which necessitates are being established of temporary interactions among fragments to hold them in a position [45]. Tg re-heating releases segments from their restraints, restoring their original shape [46–49]. No matter how much external force is applied, this entropy-driven form memory effect cannot be reversed; the SMP cannot go from its primary to secondary shape. SMPs are notable for their irreversibility, which can be beneficial in some cases (for example, self-assembling stable structures). SMP 4D printing uses this property as the polymer will become less stiff during the changeover from a glassy to a rubbery form. They are utilizing temperature as a straightforward stimulus; SMP 4D printing has been established for many applications. Solar concentrators, smart structures, jewelry, mechanical parts, electronics, and biomedical devices are just a few of the many possible applications.

SMPs are widely used today, and new 4D printable versions are consistently created. Promotional SMPs from Stratasys' Tango and the Vero series seem to be particularly well liked. Polyvinyl alcohol (PVC), polystyrene (PS), polycaprolactone (PCL), high impact polystyrene (HIPS), polyester (PE), polyurethane (PU), polyamide, acrylonitrile butadiene styrene (ABS), as well as polylactic acid (PLA), are examples of common 4D-printable polymers that have demonstrated shape memory consequences (PVA). SMPs were, however, made from a variety of acrylate monomers that had also been polymerized even during the 3D-printing process, including ethylene glycol phenyl ether acrylate, isobornyl acrylate (IBOA), tert-butyl acrylate, lauryl acrylate, acrylic acid, 2-ethyl hexyl acrylate (PEGDMA), benzyl methacrylate, methyl acrylate, along with butyl acrylate.

10.2.2 Hydrogels

In addition to hydrogels, other polymeric materials are commonly used in 4D printing. Hydrophilic as well as cross-linked polymer chains make these materials insoluble. Water absorption and swelling are essential properties of hydrogels. Water-responsive actuators are made possible by taking advantage of this property in 4D printing. By printing based on a fundamental with an active hydrogel layer that expands in H_2O, a strain incongruity can create between the swelling layer and the non-swelling layer [50,51]. As a result, the actuator bends towards the passive side due to this net actuating force. After drying the hydrogel, the polymer matrix can return to its original state. With this repairable shape change effect, it is possible to fabricate self-folding structures, unlike the point-to-point shape change in SMPs [52–54]. Hydrophilic Polyurethane Hydrogel sandwiched among Hydrophobic Polyurethane Elastomer is printed With FDM By Baker et al. [55], for instance. It is believed that water seeped into the hydrogel at every hinge, swelling it to twice its original size. Printing materials were hydrated and dried before being used to create origami structures that could be refolded.

10.2.3 Liquid Crystal Elastomers (LCEs)

To put it another way, LCEs can reversibly shift between the liquid crystal and isotropic states. Liquid crystals within a nematic state, in which the molecules (called mesogens) have decided to order in a standard orientation characterized by a director, are typically used in 4D printing. This sequence is lost above a specific temperature, causing the mesogens to shrink along the direction and expand orthogonal to it. Nematic-to-isotropic temperature (TNI) is diverse from the transition temperature of glass (Tg). Extrusion-based additive manufacturing methods like DIW allow mesogens to be brought into line along the printing path while extruded through the nozzle. Due to the subsequent curing, formations with inherent anisotropy are created. Stimuli can force the LCE to contract in a specific direction, which is how a coded shape change is achieved in 4D printing. LCEs change their shape when heated, unlike SMPs, which do not change shape reversibly. Several studies have found that increasing the temperature causes shape changes throughout 4D-printed LCE objects, which can be reverted by cooling [56–62]. The LCEs 4D print of a temperature-triggered changeable hinge is an example of this. It was done by printing on two elastomer layers, followed by another layer of LCE. EDDET chain extenders were used in the LCE ink. Joule heat treatment happened until the TNI of 42 degree Celsius, which ultimately resulted in the LCE component contracting in length when the current was applied. An imbalance in the LCE and elastomer caused the hinge to bend. An example of this hinge is a prosthesis hand that can turn and straighten its fingers in reaction to temperature changes.

10.3 POLYMER 4D-PRINTING APPLICATIONS

Many researchers have been intrigued by the potential of additive manufacturing technology's four-dimensional (4D) printing. This interest has only grown since then.

FIGURE 10.1 4D printer variants, materials, and applications [Redrawn from 64].

With 4D printing, complex geometries can be processed by giving stationary printed structures with dynamic characteristics that transform over a period. Printed components can be exposed to various stimuli to alter their shapes and features. This modification can be pre-planned but also in a well-controlled manner. It allows for creating of geometries that can be controlled and manipulated in real time.

The capabilities of 4D printing have significantly widened in the last few years due to the rapid advancements in intelligent substances and novel printing methods. 4D-printing technologies and polymer science, as well as engineering, are discussed in this article. As this technology advances, it provides a platform for novel applications thanks to advanced printer design and materials. 4D printing, because of its ability to change its state over time, has the potential to grow at an exponential pace [63,64]. (Figure 10.1).

10.4 FUTURE SCOPE

In general, the future scope is assessed based on the research articles, industry bulletins, forums of scientific communities, and patent landscape. This patent landscape has been chosen due to its immediate updating over another database. (Table 10.1).

Patent landscape analysis is carried out for the keyword "4D printing" in the "English All" category; the total count pops out at 830 (as on 2 August 2022).

The above table reveals the USA dominates in this technology, followed by China. If the researcher wants to work in this domain and is looking for research collaboration, they can look to these countries. The primary applicant is "Pure Storage Inc." with a filing count of 289 patents; none of the other players are able to compete. As an individual player, "Mr. Ronald Karr" has a patent filing count of 78. Researchers aim for this company, and this collaborator might have good scope for their research. Intentional patent classification (IPC) code plays a pivotal role for the

TABLE 10.1
4D-Printing Technology Patent Count Based on Country, IPC Code, and Year [65]

S. No.	Countries	Count	IPC Code	Count	Year	Count
1	USA	418	G06F	299	2013	2
2	China	206	B33Y	249	2014	2
3	Patent Cooperation Treaty	141	B29C	148	2015	8
4	European Patent Office	20	B22F	63	2016	16
5	Australia	15	H04L	63	2017	26
6	Republic of Korea	12	A61L	37	2018	65
7	Canada	6	C08L	31	2019	135
8	India	5	C09D	30	2020	134
9	United Kingdom	3	A61F	28	2021	284
10	Japan	2	C08F	24	2022	132

researchers to converge their search with the aid of G06F, B33Y, and B29C are dominant. Growth in this technology is depicted by the number of patents filed over the years, showing a progressive rise. In the future, it will be a promising technology for versatile applications (Table 10.2).

A patent search is carried out for the keyword "4D printing" "polymer" in the "English All" category; the total count pops out at 292 (as on 2 August 2022).

The above table reveals the USA dominates in this technology, followed by China. If the researcher wants to work in this domain and is looking for research collaboration, they can look to these countries. The primary applicant is "Harbin institute of tech" with a filing count of 12 patents; none of the other players are able to compete. As an individual player, "Mr. Leng Jingsong" has a patent filing count

TABLE 10.2
4D-Printing Technology of Polymers Patent Count Based on Country, IPC Code, and Year [65]

S. No.	Countries	Count	IPC Code	Count	Year	Count
1	USA	116	B33Y	110	2014	1
2	Patent Cooperation Treaty	89	B29C	82	2015	1
3	China	57	C09D	26	2016	13
4	Australia	12	A61L	25	2017	18
5	European Patent Office	9	C08L	17	2018	34
6	India	4	B22F	16	2019	61
7	Canada	2	C08F	16	2020	50
8	Republic of Korea	1	G01N	16	2021	77
9	United Kingdom	1	A61B	14	2022	31
10	Japan	1	C08G	14		

of 12. Researchers aim for this company, and this collaborator might have good scope for their research. Intentional patent classification (IPC) code plays a pivotal role for the researchers to converge their search with the aid of B33Y, B29C, C09D, and A61L are pretty dominant. Growth in this technology is depicted by the number of patents filed over the years, showing a progressive rise. In the future, it will be a promising technology for versatile applications.

10.4.1 Bio-Cell Printing

The stem cell dividing line can accommodate by a 4D-printed configurable culture substrate that can change shape with the stem cells themselves. During a pre-determined period in responding to a stimulus, including such a temperature, the 4D-printed culture material would include a shape memory polymer which thus could also transform from one topographical shape to another. The first topographical shape could consist of micro-wells. In contrast, the second topographical shape could include microgrooves, accommodating neural stem cells' growth and differentiation that can compensate for the development and differentiation of neural stem cells [66] (Figure 10.2).

10.4.2 Scaffold Printing

Scaffold printing was designed to have shear shape memory properties and a resorbable star PPF 4D-printed configuration. Printed structures are gyroid-structured and can be compressed at room temperature to achieve insertion into the body.

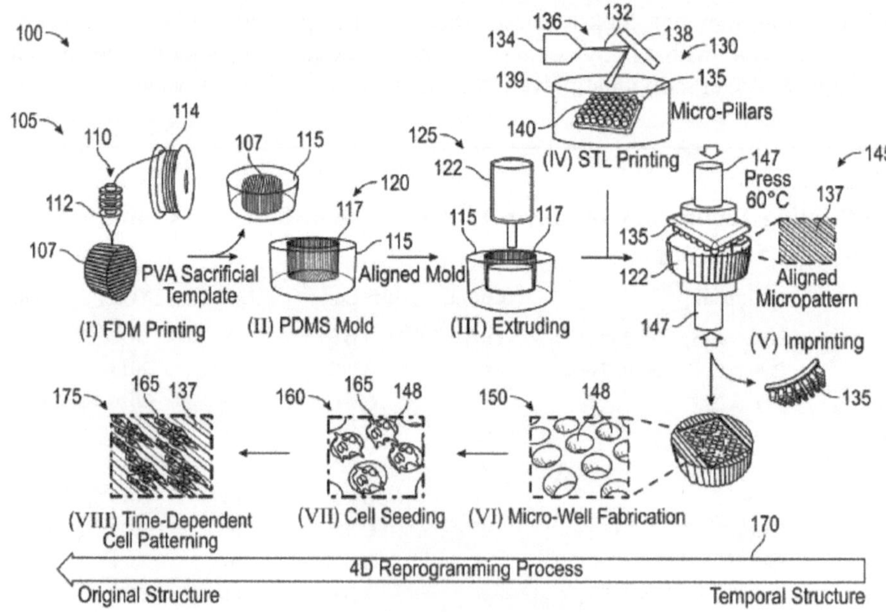

FIGURE 10.2 Process of 4D cell printing [66].

Afterward, they are heated to body temperature and expand to a desired (third) thickness. Although this behavior focuses on the compression and temperature-based augmentation of printed structures, the contraction, and expansion of such printed forms might have a profound effect on the printed system before and after insertion in the body. It is easier to insert things like bone scaffolds and stents into the body because they can be condensed before inserting them. It allows for a more precise placement of the stent. Polymeric resins based on one or more multi-arm "star" poly (propylene fumarate) (PPF) copolymers are used in one or more embodiments of the present invention to create resorbable star PPF 4D-printed structures. Diethyl fumarate (DEF) is commonly used as a solvent and a co-crosslinker in these resins to lower the viscosity to a range suitable for 3D-printing applications. These resins require less DEF to reach a printable consistency and have shorter curing times than comparable linear PPF oligomers because the complex viscosity of the multi-arm ("star") poly (propylene fumarate (PPF) copolymers used in the present invention are lower at a comparable number average molecular weight (Mn) [67] (Figure 10.3).

For use with an absorbable star, PPF 4D-printed structure with a first shape, second compressed form, and third recovered form, wherein the absorbable star PPF 4D-printed structure might very well transform from the first shape to the second compact form over a predetermined period or at a predetermined temperature when compressed. One or more embodiments of this structure include a resorbable star PPF polymer with a degree of polymerization ranging from about 40 to about 200, produced by ROCOP using a catalyst and multifunctional alcohol initiator to polymerize two cyclic anhydrides with an epoxide. One or more of those mentioned above, in which the cyclic anhydride is at least one of maleic anhydride or succinic anhydride, may be included in the resorbable star PPF 4D-printed structure. Propylene oxide may be used as the epoxide in some cases. For example, the present invention's resorbable star PPF 4D-printed structure can include any of the previously mentioned embodiments of the first aspect, in which a multi-arm PPF star polymer is formed by controlled ring-opening copolymerization (ROCOP) of maleic anhydride but also propylene oxide using a catalyst along with a multifunctional alcohol initiator, the PPF star polymer having a degree of the polity. Any

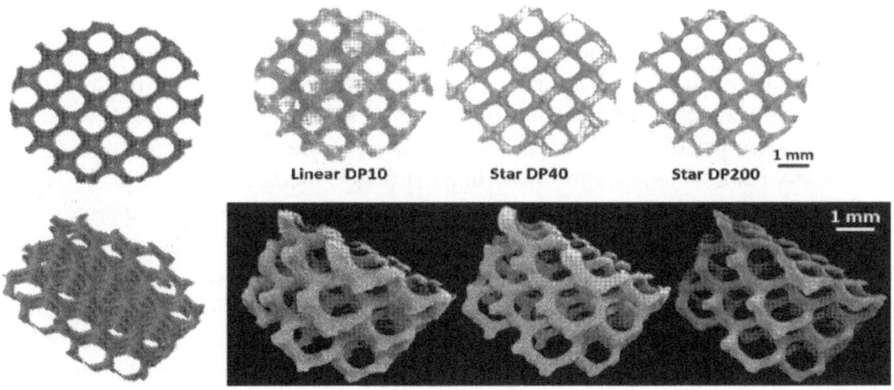

FIGURE 10.3 Method of 4D printing [67].

one or a combination of the above-mentioned embodiments of the first aspect may be used to form the resorbable star PPF 4D-printed structure, wherein a multi-arm PPF star polymer formed by controlled ring-opening copolymerization (ROCOP) of succinic anhydride as well as propylene oxide is used as a catalyst for and a multifunctional alcohol initiator for the ROCOP, the multi-arm PPF star polymer exhibiting a degree of polymerization of between 40 and 60. To illustrate this point, any one or more of the multi-arm PPF star polymer embodiments with three to five arms may be included in the present invention's resorbable star PPF 4D-printed structure. A glass transition temperature (Tg) of about 10°C to about 60°C, preferably 20°C to about 50°C, and even more preferably 30°C to about 40°C, may be used for the resorbable star PPF 4D-printed structure. For example, provides a resorbable star PPF 4D-printed structure with a compressive modulus ranging from about 1 MPa to about 60 MPa in any of the above-mentioned about 2 MPa to about 40 MPa, as well as still further preferred from about 3 MPa to about 25 MPa. It is possible to use any of the above-referenced embodiments of the resorbable star PPF 4D-printed structure in some cases where the first shape is a gyroid with a uniform pore geometry and porosity, as in the first aspect of the invention. A resorbable star PPF 4D-printed structure includes any one or more embodiments from elements of the first aspect of the present invention with a plurality struts of from about 50 microns to 1,000 microns, preferred from about 100 microns to about 500 microns, and also more preferably from about 140 microns to 280 microns in strut size. Struts may be regularly spaced in one or more embodiments of this invention's resorbable star PPF 4D-printed structure, which is described in greater detail below. The struts can be anisotropic in some cases. Where the resorbable star PPF 4D-printed structure has a porosity between about 15% and about 95%, preferably between about 30% and 92%, and even more preferably between about 50% and 90%. Resorbable star PPF 4D-printed structure includes any embodiments, for example, with pore sizes ranging from about 50 microns to about 5,000 microns, especially between 50 microns and 2,500 microns, and much more preferably between 50 microns and 1,000 microns [67].

When compressed, the present invention's resorbable star PPF 4D-printed structure can take on any of the above-mentioned embodiments of the invention's first aspect, which calls for a denser structure. When the resorbable star PPF 4D-printed structure of the present invention has been post-cured with UV irradiation after printing, post-curing of the resorbable star PPF 4D-printed structure has taken place for from about 1 minute to a maximum of 12 hours with UV radiation after printing. For example, in one or more embodiments, the present invention's resorbable star PPF 4D-printed structure includes any of the first-aspect embodiments listed above, with a time interval ranging from 1 hour to approximately 72 hours at ambient temperature. There are several examples of the resorbable star PPF 4D-printed structure, which includes a time interval of about 1 minute to about 60 minutes at ambient temperature. From about 20°C to about 50°C, a predetermined temperature for the resorbable star PPF 4D-printed structure can be used, as described above. A resorbable star PPF 4D-printed structure includes one or more of the above-mentioned embodiments wherein a predetermined temperature ranges from about 20°F to approximately 40°F. A resorbable star PPF structure is printed when the predetermined temperature ranges from about 30°C to 42°C. Some of the embodiments mentioned above of the first aspect of the present invention may be included in the present invention's resorbable star PPF 4D-printed structure, depending on whether the predetermined temperature is 30°C or more. When the resorbable lead PPF 4D-printed structure is implanted or inserted into a mammal, it may include any of those embodiments mentioned above of the first aspect, wherein the predetermined temperature is the body temperature of the mammal or human (37°C). One or more of the embodiments mentioned above of the first aspect of the present invention, which has tunable degradation and resorbability, may be included in the resorbable star PPF 4D-printed structure.

Resorbability can be controlled by varying molar mass of multi-arm PPF star polymer, which may be included in the resorbable star PPF 4D-printed structure. At least one embodiment consists of any of the embodiments mentioned above of the first aspect of this invention, where the first and third recovered shapes seem to be identical in one or more resorbable star PPF 4D-printed structures. As described above, a resorbable star PPF 4D-printed structure may include any one or more wherein the third recovered shape is between about 65% and 100% of the first shape. A star PPF 4D-printed structure may include one or more linear PPF polymers. This invention consists of any one or more of the examples mentioned above of the first aspect of the present invention, which includes a bone scaffold, vascular stent, kidney stent, urethral, colitis, esophageal, colon, and intestinal and vein structures, as well as a star PPF 4D-printed structure. As an additional aspect, the present technology is directed to the production of the previously described resorbable star PPF 4D-printed structure, which includes the steps of making a 3D-printable resin comprising a star PPF polymer, printing a 3D structure from the star PPF polymer using a suitable 3D printer, and curing the 3D-printed structure by UV irradiation after it has been produced. According to some embodiments, a multi-arm star polymer (MPF star) is formed by controlled ring-opening copolymerization (ROCOP) using multifunctional alcohol initiators and catalysts. It has from about 40° to about 200° of polymerization. Resorbable star PPF 4D-printed structure. The cyclic anhydride may be at least one of the following maleic anhydride, and the epoxide may be propylene oxide, as described above. This invention's resorbable star PPF 4D-printed structures can be made using any one or more wherein the multi-arm PPF star polymer has three to five arms linked together at a central core, where the main body comprises the residue of the multifunctional alcohol initiator. The 3D-printable resin comprises star PPF polymer from three to five arms, DEF, and photoinitiator is included in the method for making a resorbable star PPF 4D-printed structure. Resorbable star PPF 4D-printed structure using a continuous digital light processing (cDLP) 3D printer. A post-curing step includes irradiating the 3D-printed structure with UV light for about 1 minute to about 1,200 minutes, preferably 20 minutes to about 600 minutes, and most preferentially 45 minutes or more. It is possible to make the resorbable star PPF 4D-printed structure by using any of the above-mentioned embodiments, wherein the 3D structure is a gyroid with considerably uniform pore geometry and porosity. Resorbable star PPF 4D-printed structures of this invention can be made wherein the 3D structure has a number of struts that range in size from about 50 to about 1,000, ideally from about 100 to about 500, and also more preferably from about 140 to 280. A resorbable star PPF 4D-printed structure can be made using, wherein the 3D structure is a gyroid with porosity of about 15% to about 95%, especially from about 30% to about 92%, as well as more preferably from 50% to 90%. Methods for creating a resorbable star PPF 4D-printed structure with any one or more of the embodiments mentioned above of the second aspect of this invention, where the 3D structure is a gyroid with a pore size of between 50 microns and 5,000 microns, more preferably between 50 microns and 1,000 microns, and even more preferably between 50 microns and 2,500 microns [67].

10.5 CONCLUSION

4D printing is a promising technology owing to its morphing capability over time, with or without external aid. Mainly, novel applications have been developed during this last decade. Many applications are discussed but not converged only with that; it is expanding daily. The future trend is predicted with the aid of the patent landscape, which reveals that the growth of this technology is accelerating much faster in the last five years since 2017. The expansion would not be decelerated in the future because the patent-filed techniques will take some time to come into the market.

REFERENCES

[1] Hull, C.W., UVP Inc, 1986. Apparatus for production of three-dimensional objects by stereolithography. U.S. Patent 4,575,330.

[2] Prince, J.D., 2014. 3D printing: An industrial revolution. *Journal of Electronic Resources in Medical Libraries*, 11(1), pp. 39–45.

[3] Nickels, L., 2015. AM and aerospace: An ideal combination. *Metal Powder Report*, 70(6), pp. 300–303.

[4] Tetsuka, H. and Shin, S.R., 2020. Materials and technical innovations in 3D printing in biomedical applications. *Journal of Materials Chemistry B*, 8(15), pp. 2930–2950.

[5] Tay, Y.W.D., Panda, B., Paul, S.C., Noor Mohamed, N.A., Tan, M.J. and Leong, K.F., 2017. 3D printing trends in building and construction industry: A review. *Virtual and Physical Prototyping*, 12(3), pp. 261–276.

[6] Valentine, A.D., Busbee, T.A., Boley, J.W., Raney, J.R., Chortos, A. and Kotikian, A., 2017. JD 1204 Berrigan, MF Durstock, JA Lewis. *Advanced Materials*, 29(40), p. 1703817.

[7] Pasricha, A. and Greeninger, R., 2018. Exploration of 3D printing to create zero-waste sustainable fashion notions and jewelry. *Fashion and Textiles*, 5(1), pp. 1–18.

[8] Tibbits, S., 2013, February. The emergence of "4D printing". In TED Conference.

[9] Schattling, P., Jochum, F.D. and Theato, P., 2014. Multi-stimuli responsive polymers – The all-in-one talents. *Polymer Chemistry*, 5(1), pp. 25–36.

[10] https://www.lexico.com/definition/4d_printing – Accessed on 7 July 2022.

[11] Oladapo, B.I., Oshin, E.A. and Olawumi, A.M., 2020. Nanostructural computation of 4D printing carboxymethylcellulose (CMC) composite. *Nano-Structures & Nano-Objects*, 21, p. 100423.

[12] Wang, F., Yuan, C., Wang, D., Rosen, D.W. and Ge, Q., 2020. A phase evolution based constitutive model for shape memory polymer and its application in 4D printing. *Smart Materials and Structures*, 29(5), p. 055016.

[13] Oladapo, B.I., Adebiyi, A.V. and Elemure, E.I., 2021. Microstructural 4D printing investigation of ultra-sonication biocomposite polymer. *Journal of King Saud University-Engineering Sciences*, 33(1), pp. 54–60.
[14] Han, M., Yang, Y. and Li, L., 2020. Energy consumption modeling of 4D printing thermal-responsive polymers with integrated compositional design for material. *Additive Manufacturing*, 34, p. 101223.
[15] Mathews, A.S., Abraham, S., Kumaran, S.K., Fan, J. and Montemagno, C., 2017. Bio nano ink for 4D printing membrane proteins. *RSC Advances*, 7(66), pp. 41429–41434.
[16] Bagheri, A., Engel, K.E., Bainbridge, C.W.A., Xu, J., Boyer, C. and Jin, J., 2020. 3D printing of polymeric materials based on photo-RAFT polymerization. *Polymer Chemistry*, 11(3), pp. 641–647.
[17] Bagheri, A., Bainbridge, C.W.A., Engel, K.E., Qiao, G.G., Xu, J., Boyer, C. and Jin, J., 2020. Oxygen tolerant PET-RAFT facilitated 3D printing of polymeric materials under visible LEDs. *ACS Applied Polymer Materials*, 2(2), pp. 782–790.
[18] Zhang, Z., Corrigan, N., Bagheri, A., Jin, J. and Boyer, C., 2019. A versatile 3D and 4D printing system through photo controlled RAFT polymerization. *Angewandte Chemie*, 131(50), pp. 18122–18131.
[19] Zhao, Z., Tian, X. and Song, X., 2020. Engineering materials with light: Recent progress in digital light processing based 3D printing. *Journal of Materials Chemistry C*, 8(40), pp. 13896–13917.
[20] Huang, J., Qin, Q., and Wang, J., 2020. A review of stereolithography: Processes and systems. *Processes*, 8(9), p. 1138.
[21] Harinarayana, V. and Shin, Y.C., 2021. Two-photon lithography for three-dimensional fabrication in micro/nanoscale regime: A comprehensive review. *Optics & Laser Technology*, 142, p. 107180.
[22] Lewis, J.A., 2006. Direct ink writing of 3D functional materials. *Advanced Functional Materials*, 16(17), pp. 2193–2204.
[23] Fischer, F., 2015. FDM and Polyjet 3D printing. *Popular Plastics & Packaging*, 60(6), pp. 1–7.
[24] Mallakpour, S., Tabesh, F. and Hussain, C.M., 2021. 3D and 4D printing: From innovation to evolution. *Advances in Colloid and Interface Science*, 294, p.102482.
[25] Patil, A.N. and Sarje, S.H., 2021. Additive manufacturing with shape-changing/memory materials: A review on 4D printing technology. *Materials Today: Proceedings*, 44, pp. 1744–1749.
[26] Huang, J., Xia, S., Li, Z., Wu, X. and Ren, J., 2021. Applications of four-dimensional printing in emerging directions: Review and prospects. *Journal of Materials Science & Technology*, 91, pp. 105–120.
[27] Subeshan, B., Baddam, Y. and Asmatulu, E., 2021. Current progress of 4D-printing technology. *Progress in Additive Manufacturing*, 6(3), pp. 495–516.
[28] Kumar, S.B., Jeevamalar, J., Ramu, P., Suresh, G. and Senthilnathan, K., 2021. Evaluation in 4D printing–a review. *Materials Today: Proceedings*, 45, pp. 1433–1437.
[29] Mohol, S.S. and Sharma, V., 2021. Functional applications of 4D printing: A review. *Rapid Prototyping Journal*, 27(8), pp. 1501–1522.
[30] Ma, S., Zhang, Y., Wang, M., Liang, Y., Ren, L. and Ren, L., 2020. Recent progress in 4D printing of stimuli-responsive polymeric materials. *Science China Technological Sciences*, 63(4), pp. 532–544.
[31] Falahati, M., Ahmadvand, P., Safaee, S., Chang, Y.C., Lyu, Z., Chen, R., Li, L. and Lin, Y., 2020. Smart polymers and nanocomposites for 3D and 4D printing. *Materials Today*, 40, pp. 215–245.
[32] Peng, B., Yang, Y. and Cavicchi, K.A., 2020. Sequential shapeshifting 4D printing: Programming the pathway of multi-shape transformation by 3D printing stimuli-responsive polymers. *Multifunctional Materials*, 3(4), p. 042002.

[33] González-Henríquez, C.M., Sarabia-Vallejos, M.A. and Rodriguez-Hernandez, J., 2019. Polymers for additive manufacturing and 4D-printing: Materials, methodologies, and biomedical applications. *Progress in Polymer Science*, 94, pp. 57–116.
[34] Huang, X., Panahi-Sarmad, M., Dong, K., Li, R., Chen, T. and Xiao, X., 2021. Tracing evolutions in electro-activated shape memory polymer composites with 4D printing strategies: a systematic review. *Composites Part A: Applied Science and Manufacturing*, 147, p. 106444.
[35] Gauss, C., Pickering, K.L. and Muthe, L.P., 2021. The use of cellulose in bio-derived formulations for 3D/4D printing: A review. *Composites Part C: Open Access*, 4, p. 100113.
[36] Le Duigou, A., Correa, D., Ueda, M., Matsuzaki, R. and Castro, M., 2020. A review of 3D and 4D printing of natural fibre biocomposites. *Materials & Design*, 194, p. 108911.
[37] Mehrpouya, M., Vahabi, H., Janbaz, S., Darafsheh, A., Mazur, T.R. and Ramakrishna, S., 2021. 4D printing of shape memory polylactic acid (PLA). *Polymer*, 230, p. 124080.
[38] Andreu, A., Su, P.C., Kim, J.H., Ng, C.S., Kim, S., Kim, I., Lee, J., Noh, J., Subramanian, A.S. and Yoon, Y.J., 2021. 4D printing materials for vat photopolymerization. *Additive Manufacturing*, 44, p. 102024.
[39] Shie, M.Y., Shen, Y.F., Astuti, S.D., Lee, A.K.X., Lin, S.H., Dwijaksara, N.L.B. and Chen, Y.W., 2019. Review of polymeric materials in 4D printing biomedical applications. *Polymers*, 11(11), p. 1864.
[40] Champeau, M., Heinze, D.A., Viana, T.N., de Souza, E.R., Chinellato, A.C. and Titotto, S., 2020. 4D printing of hydrogels: A review. *Advanced Functional Materials*, 30(31), p. 1910606.
[41] Miao, S., Castro, N., Nowicki, M., Xia, L., Cui, H., Zhou, X., Zhu, W., Lee, S.J., Sarkar, K., Vozzi, G. and Tabata, Y., 2017. 4D printing of polymeric materials for tissue and organ regeneration. *Materials Today*, 20(10), pp. 577–591.
[42] Subash, A. and Kandasubramanian, B., 2020. 4D printing of shape memory polymers. *European Polymer Journal*, 134, p. 109771.
[43] Stutz, H., Illers, K.H. and Mertes, J., 1990. A generalized theory for the glass transition temperature of crosslinked and uncrosslinked polymers. *Journal of Polymer Science Part B: Polymer Physics*, 28(9), pp. 1483–1498.
[44] Amin, A., Sarkar, R., Moorefield, C.N. and Newkome, G.R., 2013. Preparation of different dendritic-layered silicate nanocomposites. *Polymer Engineering & Science*, 53(10), pp. 2166–2174.
[45] Behl, M. and Lendlein, A., 2007. Shape-memory polymers. *Materials Today*, 10(4), pp. 20–28.
[46] Meng, Q. and Hu, J., 2009. A review of shape memory polymer composites and blends. *Composites Part A: Applied Science and Manufacturing*, 40(11), pp. 1661–1672.
[47] Small IV, W., Singhal, P., Wilson, T.S. and Maitland, D.J., 2010. Biomedical applications of thermally activated shape memory polymers. *Journal of Materials Chemistry*, 20(17), pp. 3356–3366.
[48] Leng, J., Lan, X., Liu, Y. and Du, S., 2011. Shape-memory polymers and their composites: Stimulus methods and applications. *Progress in Materials Science*, 56(7), pp. 1077–1135.
[49] Rosales, C.A.G., Duarte, M.F.G., Kim, H., Chavez, L., Hodges, D., Mandal, P., Lin, Y. and Tseng, T.L.B., 2018. 3D printing of shape memory polymer (SMP)/carbon black (CB) nanocomposites with electro-responsive toughness enhancement. *Materials Research Express*, 5(6), p. 065704.
[50] Le Duigou, A., Castro, M., Bevan, R. and Martin, N., 2016. 3D printing of wood fibre biocomposites: From mechanical to actuation functionality. *Materials & Design*, 96, pp. 106–114.

[51] Mulakkal, M.C., Trask, R.S., Ting, V.P. and Seddon, A.M., 2018. Responsive cellulose-hydrogel composite ink for 4D printing. *Materials & Design*, 160, pp. 108–118.
[52] Bakarich, S.E., Gorkin III, R., Panhuis, M.I.H. and Spinks, G.M., 2015. 4D printing with mechanically robust, thermally actuating hydrogels. *Macromolecular Rapid Communications*, 36(12), pp. 1211–1217.
[53] Yuan, C., Wang, F. and Ge, Q., 2021. Multimaterial direct 4D printing of high stiffness structures with large bending curvature. *Extreme Mechanics Letters*, 42, p. 101122.
[54] Raviv, D., Zhao, W., McKnelly, C., Papadopoulou, A., Kadambi, A., Shi, B., Hirsch, S., Dikovsky, D., Zyracki, M., Olguin, C. and Raskar, R., 2014. Active printed materials for complex self-evolving deformations. *Scientific Reports*, 4(1), pp. 1–8.
[55] Baker, A.B., Bates, S.R., Llewellyn-Jones, T.M., Valori, L.P., Dicker, M.P. and Trask, R.S., 2019. 4D printing with robust thermoplastic polyurethane hydrogel-elastomer trilayers. *Materials & Design*, 163, p. 107544.
[56] López-Valdeolivas, M., Liu, D., Broer, D.J. and Sánchez-Somolinos, C., 2018. 4D-printed actuators with soft-robotic functions. *Macromolecular Rapid Communications*, 39(5), p. 1700710.
[57] Kotikian, A., Truby, R.L., Boley, J.W., White, T.J. and Lewis, J.A., 2018. 3D printing of liquid crystal elastomeric actuators with spatially programed nematic order. *Advanced Materials*, 30(10), p. 1706164.
[58] Ambulo, C.P., Burroughs, J.J., Boothby, J.M., Kim, H., Shankar, M.R. and Ware, T.H., 2017. Four-dimensional printing of liquid crystal elastomers. *ACS Applied Materials & Interfaces*, 9(42), pp. 37332–37339.
[59] Saed, M.O., Ambulo, C.P., Kim, H., De, R., Raval, V., Searles, K., Siddiqui, D.A., Cue, J.M.O., Stefan, M.C., Shankar, M.R. and Ware, T.H., 2019. Molecularly-engineered, 4D-printed liquid crystal elastomer actuators. *Advanced Functional Materials*, 29(3), p. 1806412.
[60] Barnes, M., Sajadi, S.M., Parekh, S., Rahman, M.M., Ajayan, P.M. and Verduzco, R., 2020. Reactive 3D printing of shape-programmable liquid crystal elastomer actuators. *ACS Applied Materials & Interfaces*, 12(25), pp. 28692–28699.
[61] Zhang, C., Lu, X., Fei, G., Wang, Z., Xia, H. and Zhao, Y., 2019. 4D printing of a liquid crystal elastomer with a controllable orientation gradient. *ACS Applied Materials & Interfaces*, 11(47), pp. 44774–44782.
[62] Yuan, C., Roach, D.J., Dunn, C.K., Mu, Q., Kuang, X., Yakacki, C.M., Wang, T.J., Yu, K. and Qi, H.J., 2017. 3D printed reversible shape changing soft actuators assisted by liquid crystal elastomers. *Soft Matterials*, 13(33), pp. 5558–5568.
[63] Roach, D.J., Kuang, X., Yuan, C., Chen, K. and Qi, H.J., 2018. Novel ink for ambient condition printing of liquid crystal elastomers for 4D printing. *Smart Materials and Structures*, 27(12), p. 125011.
[64] Fu, P., Li, H., Gong, J., Fan, Z., Smith, A.T., Shen, K., Khalfalla, T.O., Huang, H., Qian, X., McCutcheon, J.R. and Sun, L., 2022. 4D printing of polymeric materials: Techniques, materials, and prospects. *Progress in Polymer Science*, p. 101506.
[65] https://patentscope.wipo.int/search/en/result.jsf?_vid=P21-L6BXPJ-55799. Accessed on 2 August 2022.
[66] "The George Washington University", 2022. "US20220204927" – 4D printing smart culture substrate for cell growth.
[67] "The University of Akron", 2022. "WO2022055558" "Resorbable complex shape memory poly (propylene fumarate) star scaffolds for 4d printing applications".

11 Investigating the Work Generation Potential of SMA Wire Actuators

Nisha Bhatt
Thapar Institute of Engineering and Technology, Patiala, Punjab, India

Sanjeev Soni
Central Scientific Instruments Organization (CSIR-CSIO), Chandigarh, India

Ashish Singla
Thapar Institute of Engineering and Technology, Patiala, Punjab, India

CONTENTS

11.1 INTRODUCTION

Shape memory alloys are distinctive metallic alloys that tend to store their forms under mechanical or thermal loadings. They showcase their fascinating properties of recovering permanent deformations up to 8% or more. As metallic alloys, they exhibit metallic properties like corrosion resistance, wear resistance, workability, stiffness, and so on. Apart from these properties, they exhibit excellent biocompatibility. These materials tend to deform at low temperatures or even at room temperatures when sufficient force is applied to cause the deformation. This deformation can be fully or partially recovered under the application of thermal

DOI: 10.1201/9781003333760-11

stresses. This is known as the shape memory effect. There are three different crystal structures called twin martensite, detwinned martensite, and austenite, associated with these materials. These structures are stable in two separate phases: a high-temperature phase (austenite) and a low-temperature phase (twinned and detwinned martensite). Naturally, SMAs exist in their twin martensite structure, and to use them as actuators, they need to be deformed. The deformed structure is called detwinned martensite. On heating, this deformation can be recovered partially or fully which entirely depends on whether the SMA is constrained or not. The purpose of an actuator is to generate force and thus, SMAs can be utilized as an actuator under constrained heating. Deformation of an unconstrained SMA can be recovered fully as no resistance will be offered to the recovery process. However, a constrained SMA offers resistance to the recovery process and thus, there is only partial recovery of the deformation. This resistance to the recovery process generates a force that can be utilized for actuation purposes. There are two categories of SMAs called one-way and two-way. In one-way, they memorized the "low-temperature state," whereas in two-way they memorized both "high temperature and low-temperature states." However, an intensive requirement of training limits the use of two-way SMAs. Thus, the usage of one-way SMAs is more prevalent in industries. In both categories, two starts and two finish temperatures are accounted for martensite and austenite crystal structure: martensite start temperature (M_s), martensite finish temperature (M_f), austenite start temperature (A_s), and austenite finish temperature (A_f). During the shape recovery process, detwinned SMA begins to contract at A_s and the recovery process completes when the transformation temperature reaches A_f. On cooling the material, martensite transformation starts when the material temperature reaches M_s and it completes when the temperature reaches M_f.

Researchers preferred thermal shape memory alloys for actuation purposes. In assembly lines, they are used as grippers or end effectors to move/manipulate the products. Compared to conventional end effectors, they bring more flexibility and cost-saving to production lines. Moreover, as end effectors, their adaptability for frequent configuration to accommodate changes in the product type makes them more favorable in assembly lines [1]. In medical industries, these actuators are preferred due to their decent biocompatibility, good ductility, and excellent wear and corrosion-resistant properties [2]. Apart from these applications, these actuators are also used to develop continuum manipulators for minimally invasive surgeries and other medical applications requiring miniature robots [3]. Automotive industries which are currently facing the technological push toward "smart" systems are now preferring these actuators over conventional electro-mechanical motors due to their lighter weight, volume, and reliability [4].

SMAs in the form of wires are the most commonly used actuators as they tend to generate large forces [5]. One-way SMAs are more radially available as compared to two-way SMAs. A one-way SMA wire requires restoring mechanism to reset the heated SMA to its initial position. The most commonly used biasing elements are a normal steel spring or another SMA wire. To use these materials as actuators, one needs to mathematically model the dynamics of the material along with its interaction with the biasing mechanism. Since the material exhibits thermal hysteresis, a high nonlinearity is associated with the phase transformation process. This nonlinearity needs to be accurately modeled along with the stress-strain relationship of the material. In literature, many mathematical models are available that model the non-linearity of the phase transformation process through different mathematical functions [6–8]. Despite a large availability of these mathematical models in the literature, only a few studies [9–12] discuss the numerical approaches adopted to determine the state vectors of the presented mathematics. Literature suggests the finite element method (FEM) as the most commonly used numerical approach to determine the state vectors of SMAs models [13–15]. Much commercial software implements FEM-based solutions for SMA material. These FEM-based solutions are computationally expensive and thus, impractical to use when complex thermo-mechanical dynamics are involved. Integrated SMA systems prefer MATLAB/SIMULINK software to model the dynamics of the actuator [16–19]. This software has a large number of inbuilt differential equation solvers that can easily take into account the nonlinearity of the phase kinetics. Moreover, this software can easily consider the history of the traveled path of an SMA material, which plays an important role in replicating the phase kinetics on simulation software. In addition, MATLAB/SIMULINK provides a platform to integrate inbuilt nonlinear controllers into the system dynamics. The advantages offered by this software make it a favorable choice for solving the non-linear dynamics of SMA systems.

It has been observed from the literature that there is limited research into the effect of particular parameters on the performance of SMA-based actuators. The selection of any actuator for a particular application depends upon its mechanical, thermal, electrical, and other parameters. Similarly, the performance of SMA-based actuators depends upon the parameters associated with its heat transfer model, constitutive model, and other geometrical parameters. In literature, a study reports performance variations of an SMA wire actuator when the parameters associated with the heat transfer model of the SMA actuator are varied. Low heating speed and high thermal conductivity of the SMA material were found to increase the mean wire temperature. In addition, thermal convection lowers the mean wire temperature [14]. Another study reports the effect of resistivity, cross-sectional area, and length of wire on the electrical resistance of the SMA wire actuator [20]. Literature also reports the influence of pre-tension and input current on the output characteristics of the SMA actuator [21]. The selection of an SMA actuator for any particular application depends upon these parameters. However, the influence of geometric parameters, materials, and other external parameters on the performance of the SMA actuator remains unknown.

The current study addresses the dynamics of the SMA wire actuator by biasing it first with a normal spring and then replacing the normal spring with another SMA wire in an antagonistic configuration. An existing phenomenological model is implemented in SIMULINK to determine the work generation potential in terms of generated force and achieved displacement. The methodology to simulate the dynamics is discussed in detail. Different geometrical, material, and other external parameters are varied in simulations to investigate their effect on the work generation potential of the SMA wire actuator in both biasing methods, which lead to some new findings. The current study concludes with a comparative investigation of the discussed bias methods. The results help in selecting an appropriate SMA wire actuator working under thermal and mechanical loading conditions.

11.2 THEORY AND METHODS

At first, SMA wire is pre-strained to some amount and kept under the biased condition with its biasing element. One end of the SMA wire is fixed and the other end is connected to the mobile unit located at the junction of the SMA wire and the bias element. Similarly, one end of the bias element is connected to the one end of the mobile unit and the other end of the bias element is kept fixed. Thus, the overall length of the SMA actuator (SMA wire and bias element) is geometrically constrained and remains constant throughout the thermal-mechanical loading. The movement of the mobile unit (moving block) during the activation and deactivation of the wire will determine the overall achievable displacement of the actuator. A schematic of the above arrangement is presented in Figure 11.1 with two different biasing elements: (a) a normal spring and (b) another SMA wire.

A basic block diagram describing the mathematics of an SMA wire actuator comprises three sub-models: heat transfer model, constitutive model, and phase kinetics model. Input to the system is an external voltage supply that heats the wire through *Joule* heating. A heat transfer model describes the relationship of the input voltage to the average temperature of the wire under heating and cooling conditions. It is assumed that the wire temperature varies uniformly throughout its length. A heat transfer model outputs the average temperature of the wire at a given time through Eq. 11.1 [11].

$$C_v \frac{\partial T_w}{\partial t} - \mathcal{L}\frac{\partial \xi}{\partial t} - \frac{V^2}{\beta_r L_w^2} + \frac{4h}{d_w}(T - T_a) = 0 \quad (11.1)$$

The meaning and values of all symbols are provided in Table 11.1. The first term denotes the temperature variation due to heat absorbed by the SMA material. The second term represents the latent warmth of SMA necessary to complete the phase transformation process. The third term is the power input to the SMA material through a voltage supply. The last term shows convective heat transfer to the ambient temperature.

Another sub-model is the constitutive model, which represents the relationship between stress and strain of the material. However, in addition to strain, stress depends on temperature and martensite fraction. Strain represents the interaction of biasing element with the SMA wire and it could be represented in the form of stress only. At a given moment in the constitutive model, there are two unknowns: stress and martensite fraction. A martensite fraction is a non-dimensional quantity that

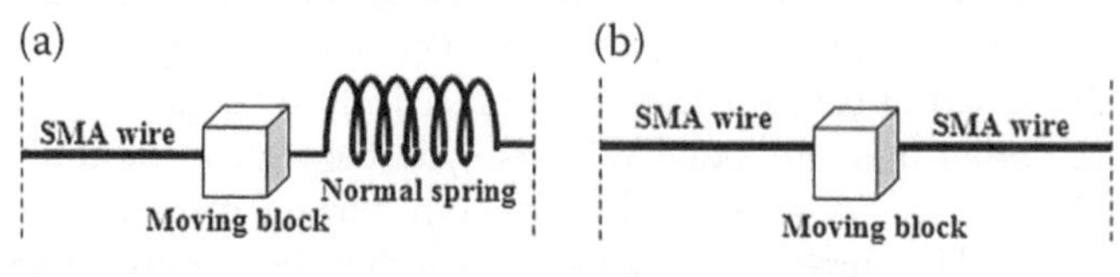

FIGURE 11.1 Arrangement of SMA wire with (a) bias spring (adapted from [12]); (b) another SMA wire.

TABLE 11.1
Properties of NiTi Wire and Steel Spring

Parameters	Value
Diameter of SMA wire (d_w)	125 μmm
Unstretched length of SMA wire (L_{ow})	250 mm
Stiffness of steel spring (K)	0.24 N/mm
Ambient temperature (T_a)	26°C
Density of NiTi (50:50) (ρ)	6,450 × 10^{-9} Kg/m^3
Resistivity (β_r)	0.009 Ohm-mm
Specific heat capacity (C_v)	0.0052 J/K-mm^3
Stress-temperature slope in austenite (C_a)	12 MPa/°C
Stress-temperature slope in martensite (C_m)	10 MPa/°C
Start temperature of martensite phase (M_s)	58°C
Finish temperature of martensite phase (M_f)	45°C
Start temperature of austenite phase (A_s)	60°C
Finish temperature of austenite phase (A_f)	75°C
Young's modulus (E)	51,500 N/mm^2

describes the volume fraction of martensite in the material. The value of the martensite fraction is 1 when the material structure is fully martensite and 0 when the structure is austenite. The constitutive model outputs the value of stress or force at a given moment if temperature and martensite fraction is known at that instant. Since the temperature is known from the heat transfer model, the other unknown-martensite fraction is evaluated from the phase kinetics model. In the following paragraph, the stress-strain relationship between the two different biasing cases is discussed.

11.2.1 Case 1: SMA Wire with Normal Spring

One of the easiest methods of resetting the SMA wire is using a linear non-SMA spring. The spring offers resistance force to the wire and thereby, constrained the free recovery of SMA's deformation during heating. Moreover, the same spring pulls back the SMA when the material transforms from austenite to martensite during cooling. It is easy to deform the martensite crystal structure as Young's modulus at this stage is much lower than that in the austenite phase. When a pre-stressed SMA wire is heated by the Joule effect, it starts to contract as soon as the transformation temperature reaches A_s. This contraction of the wire is resisted by the linear spring which ultimately generates recovery stress in the wire. This recovery stress is large enough to deform the linear spring and it shifts the junction towards the wire side. This displacement represents the work generation potential of the SMA wire actuator. Once cooled, the transformation of the austenitic phase to the martensite phase occurs and the recovery stress starts to decline. Soon, the spring resists its deformation and starts to come back to its initial position. In doing so, it pulls back the wire to its initial position. The stress-strain relationship of this configuration is depicted in Eq. 11.2:

$$\Delta = \frac{F_s}{K} \tag{11.2}$$

where Δ is the deflection of spring and K represents the stiffness of spring. F_s is the force acting on the spring due to the deflection. This force is equal to the restoring force of SMA, as represented in Eq. 11.3:

$$F_s = \sigma \mathrm{A} \tag{11.3}$$

Here, σ represents restoring stress and A is the cross-sectional area of the wire. Again, the kinematic relation of the SMA wire actuator provides Eq. 11.4 [10].

$$\Delta = L_{w_o}(1 + \varepsilon_o) - L_{w_o}(1 + \varepsilon) \tag{11.4}$$

Therefore, strain can be represented by Eq. 11.5:

$$\varepsilon = \varepsilon_o - \frac{\sigma \mathrm{A}}{K} \tag{11.5}$$

Once the martensite fraction is known, the stress in the wire can be calculated by Eq. 11.6:

$$\frac{d\sigma}{dt} = E\frac{d\varepsilon}{dt} - E\varepsilon_l \frac{d\xi_s}{dt} \tag{11.6}$$

Here, the first term on the right is the stress caused by the axial deformation and the second is the stress caused by the phase transformation [8].

11.2.2 Case 2: SMA Wire in an Antagonistic Configuration

This case represents biasing of an SMA wire with an identical SMA wire. Such a configuration is known as an antagonistic SMA wire configuration. In this configuration, like a normal spring, the other SMA wire provides biasing force to the heated SMA wire. However, unlike in Case 1, here the moving block can be displaced in both extreme positions. When the first wire is heated, the other wire serves as bias and, similarly, when the second wire is heated, the first will serve as bias. Interestingly, the two wires are mechanically linked but electrically isolated from one another. Under static loading, the junction is stable at its neutral position and the force exerted by both wires is equal.

$$F_1 = F_2 \tag{11.7}$$

where subscripts 1 and 2 represent SMA wire 1 and SMA wire 2, respectively. The displacement of the block can be represented by Eq. 11.8:

$$\Delta = L_{W_o}\left(1 + \varepsilon_{o_1}\right) - L_{W_o}(1 + \varepsilon_1) = L_{W_o}(1 + \varepsilon_2) - L_{W_o}\left(1 + \varepsilon_{o_2}\right) \tag{11.8}$$

When equally pre-strained,

$$\varepsilon_{o_1} = \varepsilon_{o_2} \tag{11.9}$$

Again, the stress-strain relation remains the same as depicted by Eq. 11.6:

$$\frac{d\sigma_1}{dt} = E_1\frac{d\varepsilon_1}{dt} - E_1\varepsilon_l\frac{d\xi_{s_1}}{dt} \tag{11.10}$$

$$\frac{d\sigma_2}{dt} = E_2\frac{d\varepsilon_2}{dt} - E_2\varepsilon_l\frac{d\xi_{s_2}}{dt} \tag{11.11}$$

A phase kinetics model represents the phase kinetics involved in two phase transformations: martensite to austenite and vice versa. There are numerous mathematical models which represent the martensite fraction relationship with temperature and stress. In the current study, an arbitrary loading-based model that incorporates the reflection of traveled path history in the phase transformation process is adapted and simulated in a SIMULINK environment. Originally the adapted model is based on implicit differential solver *ode15i*, which is highly sensitive to the selected tolerances. Moreover, the implicit scheme is difficult to be clubbed with control algorithms because control algorithms generally work on Laplace transforms, whereas implicit solvers are time domain solvers and are difficult to convert into the Laplace transforms. The phase kinetics model takes inputs as temperature and stress and provides martensite volume fraction as output. The martensite fraction is a combination of stress-induced and temperature-induced martensite fractions. Thus, the stress-induced martensite fraction can also be obtained as the output from the phase kinetic model. The martensite and stress-induced martensite fractions can be represented by Eqs. 11.12 and 11.13 [9]:

$$\frac{\partial\xi}{\partial t} - \eta_1\frac{\partial T}{\partial t} - \eta_2\frac{\partial\sigma}{\partial t} = 0 \tag{11.12}$$

$$\frac{\partial\xi_s}{\partial t} - \eta_3\frac{\partial T}{\partial t} - \eta_4\frac{\partial\sigma}{\partial t} = 0 \tag{11.13}$$

where η_{1-4} function values have been discussed in the arbitrary loading model [9]. The interaction among inputs and outputs of all three sub-models is shown in Figure 11.2.

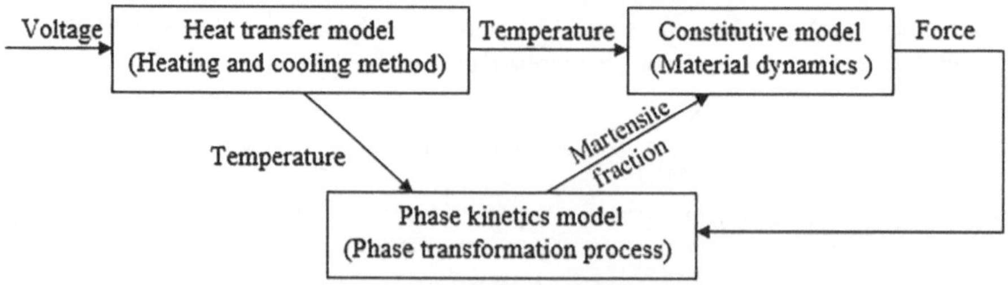

FIGURE 11.2 Interaction among the submodels of SMA wire actuator (adapted from [12]).

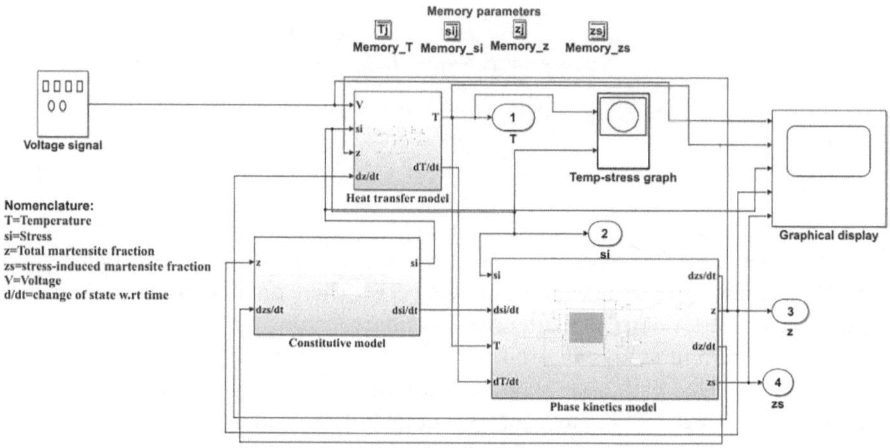

FIGURE 11.3 SIMULINK model of SMA wire actuator with normal spring as bias.

Once, the mathematics is formulated, a block diagram representation like Figure 11.2 is adapted in the SIMULINK platform, as shown in Figure 11.3, and all properties of the model are saved in the base workspace. Table 11.1 shows the properties of the selected SMA actuator.

A voltage signal is generated through the signal generator and three sub-models are formed. Eqs. 11.12 and 11.13 are called through MATLAB function and memory parameters that define the earlier traveled path of SMA material during phase transformation are saved as data store memory blocks. Once the whole model is formed, a fixed step solver called Backward Euler (*ode 1be*) is called with a step size of 0.001. After inserting the simulation start and end time, the model is made to run and a temperature-stress graph pops up that simultaneously plots the stress values on the y-axis with temperature values on the x-axis. All state values such as the temperature of the wire, the stress in the wire, the total martensite fraction, and the stress-induced martensite fraction are saved to the workspace. Although the dynamics can be solved in MATLAB software, a SIMULINK model like Figure 11.3 is helpful in scenarios where control architecture needs to be integrated with the system. There are different inbuilt solvers available in the software but for the particular model, Backward Euler is found to be the most suitable one.

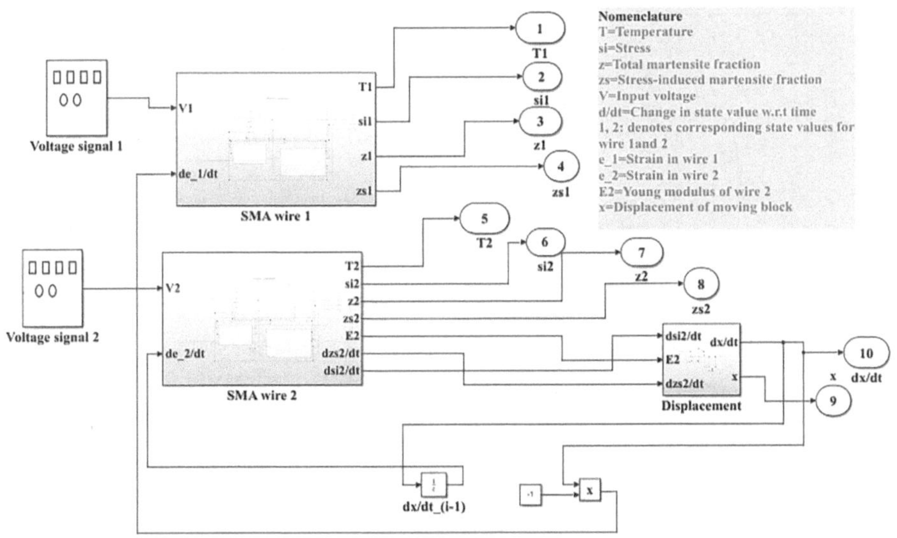

FIGURE 11.4 SIMULINK model of antagonistic SMA wire actuator.

Likewise, the block diagram of the antagonistic SMA wire actuator is developed in the SIMULINK platform and identical SMA wires are selected with the properties presented in Table 11.1. A schematic of the same is shown in Figure 11.4.

Again, Backward Euler solver is used to simulate the model as it gives the best results when compared with other solvers.

Once the model is built and run on the software, parametric variations are performed under the same conditions to find out their effect on the actuator's output: force and displacement. Parameters of SMA wire like diameter/length, maximum recoverable strain, stress-temperature slopes, initial pre-tension, and other external parameters like input voltage and ambient temperature are varied to identify the most influencing parameter. Simulation results for these variations are important during the design and selection process of SMA wire actuators.

11.3 RESULTS AND DISCUSSIONS

This section presents the results obtained by simulating both configurations of the SMA wire actuator within the SIMULINK platform.

11.3.1 Case 1: SMA Wire with Normal Spring

Initially, a SMA wire is pre-tensed to 33% of its length with a force of 0.98 N. When the SMA wire is subjected to a square wave voltage pattern as shown in the inset picture of Figure 11.5(a), the maximum temperature reaches 105°C. During heating, when the voltage is increased from 0 V to 7 V, a transformation from martensite to austenite takes place, as marked by a red circle in Figure 11.5(a). Similarly, when the voltage drops from 7 V to 0 V, the cooling of the wire initiates

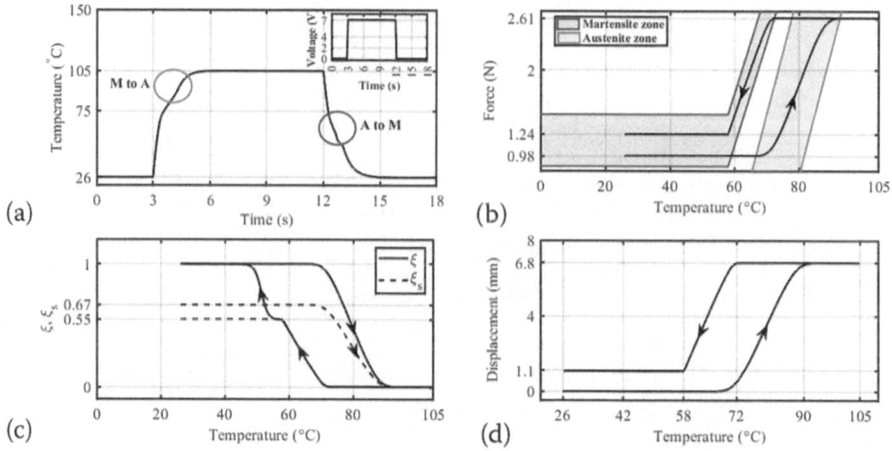

FIGURE 11.5 Results obtained when SMA-bias spring actuator subjected to a square wave (inset): (a) temporal variation of temperature; (b) variation of force; (c) variation of martensite fraction; and (d) variation of displacement with temperature.

the transformation from austenite to martensite phase as indicated by a blue circle in Figure 11.5(a). As the SMA wire is constrained by the bias spring, free recovery of initial strain in the wire is resisted by the spring. Thus, a maximum of 2.61 N of force is achieved on completion of austenite transformation. Heating further from here doesn't increase the maximum force achievable and, thus, the force retains its maximum value of 2.61 N until 105°C, as shown in Figure 11.5(b). In the course of cooling, when the temperature of the wire reaches the martensite zone (indicated by the blue shaded area), the recovery force starts to decrease and finally settles at 1.24 N.

A full transformation of martensite into austenite and vice versa is represented by the value of the total martensite fraction (ξ). Originally, $\xi = 1$, when the material is in its martensite phase and as the temperature of wire reaches the austenite start temperature (A_s), ξ begins to decrease. When ξ reaches 0 near A_f, a full conversion from martensite to austenite takes place. Similarly, during cooling, austenite to martensite transformation starts when the wire temperature cools down to M_s and ξ starts to increase from 0. A full transformation from austenite to martensite is marked when the temperature of the wire reaches M_f and $\xi = 1$, as indicated by Figure 11.5(c).

Likewise, ξ_s varies with the temperature of the wire. An initial value of 0.67 indicates the percentage fraction of total martensite due to initial pre-tensioning in the wire. As the wire temperature increases, ξ_s decreases and it becomes 0 when the material crystal structure changes from martensite to austenite. Thereafter, the value of ξ_s increases when the biasing spring starts to bias the cooling SMA. The final values of ξ_s depend upon the stiffness of the biasing spring. Here, it settles to 0.55, as shown in Figure 11.5(c).

The ability of the wire actuator to generate work depends on the maximum achievable force and the maximum displacement of the moving unit placed between the junction of the SMA wire and the bias spring. The neutral position of the moving

TABLE 11.2
Comparison of the Current Study with Published Literature [11]

Parameters	Published Results	Current Study
Input voltage	Sinusoidal (7V)	Square (7V)
Maximum value of stress generated (force)	166 MPa (2.037 N)	212.68 (2.61 N)
Final value of stress-induced martensite fraction	0.5	0.55
Maximum temperature	108°C	105°C

block is indicated by 0 mm in Figure 11.5(d) and its maximum value is 6.8 mm. After complete cooling, the neutral position is shifted to 1.1 mm toward the positive direction. Here, positive direction indicates the side where the SMA wire is attached to the moving block i.e., left (as shown in Figure 11.1(a)) and the negative direction is towards the side where the bias spring is attached to the block. The findings are compared with the published results [11] on the same configuration and are shown in Table 11.2.

Though in the current study, the force obtained is a little overestimated but the simplified model in SIMULINK and easiness to connect the plant model with the controller makes it justifiable.

11.3.2 Case 2: SMA Wire in an Antagonistic Configuration

An SMA wire can also be biased with another SMA wire, attached antagonistically. Here, identical SMA wires are taken and are equally pre-stressed to 33% of their lengths. Like Case 1, the left SMA wire is heated first and subjected to a heating-cooling cycle which depends upon the voltage pattern as indicated by a black solid line in Figure 11.6(a). The corresponding variation in temperature is reported by T_1 in Figure 11.6(b).

Initially, there exists a force of 0.98 N due to pre-tensioning in both wires. During the heating of wire 1, this force increases and reaches a maximum of 5.18 N. It finally settles to 0.98 N during cooling. When the first wire contracts at a high temperature, the second wire extends at room temperature. During this process, there is an increase in the force of the second wire from 0.98 N to 5.18 N. However, during the cooling of the first wire, the force in the second wire as well as the first wire decreases. Thereafter, heating the second wire results in a maximum force of 9.48 N. Although both the wires are subjected to the same voltage and same highest temperature the maximum value of force is more in the second wire as compared to the first. It is because the second wire is more stressed as compared to the first one. Despite equal pre-tensioning, the second wire is subjected to higher detwinning during the heating cycle of the first wire. At this stage, the first wire detwins due to the heating force of the second wire, and its force reaches 9.48 N at room temperature. This force value finally settles to 1.25 N during the cooling of the second wire, as depicted in Figure 11.6(c). The movement of the block at the junction is illustrated in Figure 11.6(d). A positive displacement shows the movement of the

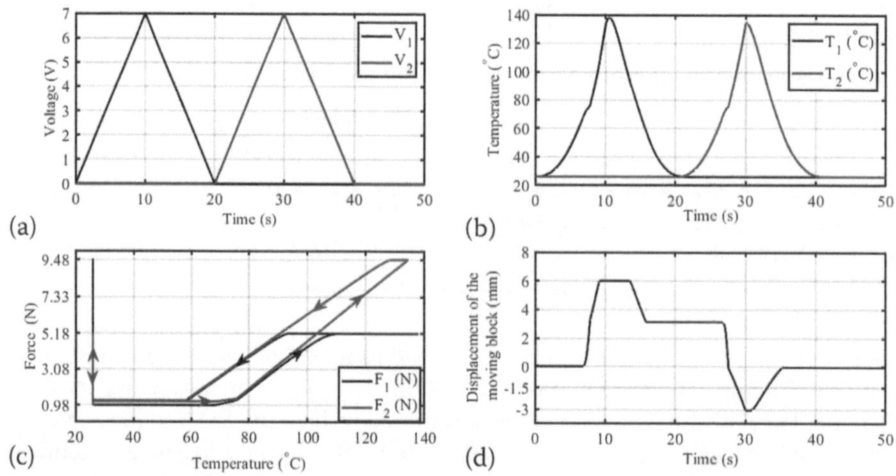

FIGURE 11.6 Results obtained when an antagonistic SMA wire actuator is subjected to thermal and mechanical loading: (a) voltage pattern; (b) temporal variation of temperature; (c) change in force with change in temperature of wire 1 and wire 2; and (d) temporal variation of displacement.

junction from the neutral position (0 mm) towards the left side and a negative displacement indicates the movement of the junction towards the right side. A maximum of 6 mm positive displacement is observed, which settles down to 3 mm towards the left during the full heating and cooling cycle of the first wire. This is indicated by a constant solid black line at 3 mm in Figure 11.6(d). Thereafter, the heating cycle of the second wire shifts the block from a positive 3 mm to a negative 3 mm. On cooling the second wire, the moving block finally shifts back to its initial neutral position of 0 mm. The obtained results follow the characteristic profile of antagonistic wires, as presented in the published literature [9].

Once the results are obtained under thermo-mechanical loadings, material, geometrical, and other external parameters are varied to study their effect on the SMA wire actuator's performance.

11.3.3 Parametric Variations

An important geometrical parameter of the SMA wire actuator is its diameter. If the wires are kept under the same stressed conditions and the total volume of the actuator is kept constant, then the wire with a lower diameter or longer length will generate lesser force as compared to the wire with a larger diameter or smaller length. The same statement can be verified in Figure 11.7(a).

Here, a wire with 100 μmm generates a force of 2.07 N, and a wire with 200 μmm generates 4.24 N. Likewise, the output displacement keeps on increasing with an increase in the diameter of the wire as depicted in Figure 11.7(b). It is 6 mm for 100 μmm and 7.2 mm for 200 μmm. This behavior, in general, is obvious too, because if the stress conditions are constant, the area is directly proportional to the force.

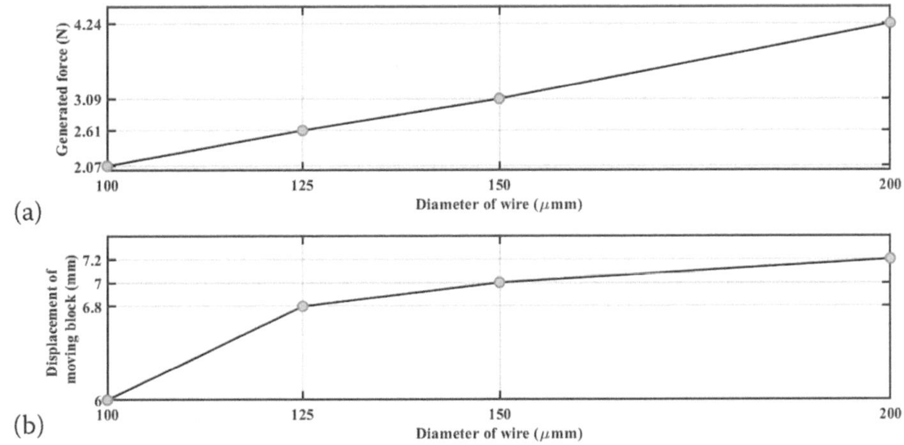

FIGURE 11.7 Effect of SMA wire's diameter (μmm) on (a) generated force (N), and (b) displacement of moving block (mm).

Similarly, as per Eq. 11.5, the output displacement increases with an increase in the area.

The sensitivity parameter can be calculated as $s = \frac{Change\ in\ output}{change\ in\ input}$

Thus, $s = 0.0217N/\mu mm$ for the generated force and it is 0.012 mm/μmm for output displacement.

Apart from the length or diameter of the wire, parameters like maximum recoverable normal strain and stress-temperature slopes in martensite and austenite have a significant role in the phenomenological modeling of the actuator. The stress-temperature slopes decide the wideness of the transformation zones and the maximum value of recoverable normal strain plays an important role in deciding the last limit of recovery of the deformation through the shape memory effect. Over and above that, there will be plastic deformation in the material that cannot be recovered by heating. These two parameters are the material parameters and depend upon the composition of the shape memory alloy.

There is not any predefined range of stress-temperature slopes. They can have any numeric values which depend on their transformation temperatures and yield strength. Therefore, in this study, instead of manipulating their numeric values, three possible cases are considered to analyze their effect on the performance of the actuator.

1. $C_a > C_m$,
2. $C_a = C_m$, and
3. $C_a < C_m$

The simulation results in Figures 11.8(a) and 11.8(b) suggest that even though the values of stress-temperature slopes decide the width of the transformation zones, they don't much affect the output displacement and force generation capability of the wire actuator. In all three cases, the force is around 2.5–2.6 N and the output displacement is 6.8 mm.

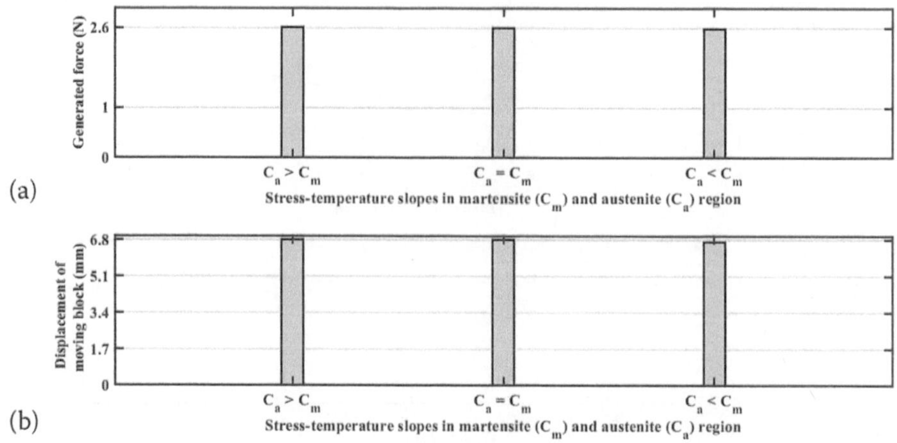

FIGURE 11.8 Effect of C_a and C_m (MPa/°C) on (a) generated force (N) and (b) displacement of moving block (mm).

Another considerable parameter is the maximum value of recoverable normal strain in the material. Generally, the maximum value of recoverable strain in the SMA wire is up to 8% of its total length. The simulation results show that the maximum recoverable strain does not affect the force generation capability and output displacement of the actuator. Here, the maximum force is limited to 2.61 N and output displacement is 6.81 mm for all values of maximum recoverable strain, as shown in Figures 11.9(a) and 11.9(b), respectively.

Next, the impact of the initial pre-stressing on the production potential of the actuator is analyzed. It is observed that the larger the initial pre-tension in the wire, the higher will be the force generation capability of the actuator, as seen in Figure 11.10(a). Also, it is observed from Figure 11.10(b) that the output displacement tends to increase

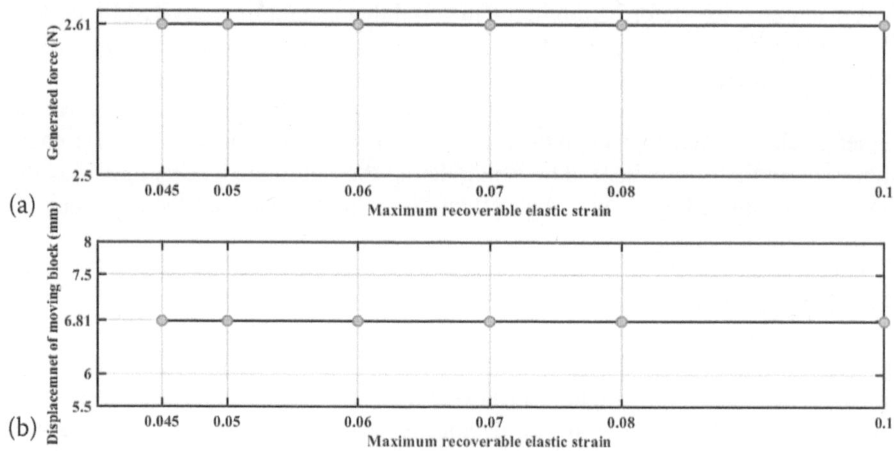

FIGURE 11.9 Effect of maximum recoverable elastic strain on (a) generated force (N) and (b) displacement of moving block (mm).

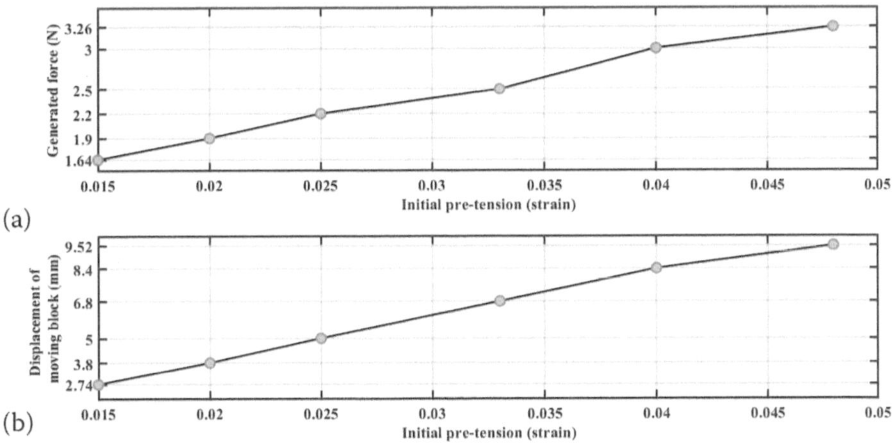

FIGURE 11.10 Effect of initial pre-tension on (a) generated force (N) and (b) displacement of moving block (mm).

with a rise in the pre-tensioning of the wire. It is to be noted that the pre-tensioning in the wire needs to be varied in the limit. This parameter directly affects the initial value of ξ_s, which is between 0 and 1. Again, the sensitivity parameter for output force and displacement is 49.09 N and 204.45 mm.

Apart from the discussed parameters, external factors like ambient temperature directly affect the average temperature rise in the wire due to the applied voltage. The higher the ambient temperature, the lesser will be the heat loss to the environment due to convection. Thus, more will be the average temperature of the wire. However, increased wire temperature doesn't much affect the performance of the actuator. It can be observed from Figure 11.11(a) that the generated force is 2.62 N below 26°C temperature and it is 2.61 N above 26°C.

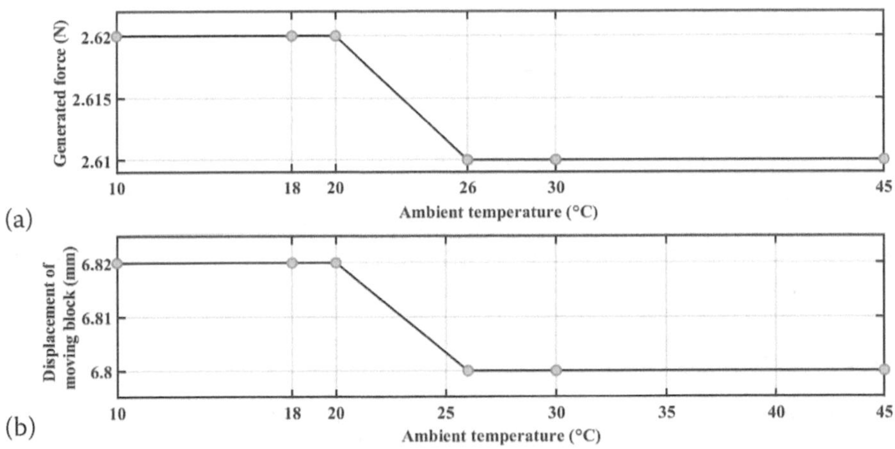

FIGURE 11.11 Effect of ambient temperature (°C) on (a) generated force (N) and (b) displacement of moving block (mm).

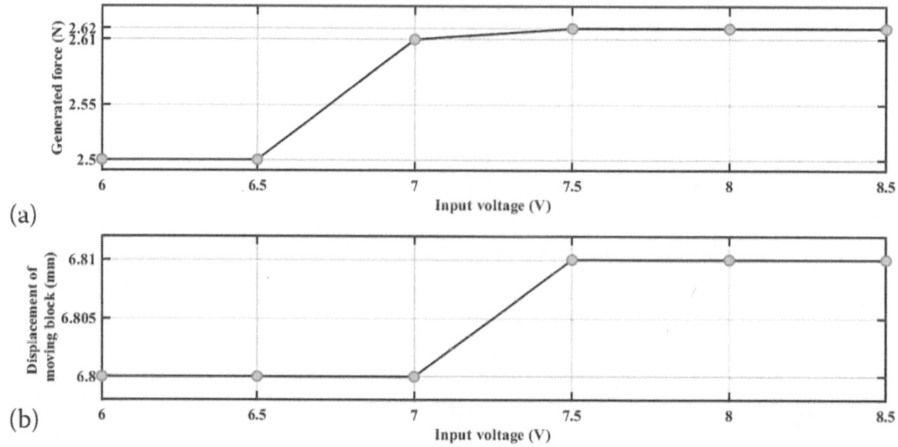

FIGURE 11.12 Effect of input voltage (V) on (a) generated force (N) and (b) displacement of moving block (mm).

Similarly, a change in the ambient temperature doesn't influence the displacement of the moving block. It is 6.8 mm above and at 26°C, whereas it is 6.82 mm below 26°C, as shown in Figure 11.11(b). These small variations in output force and displacement are negligible. Moreover, it is found from the simulation that if the ambient temperature is greater than M_f, it will result in an incomplete transformation from austenite to martensite. Thus, in Figure 11.11, the highest ambient temperature is equal to M_f.

Another external parameter is the input voltage supplied to the wire. Variations in the input voltage must be within limits. Consistent application of high voltages can permanently change the properties of SMA materials. However, low voltages don't interfere with the properties of the material but voltages below the prescribed limit don't induce austenite to martensite transformation. Figure 11.12(a) shows the variation of output force with the magnitude of input voltage. Here, the magnitude of input voltage varies from 6 V to 8.5 V. At first, with an increase in the input voltage, generated force increases and, after reaching a limit, it saturates. The maximum value of generated force is 2.62 N for 7.5 V and above. A similar observation from Figure 11.12(b) shows the variation of output displacement with input voltage. The output displacement is 6.8 mm until the voltage is 7 V and, after that, it attains 6.81 mm for voltages above 7 V. Thus, it can be concluded from Figure 11.12 that with an increase in the voltage supply, force tends to increase in a limit, whereas an increase in the output displacement remains negligible.

11.4 CONCLUSIONS

This research presented a mathematical model of an SMA wire actuator based on existing phase kinetics that incorporates the traveled path history of SMA. In the literature the existing phase kinetics has been implemented through an implicit solver which was found to be quite sensitive towards the relative tolerances and was difficult to combine with the controller's mathematics. Hence, the current study has presented a SIMULINK-based approach to model the same and it has been found that the obtained results follow the typical characteristic profile of an SMA actuator. Further, the biasing of SMA wire via antagonistic configuration has more work generation potential compared to biasing via a normal spring. A maximum force of 9.48 N has been generated by an antagonistic configuration compared with 2.61 N force via biasing with a normal spring. Also, the antagonistic configuration generates bidirectional output displacement (+ 6 mm and −3 mm) compared with the unidirectional output displacement (+ 6.8 mm) of the SMA wire-bias spring configuration. Subsequently, a parametric study has been performed by varying one parameter at once and maintaining all other parameters as constants. It has been found that among all the considered parameters, the diameter of the wire and initial pre-tension has a significant effect on the work generation potential of the wire. Increased pre-tension and wire diameter led to higher work generation potential i.e., higher force generation capability and higher output displacement. Initial pre-tension in the wire has been considered the most sensitive parameter among the diameter of the wire, maximum recoverable elastic strain, input voltage, stress-temperature slopes, and ambient temperature. The simplified model presented in this paper will help novice researchers of SMAs to build mathematical models as per their applications. Moreover, the results of the performed parametric study will help in the selection of design parameters for the SMA wire actuator.

REFERENCES

[1] Motzki, P., and Seelecke, S. 2022. Industrial applications for shape memory alloys, In *Encyclopedia of Smart Materials*, ed. Olabi, A. G., 254–266. Elsevier. 10.1016/B978-0-12-803581-8.11723-0

[2] Patel, S. K., Behera, B., Swain, B., Roshan, R., Sahoo, D., and Behera, A. 2020. A review on NiTi alloys for biomedical applications and their biocompatibility. *Materials Today: Proceedings* 33, No. 8: 5548–5551. 10.1016/j.matpr.2020.03.538

[3] Goergen, Y., Chadda, R., Britz, R., Scholtes, D. et al. 2019. Shape memory alloys in continuum and soft robotic applications. In *Proceedings of the ASME 2019 Conference on Smart Materials, Adaptive Structures, and Intelligent Systems*. Louisville, Kentucky, USA. 9–11 September, pp. 1–6. 10.1115/SMASIS2019-5610

[4] Riccio, A., Sellitto, A., Ameduri, S., Concilio, A., and Arena, M. 2021. Shape memory alloys (SMA) for automotive applications and challenges, In *Shape Memory Alloy Engineering* (Second edition), ed. Concilio, A., Antonucci, V., Auricchio, F., Lecce, L., and Sacco, E., 785–808. Butterworth-Heinemann. 10.1016/B978-0-12-819264-1.00024-8

[5] Materials. 2018. Shape memory alloy achieve high force and precision. https://www.techbriefs.com/component/content/article/tb/pub/features/articles/29019

[6] Tanaka, K. 1986. A thermo-mechanical sketch of shape memory effect: One-dimensional tensile behavior. *Res Mechanica* 18, 251–263.

[7] Liang, C., and Rogers, C. A. 1990. One dimensional thermomechanical constitutive relations for shape memory materials. *Journal of Intelligent Material Systems and Structures* 1, 207–234. 10.1177/1045389X9000100205

[8] Brinson, L. C. 1993. One-dimensional constitutive behavior of shape memory alloys: Thermo mechanical derivation with non-constant material functions and redefined martensite internal variable. *Journal of Intelligent Material Systems and Structures* 4, No. 2: 229–242. 10.1177%2F1045389X9300400213

[9] Banerjee, A. 2012. Simulation of shape memory alloy wire actuator behavior under arbitrary thermo-mechanical loading. *Smart Materials and Structures* 21, No. 125018: 1–13. 10.1088/0964-1726/21/12/125018

[10] Gurung, H. 2017. Self-sensing shape memory alloy wire actuators using Kalman filters, (accession no. TH-1636) Ph.D. diss., Indian Institute of Technology Guwahati, India. http://gyan.iitg.ernet.in/handle/123456789886

[11] Bhatt, N., Soni, S., and Singla, A. 2021. Comparative analysis of numerical methods for constitutive modeling of shape memory alloys. *Modeling and Simulation in Materials Science and Engineering* 29, No. 8: 1–23. 10.1088/1361-651X/ac3052

[12] Bhatt, N., Soni, S., and Singla, A. 2022. Analyzing the effect of parametric variations on the performance of antagonistic SMA spring actuator, *Materials Today Communications* 31, No. 103728: 1–13.10.1016/j.mtcomm.2022.103728

[13] Gao, X., Qiao, R., and Brinson, L. C., 2007. Phase diagram kinetics for shape memory alloys: A robust finite element implementation. *Smart Materials and Structures* 16, 2102–2115. 10.1088/0964-1726/16/6/013

[14] Alipour, A. Kadkhodaei, M., and Ghaei, A. 2015. Finite element simulation of shape memory alloy wires using a user material subroutine: Parametric study on heating rate, conductivity, and heat convection. *Journal of Intelligent Material Systems and Structures* 26, No. 5: 554–572. 10.1177%2F1045389X14533431

[15] Zhou, B., Kang, Z., Wang, Z., and Xue, S. 2019. Finite element method on shape memory structure and its applications, *Chinese Journal of Mechanical Engineering* 32, No. 84: 1–11. 10.1186/s10033-019-0401-3

[16] Copaci, D., Blanco, D., Clemente, A. M., and Moreno, L. 2020. Flexible shape memory alloy actuators for soft robotics: Modelling and control, *International Journal of Advanced Robotic Systems* 17(1), 1–15. 10.1177%2F1729881419886747

[17] Pillai, R. R., Murali, G., and Gopal, M. 2018. Modeling and simulation of a shape memory alloy spring actuated flexible parallel manipulator, *Procedia Computer Science* 133, 895–904. 10.1016/j.procs.2018.07.104

[18] Zhang, D., Zhao, X., Han, J., Li, X., and Zhang, B. 2019. Active modeling and control for shape memory alloy actuators, *IEEE Access* 7, 162549–162558. 10.1109/ACCESS.2019.2936256

[19] Hadi, A., Akbari, H., Tarvirdizadeh, B., and Alipour, K. 2016. Developing a novel continuum module actuated by shape memory alloys, *Sensors and Actuators A: Physical* 243, 90–102. 10.1016/j.sna.2016.03.019

[20] Lu, Y., Zhang, R., Xu, Y., Wang, L., and Yue, H., Resistance characteristics of SMA actuator based on the variable speed phase transformation constitutive model. *Materials* 13, No. 6: 1–11. 10.3390/ma13061479

[21] Todorov, T., Mitrev, R., Yatchev, I., Fursov, A., Ilin, A., and Fomichev, V. 2020. A parametric study of an electrothermal oscillator based on shape memory alloys. In *21st International Symposium on Electrical Apparatus & Technologies (SIELA)*, Bourgas, Bulgaria. 3–6 June 2020, pp. 1–4. 10.1109/SIELA49118.2020.9167127

12 Troubleshooting on the Sample Preparation during Fused Deposition Modeling

Pradeep Singh, Ravindra Mohan, and J.P. Shakya
Department of Mechanical Engineering, Samrat Ashok Technological Institute, (Engineering College), Vidisha (MP), India

CONTENTS

DOI: 10.1201/9781003333760-12

12.1 INTRODUCTION

Additive manufacturing or 3D printing renders the CAD designs from the desktop world to actual real product by wasting material comparatively negligible as in the conventional manufacturing process. AM was primarily introduced in the 1980s to fulfill the demand for customized product manufacturing [1]. Nowadays, AM is promptly full-blown and became the center of attention in different sectors of manufacturing techniques such as consumer products, aerospace, automobile, medical, drug delivery systems, tissue engineering, jewelery etc. and is capable of producing complicated contoured parts in a small lead time and cheaper as compared to the classic manufacturing processes [2]. Mostly three types i.e., polymer resins, metal powders, and ceramics are extensively used in AM to produce customized objects and multifunctional object applications [3].

Additive manufacturing is a unique technique of manufacturing objects just from digital CAD design to physical three-dimensional objects by ultrafine layer upon layer of material accrue manner [4]. It is a tool-less manufacturing method that can be able to produce compact objects with high accuracy. The liberty of object design, part intricacy, lightweight, reinforcement of part, and functionality are some extraordinary features of additive manufacturing that magnetize the attention of researchers and manufacturing industries of automobiles, marine, oil and gas, aerospace, and biomedical applications [5].

General steps in the additive manufacturing process are as follows:

i. Formulation and CAD modeling,
ii. Export file in STL format,
iii. Check the STL errors and remove it,
iv. Slice the corrected STL file,
v. Give the machine parameters,
vi. Allot the material to machine and check the material availability,
vii. Upload file to the machine,
viii. Set up the machine,
ix. Print the part,
x. Remove the part,
xi. Post-process the part,
xii. The final product is ready to use.

This article is focused on describing practical challenges faced during the three-dimensional printing of different materials and their most appropriate solution based on experience and the study of different parameters and factors throughout the printing process.

12.2 CLASSIFICATION OF ADDITIVE MANUFACTURING TECHNIQUES AS PER ASTM STANDARDS

According to the ASTM international committee, different AM processes are classified as follows and represented in Figure 12.1.

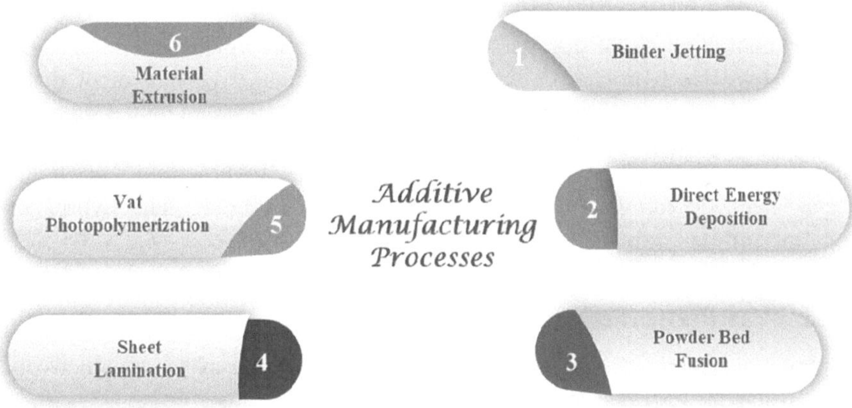

FIGURE 12.1 Classification of additive manufacturing processes according to ASTM.

12.2.1 Binder Jetting (BJ)

Binder jetting is one of the types of AM processes in which material powder particles are joined together by selectively deposited power material and three-dimensional part are created by fusing the particles together. In a binder jetting machine, the print head is designed in such a way that the bonding agent is dropped onto the powder and the build platform is lowered down to deposit an upcoming layer of powder material. The benefit of the binder jetting process of AM is huge build volume and intense printing speed with no need for support and design freedom. There are a wide variety of materials available in BJ including ceramics, glass, polymers, stainless steel, chromite, zircon, soda lime, tungsten carbide, etc. [6].

12.2.2 Direct Energy Deposition (DED)

DED is one of the processes in which high energy is focused into a small area to heat a substrate and melt it that is being deposited on a continuous stream of metal powder. Usually, a high-intensity laser beam or electron beam is used for fabrication of three-dimensional parts using metal powder. Stainless steel, titanium, nickel, cobalt, copper, tin, Inconel metal, etc. are usually used as materials for the DED AM process [7].

12.2.3 Powder Bed Fusion (PBF)

PBF is one of the types of AM processes in which a laser/electron beam is used as a thermal source for partial or full fusion of metal powder particles of a powder bed and a smooth layer of powder is recoated by a roller or blade. Laser sintering, which is observed as partial melting, is a principal binding mechanism of PBF. A thermal heating process using extremely energized electrons or a laser beam is a basic process of powder bed fusion [8].

12.2.4 Sheet Lamination (SL)

Sheet lamination is one of the types of AM processes in which material sheets are either cut by a laser process or integrated by ultrasound. Every layer of material sheet is considered an individual cross-sectional layer of the solid model. Laminated object modeling (LOM) and ultrasonic additive manufacturing (UAM) processes mainly fall into this category. Thermal bonding, adhesive bonding, and clamping are classified according to the bonding mechanism in LOM. The hybrid sheet lamination process through UAM falls into this category [9].

12.2.5 Vat Photo Polymerization (VP)

Vat photo polymerization is also known as stereolithography, in which photo curable resins are exposed and cured in the UV range by laser or by UV light and undergo a chemical reaction to convert it into a solid from a liquid. This chemical reaction is known as photopolymerization. The photopolymerization is processed when small monomers are cross-linked into a chain-structured polymer. The photoinitiator, additives, etc. are used as a catalyst to increase the polymerization rate. During the polymerization, it should be taken care that the prepared polymer has enough cross-linked, otherwise polymerized, molecules that will get re-dissolved within the liquid monomers. Stereo lithograph apparatus (SLA), direct light processing (DLP) and continuous liquid interface productions are the most prominent examples of VP [10].

12.2.6 Material Extrusion (ME)

3D material extrusion is one of the widespread AM processes. Layer-by-layer filling of liquified and semi-liquified thermoplastic polymers and a solution of polymers and pastes done by computer monitored deposition on a heated bed use a movable extrusion nozzle serving as the print head. After deposition of the first layer, either the extrusion head goes up or the heating bed goes down for depositing the upcoming layers. Fused deposition modeling (FDM) also known as fused filament fabrication (FFF), three-dimensional dispersing, three-dimensional microfiber extrusion, three-dimensional fiber deposition, three-dimensional micro-extrusion fluid dosing and deposition, and three-dimensional plotting fall into this category [2,11].

12.3 FUSED DEPOSITION MODELING

An additive manufacturing process that is enormously distinguished and governed by customers and industries possessing good integrity, economical, and user friendly is the FDM technique. Scott Crump invented and patented fused deposition modeling (FDM) in 1989 and soon thereafter founded the first FDM 3D printer company i.e., Stratasys Commercialization Limited in the USA and made FDM trademark of the company. As well, the Rep rap community used the similar term "fused filament fabrication." FDM has a mechanical extrusion nozzle in which thermoplastic material is fed in the shape of thin filaments, generally a diameter in the range of 1.75–3 mm from a filament wound in the spool to the extrusion head at a constant pressure and extruded according to a spatial coordinate. The heated nozzle melts the polymer filament above the melting temperature for a semi-crystalline type polymer and above the glass transition temperature for an amorphous polymer filament. FDM is one the cheapest 3D printers among all types of 3D printers and useful in the home and office environment [9,12].

12.3.1 Basic Principle of FDM

FDM has three basic parts: a) printing bed on which parts to be printed, b) a spool of filament (coil ring) that acts as printing material, and c) an extruder with a nozzle that pushes out melted filament to construct a model layer by layer on the printing bed (Figure 12.2). The extruder nozzle moves on a printing bed in X, Y, Z axes to draw the cross-section of the model on the printing bed. This deposits a fine layer of polymer that cools and hardens and instantly joins over an underneath layer. After completion of one layer, either the printing head moves up or the printed bed moves down by 100 microns, depending upon the machine configuration to deposit the next layer [13].

The printing steps start with designing the model of the product using any CAD software and then file of a model is exported in STL format, which is the language of a 3D printer. After exporting, it removes STL export errors and imports to slicing software. Here, an open-source slicer is available, like Cura, Silcer3r, Siimplify3D, Kisslicer, Fusion 360, etc. The model is sliced in a number of layers and the 3D printing machine print parameters are given. Thereafter, a G-code file is uploaded to a 3D-printer machine. The FDM machine starts printing after achieving a temperature of nearly 180°C–210°C (depending upon the type of polymer filament used) because this temperature is required for melting the polymer. When this temperature is achieved, the filament is fed into the hot extruder for printing the model.

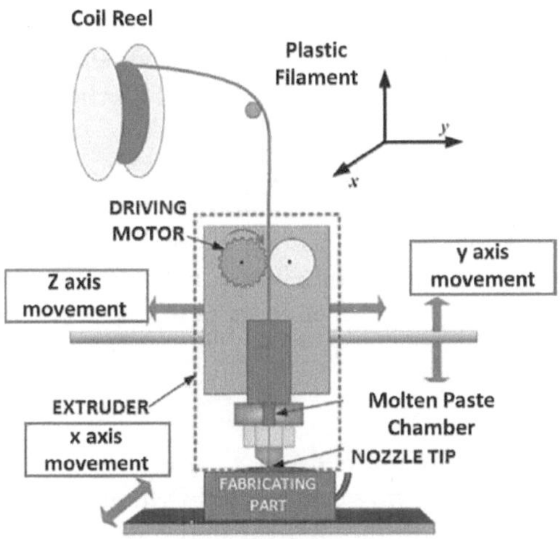

FIGURE 12.2 Typical fused deposition modeling machine parts [14].

Two types of materials are used in FDM printers. One is a base material from which the model is created and the other is support material that supports the extended parts of a model and serve as a scaffold for a model to be printed. This support is made of the same base material of the model or from a different material that can be easily breakable/water-soluble detergent or chemical soluble or limonene or other solution. Also, a model is finished by sandpaper/filed/ milled/painted after printing to enhance aesthetics and functionality.

The build time is dependent upon various functions like layer thickness, infill density, part size, print speed, build orientation, air gap, shell width, number of shells, complexity of geometry, etc. These parameters may increase or decrease the print timing of the part.

The FDM process of AM is comparatively slower than other techniques of AM.

The equation for the specific energy of the incompressible melt is given by Eq. 12.1 [3]:

$$\rho c_p\left[\frac{dT}{dt} + (\vec{v}\,.\,\nabla)T\right] = -\nabla\vec{q} - (\vec{\tau} : \nabla\vec{v}) + \varnothing \qquad (12.1)$$

where ρ signifies the density of the melt, p denotes the applied pressure, surface force is represented by τ, dT is the temperature difference, v is the flow velocity, change in internal energy is denoted by ϕ, and ∇q is the change in heat conducted energy for unit volume and unit time [3,7].

12.3.2 Materials Available for FDM

The polymer plastic material is utilized by FDM printers. It is available in the shape of a plastic filament. These filaments are generally available in spools. It is manufactured and sold by machine suppliers.

The following FDM materials filaments are very popular: polylectic acid (PLA), acrylonitrile butadiene styrene (ABS), high impact polystyrene (HIPS), polycarbonate (PC), polyethylene terephthalate (PET), polyethylene terephthalate glycol (PETG), nylon, thermoplastic polyurethane (TPU), polyvinyl alcohol (PVA), Ultem, etc.

12.4 CHALLENGES DURING PRINTING

During the printing of a job in FDM, there are lots of common issues and problems that have to be faced. Some of the problems are listed below with their best possible solution.

12.4.1 Warping

Warping in a product print by the FDM process is shown in Figure 12.3. It is the most common problem seen during printing in FDM, especially with ABS, PC, PC ABS, and PETG. It is due to shrinkage of material while printing by which it lifts the corners of prints and detaches from the bedplate. During printing, materials

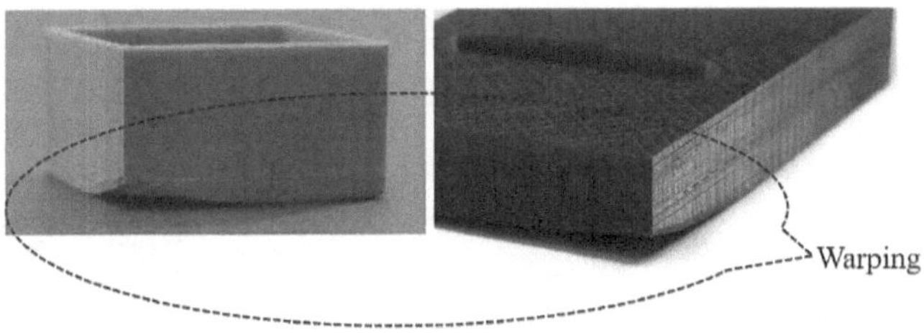

FIGURE 12.3 Warping of printed part in fused deposition modeling [15].

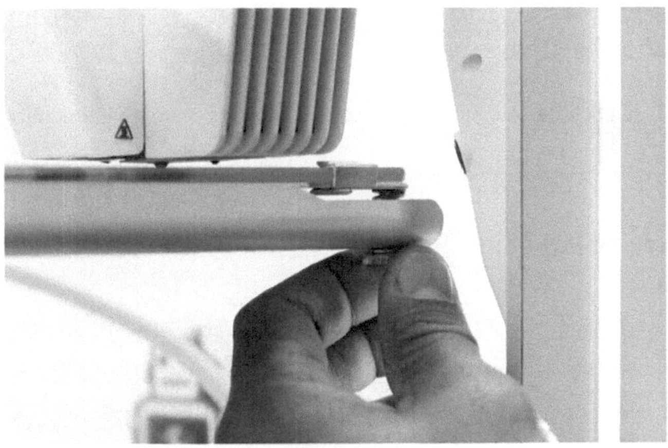

FIGURE 12.4 Manual bed leveling by adjusting leveling screw at bottom of the bed [15].

expand slightly when they come out from the extruder, but contract when they cool down. More shrinking bends up the build plate also.

Solution:

i. Bed Leveling: Bed level should be flat to avoid the warping because the bumpy bed will have uneven nozzle distance, which increases the tendency of warping. If the printing bed has a manual adjustment or auto adjustment, make sure the bed is leveled properly before the start of printing (Figure 12.4).
ii. Nozzle Tip to Bed Distance: Z height of nozzle should be ~1 mm; otherwise, it will also create the problem of warping. Figure 12.5 shows the leveling of a nozzle by a sticky note. The occurence of warping (yes/no) at different heights of the nozzle is tabulated in Table 12.1.
iii. Build Plate Temperature: A heated printing bed aids to improve bed adhesion of the first layer. The best way to prevent warping is with a good heated build platform at nearly 65°C–85°C. Always make sure that the door of the printer is closed while printing.

FIGURE 12.5 Check the Z distance of nozzle by putting a sticky note double folded [15].

TABLE 12.1
Different Nozzle Positions

Nozzle Height	View	Yes/No
Too High	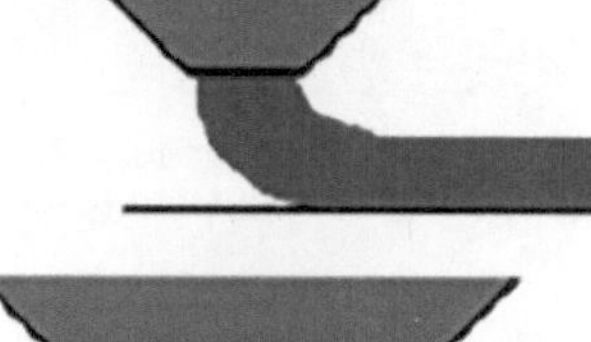	NO
Perfect		YES
Too Low	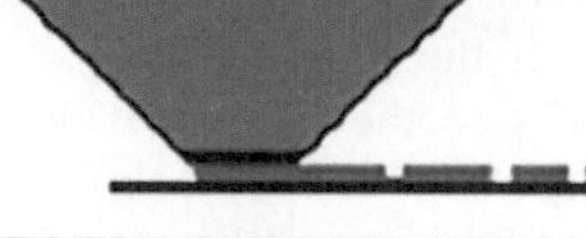	NO

iv. Slow nozzle speed at nearly 50%–60% of full speed with a minimum feed rate ~20–30 mm/s is recommended to avoid warping of the first layer, which gives more time to bond adhesion with a bed.

v. Fan Speed: Initially, the fan should be off for initiating layer printing. After printing some layers, the fan should start with regular speed. At the middle of printing, the full speed of the fan may be used.

Glue Stick **Kapton Tape**

FIGURE 12.6 Print bed adhesives.

vi. Print Bed Adhesive: To fulfill the proper adhesion, the print bed should be neat, smooth, and clean. It should be free from the oily and slippery surface. The adhesives are used to improve surface adhesion to stick the filament. Glues, Kapton tape, adhesive sheet, and some spray are recommended for increasing the stickiness of the surface according to the type of filament material (Figure 12.6). Regular cleaning of the print bed is also suggested.

vii. Brims and Rafts: Brim and raft increases the surface area of the first layer of the 3D-printed part and also increases the adhesion of the layer. Therefore, they reduce the warping with reducing the use of the adhesive as well. The brim is a slice of filament that is made around the 3D-printed part and connected to the first slice of the part. The raft is similar to the brim, but it is a thick grid that sticks before part printing. Both can be removed after printing of the final part.

12.4.2 Leaning Prints/Shifted Layers

When layers upon layers of three-dimensional prints are not properly aligned and leave behind a staggered, staircase appearance, then it known as leaning prints or shifting layers, as shown in Figure 12.7 [16]. It seems to affect some specific areas, but it makes the whole part unproductive.

Solution:

i. Examine the Belts and Pulley: The most common problem for this is either the belt loosens its tension or the pulleys wear out on the X/Y axis motors. Try tightening the belt by loosening the strips on the printer head and then again pulling the belt as forcefully as you can and tightening these strips in place again to gain maximum belt tension. Alternatively, try tightening the grub screws within the pulley as tight as you can until the Allen key starts to flex; if this doesn't solve the issue, you might need to change your pulleys or the belt.

ii. Avoid Collision: This is another regular cause of shifted layers. The 3D-printer extruder collides with the model that is being printing. It can be solved by reducing the travel speed and adding some Z hoop.

FIGURE 12.7 Leaning prints/shifted layer effect [16].

iii. Stand-Off the Electronics: Overheating of electronics components of the 3D printer may cause the shifting of layers problem, particularly overheating of stepper drives that control the movement of the motor that results in the hiccups in the printer movements. It can be resolved by providing the proper ventilation, heat sinks, etc. to printer motherboards, which provide sufficient cooling and avoid shifting of layers.

12.4.3 Stringing Effect

The stringing effect is undesirable threads of polymer filaments, as shown in Figure 12.8 [16]. When the extruder travels from one place to another place, occasionally it extrudes melted plastic, which sticks to parts and solidifies, known as the stringing effect.

Solution:

i. Retraction: Use of retraction plays a significant role to reduce the stringing effect. It retracts the filament a little bit by an extruder so that it can't come out from the nozzle during traveling between two or more printed parts, as shown in Figure 12.9 [17].
ii. Nozzle Temperature: By reducing the nozzle temperature for some degree, this can reduce the stringing effect. In Figure 12.10, it can be seen that with nozzle temperature reduction, undesirable threads become shortened.

It shows a pellucid effect by lowering the temperature, which shows a very positive effect on the stringing. Note that the temperatures shown in this image is for PLA. For other materials, the lowering of temperature may be comparatively smaller.

FIGURE 12.8 Stringing effect [16].

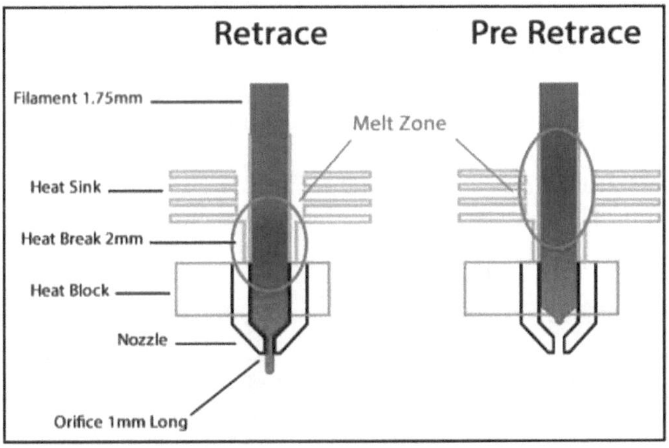

FIGURE 12.9 Retraction setting [17].

iii. Print Speed: Reducing the print speed gives significant results; ~20 mm/s will give good results for PLA for an extrusion temperature of 180°C.
iv. Travel Speed: Increasing the print speed minimizes the stringing effects due to less time to drip the material from a nozzle. About ~200 mm/sec speed gives fine prints.

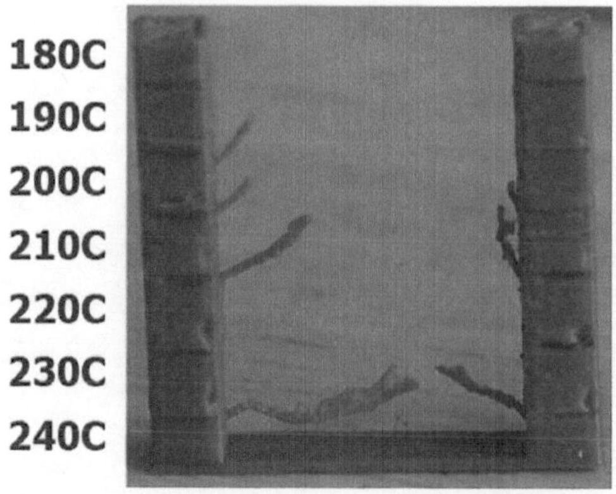

FIGURE 12.10 Effect of nozzle temperature on stringing [16].

12.4.4 Pillowing

Pillowing is the open or closed or combination of both bumps. Rough holes can be seen on the top surface of prints (Figure 12.11) [17]. It is due to an error of print settings and terrible print bed cooling.

Solutions:

i. Upper/Lower Layer Thickness: For good quality print, the lower and uppermost surface should have a quite large thickness. To overcome this, it is recommended that the minimum upper and lower thickness should be six times the layer thickness.
ii. Cooling: The suitable cooling of the uppermost layer is very crucial. At this time, the fan speed should be maximum. Otherwise, it may lead to curling of infill layers and lines.
iii. Temperature: Decreasing the print temperature is recommended to avoid pillowing. If the same problem is found in extruding, then a lower print speed is also recommended.

FIGURE 12.11 Pillowing defect in the print enclosed by the red circle [17].

FIGURE 12.12 Under extrusion [18,19].

12.4.5 Under Extrusion

The under extrusion problem is caused when the amount of material is not fed properly, as shown in Figure 12.12 [18]. It is very usual and takes place at a very high speed. The maximum speed can be calculated by multiplying the nozzle diameter by the layer height and speed of printing. If the speed is increased, the risk of clogging of the nozzle also increases.

Solutions:

i. Material and Material Settings: Use the materials that are best matches for the printer profile. Use the correct size of the nozzle can also resolve this problem.
ii. Filament Feeder: Feeder tension should be correct for proper feeding of the filament.

12.4.6 Skipped Layer/Bed Drop

A skipped layer/bed drop happens due to the absence of lubrication in the Z-axis of the lead screw. To avoid this, proper lubrication should be applied every time before the start of the print.

12.4.7 Elephant Foot

Elephant foot is very prevalent in the first couple of layers of a print that is wider than the rest of the part layers, as shown in Figure 12.13 [20]. For the manufacturing of smaller product, this defect is unnoticeable. But it may be the cause of fitting

FIGURE 12.13 Elephant foot [20].

problems during assembly, when heavier products are made by 3D printing because it decreases the tolerance significantly. The reason of generation of this defect is the bulging of the first layer due to the weight of the upper layers.

Solution:

i. Bed Leveling: Bed should be properly leveled; otherwise the layer will be crushed down.
ii. Use Chamfer: For this, simply integrate a diminutive 45° chamfer on the bottom edge of the print. This would highly be auxiliary in clearing the quandary of a 3D-printing elephant foot.
iii. Bed Temperature: Too high bed temperature also create the elephant effect issue. By lowering the bed temperature, it can be resolved.

12.5 CONCLUSION

There are numerous factors that affect the quality of the product made through FDM printing. Some of the defects can be eliminated by the experienced operator. Also, there may be some differences in use of the FDM machines as well as slicing software that influence the quality of the printed product. Therefore, some measures are required during the printing, particularly selection of print parameters, machine settings, maintenance, proper selection of slicing parameters, and apex section of the attention. Also, the position of the table of the printer may be the reason for the defective print. Therefore, it should be flat. The use of quality filament also prevents the poor quality of the print. Besides machine parmaters, product material is also responsible for the defective print. Fluidity of the melt, thermal conductivity, etc. affect the quality of the product.

REFERENCES

[1] Gibson, Ian, David W Rosen, Brent Stucker, Mahyar Khorasani, and David Rosen. 2021. *Additive manufacturing technologies*. Vol. 17: Springer.

[2] Petrovic, Vojislav, Juan Vicente Haro Gonzalez, Olga Jordá Ferrando, Javier Delgado Gordillo, Jose Ramón Blasco Puchades, and Luis Portolés Griñan. 2011. Additive layered manufacturing: Sectors of industrial application shown through case studies. *International Journal of Production Research* 49 (4):1061–1079.
[3] Chua, CK, SM Chou, and TS Wong. 1998. A study of the state-of-the-art rapid prototyping technologies. *The International Journal of Advanced Manufacturing Technology* 14 (2):146–152.
[4] Jamieson, Ron, and Herbert Hacker. 1995. Direct slicing of CAD models for rapid prototyping. *Rapid Prototyping Journal* 1(2):4–12.
[5] Upcraft, Steve, and Richard Fletcher. 2003. The rapid prototyping technologies. *Assembly Automation* 23(4):318–330.
[6] Lee, Jian-Yuan, Jia An, and Chee Kai Chua. 2017. Fundamentals and applications of 3D printing for novel materials. *Applied Materials Today* 7:120–133.
[7] Chua, C. K. and K. F. Leong. 2005. *Rapid Prototyping: Principles and Applications in Manufacturing*. John Willey, p. 400, 2005.
[8] Dizon, John Ryan C, Alejandro H Espera Jr, Qiyi Chen, and Rigoberto C Advincula. 2018. Mechanical characterization of 3D-printed polymers. *Additive Manufacturing* 20:44–67.
[9] Vaezi, Mohammad, Hermann Seitz, and Shoufeng Yang. 2013. A review on 3D micro-additive manufacturing technologies. *The International Journal of Advanced Manufacturing Technology* 67 (5):1721–1754.
[10] Dulieu-Barton, JM, and MC Fulton. 2000. Mechanical properties of a typical stereolithography resin. *Strain* 36 (2):81–87.
[11] Letcher, Todd, and Megan Waytashek. 2014. Material property testing of 3D-printed specimen in PLA on an entry-level 3D printer. Paper read at ASME International Mechanical Engineering Congress and Exposition.
[12] Melenka, Garrett W, Benjamin KO Cheung, Jonathon S Schofield, Michael R Dawson, and Jason P Carey. 2016. Evaluation and prediction of the tensile properties of continuous fiber-reinforced 3D printed structures. *Composite Structures* 153:866–875.
[13] De Leon, Al C, Qiyi Chen, Napolabel B Palaganas, Jerome O Palaganas, Jill Manapat, and Rigoberto C Advincula. 2016. High performance polymer nanocomposites for additive manufacturing applications. *Reactive and Functional Polymers* 103:141–155.
[14] Jin, Yu-an, Hui Li, Yong He, and Jian-zhong Fu. 2015. Quantitative analysis of surface profile in fused deposition modelling. *Additive Manufacturing* 8:142–148.
[15] Ultimaker, "https://support.ultimaker.com/hc/en-us/articles/360012113239-How-to-fix-warping," [Online]. Available: https://support.ultimaker.com/hc/en-us/articles/360012113239-How-to-fix-warping
[16] Ultimaker, "https://support.ultimaker.com/hc/en-us/articles/360012016280-How-to-fix-stringing," [Online]. Available: https://support.ultimaker.com/hc/en-us/articles/360012016280-How-to-fix-stringing
[17] All3DP, "https://all3dp.com/2/3d-print-stringing-easy-ways-to-prevent-it/," [Online]. Available: https://all3dp.com/2/3d-print-stringing-easy-ways-to-prevent-it/
[18] Ultimaker, "https://support.ultimaker.com/hc/en-us/articles/360012112859-How-to-fix-under-extrusion," [Online]. Available: https://support.ultimaker.com/hc/en-us/articles/360012112859-How-to-fix-under-extrusion
[19] All3DP, "https://all3dp.com/2/under-extrusion-3d-printing-all-you-need-to-know/," [Online]. Available: https://all3dp.com/2/under-extrusion-3d-printing-all-you-need-to-know/
[20] ALL3DP, "https://all3dp.com/2/elephant-s-foot-3d-printing-problem-easy-fixes/," [Online]. Available: https://all3dp.com/2/elephant-s-foot-3d-printing-problem-easy-fixes/

13 Hybrid Additive Manufacturing Technologies

M. Kumaran
Department of Production Engineering, National Institute of Technology Tiruchirappalli, Tamil Nadu, India

CONTENTS

13.1 INTRODUCTION

In the aim of research is manufacture the metal additive manufactured components using PBF and DED. The PBF and DED both are using powder material and using a laser energy source. More recently, hybrid technology of additive manufacturing has been proposed to fabricate/repair components or produce welding joints [1]. Additive manufacturing (AM) techniques used in hybrid technology so far are classified as additive manufacturing and machining, combined PBF and DED processes, and hybrid material used in the PBF process [2].

DOI: 10.1201/9781003333760-13

The following author's studies with different hybrid technology are discussed in brief. Noemie Martin et al. [3] used a hybrid process of PBF and DED fabrication methods that was investigated. The authors used Inconel 625 material for PBF and DED. The purpose of this study was to fabricate DED components on PBF substrates. This concept is used to rework or repair PBF components. Furthermore, the authors examine material connections in the interfacial region with the help of mechanical and microstructural studies. E. Cyr et al. [4] used the fracture and tensile properties of additively manufactured hybrid high-carbon steel MS1-H13 were investigated. The specimens were prepared using the PBF technique to add MS1 powder to the received H13 tool round steel rod. The specimens were then subjected to four unique heat treatments and the tensile behavior was assessed. According to a heat treatment effect study on the mechanical properties and fracture of H13 blending and additive manufacturing MS1, the best performance was obtained with H13 heat treatment alone, making it a faster and more cost-effective option for the tooling industry, making it more appealing, rather than using two heat treatment processes. Sven Muller et al. [5] studied aluminium samples that were fabricated on titanium substrates using a commercial PBF process, with a layer thickness of (20 μm) along constant energy density direction, which contradicts the prior art based on laser joining aluminium and titanium sheets in butted or overlapping configurations. The laser radiation can be transmitted to the titanium substrate through the thin first layer of aluminium powder. Because of titanium's high absorption coefficient, melting and excess formation of intermetallic compounds are unavoidable. Titanium powders produced on aluminium substrates have a high strength potential because less intermetallic compounds can form at the interface when a dense volume of titanium is formed. Pengfei Li et al. [6] investigated functional bimetallic material (FBM) that was created using a hybrid DED and thermal milling process (Inconel steel). SEM, XRD, EDX, tensile testing, and Vickers microhardness were used to characterize the FBM. Inconel-Steel FBM is manufactured using a new process. Thermal milling between the IN718 and SS316L deposits keeps the components warm while providing an excellent substrate surface for the DED. When depositing various materials, there are different interface properties. Stainless-steel powder 316L is deposited on the milled surface of IN718 with no diffusion layer and no clear interface. However, the same methods and materials produce different results. There is a 450-m wide diffusion layer. Although SS316L is the main component of this layer, the microhardness is much higher than that of SS316L. Zhanqi Liu et al. [7] used laser additive manufacturing (LAM) technology and successfully used it to produce TiAl/TC4 bimetallic (BS) structures free of metallurgical defects, effectively reducing the formation of brittle A-2 phases.

The current study aims to demonstrate recent advances in various hybrid additive manufacturing technology approaches. All hybrid techniques will examine the results of interfacial regions, material bonding, tensile testing, and microstructural characterization.

13.2 OVERVIEW OF ADDITIVE MANUFACTURING

As manufacturing processes continue to improve and evolve, the need for faster and cheaper manufacturing processes has led to the development of many rapid prototyping (RP) processes. RP technology is not limited by specially designed tools and fixtures. As a result, almost any geometry of varying size and complexity can be fabricated with high precision. RP technology is capable of generating complex 3D geometries using additive manufacturing (AM). Unlike the case of subtractive manufacturing (SM), where material is removed from the raw material to get the desired shape, material is added in specially shaped layers to build the geometry in the case of AM [8,9]. Depending on the process used, each successive layer of material is bonded to the previous layer by some type of controlled heat application. Initially, additive manufacturing techniques were used to create prototypes or low-volume parts. Although the current use of AM is not limited to prototyping or as a transitional stage in design, it encompasses many applications of the technology, including modeling, pattern making, tooling, and mass production of end-use parts. These part property profiles typically include the following property groups:

i. Geometric properties such as minimum wall thickness and surface roughness.
ii. Mechanical properties such as yield point, elongation at break, tensile strength, elastic modulus, hardness, and where appropriate, dynamic fatigue life.
iii. Thermal properties include specific heat, thermal conductivity, and coefficient of thermal expansion.

The purpose of this research is to compare the mechanical properties of 316L stainless steel components manufactured by two AM processes (PBF and DED). The mechanical properties of these materials are based on tensile tests. Tensile testing is done by pulling on the material, which usually results in failure. This test produces a tension curve that shows how the material responds to an applied force. This experiment examines stress-strain, tensile strength, yield strength, and Young's modulus. Stress is the internal resistance of a material to torsional action of an external load or force [10,11].

Additionally, additive manufacturing was developed as part of a new industrial movement known as Industry 4.0, which is bringing significant changes to the way products are made through digitalization [12].

13.3 ADDITIVE MANUFACTURING PROCESS CHAIN

Figure 13.1 shows the process flow of additive manufacturing and machines for additive manufacturing. Additive manufacturing technology creates solid

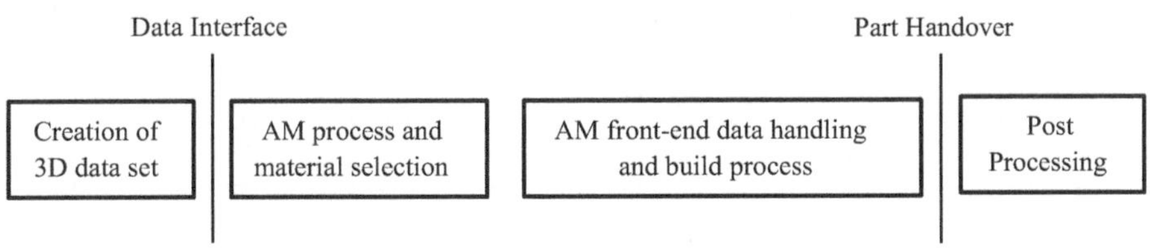

FIGURE 13.1 AM process chain.

components through a layer-by-layer powder melting or powder feeding process. There are many 3D-printing technologies in the market that use specific machines for the job required. In the AM concept, the following general flow is used from design to final assembly. AM processes are characterized by a process chain, as shown in Figure 13.1. The process starts with a (virtual) 3D CAD file representing the part to be produced. During the design phase, data files are typically generated using 3D CAD design, scanning, or imaging techniques such as computed tomography (CT scan). Regardless of how the 3D data set is generated, it is first decomposed into layers or layers using special software. Therefore, a data set containing contour data (xy), thickness data (dz), and the number of layers (or z-coordinates) per layer is then sent to a machine that performs two basic processing steps for each layer to generate a part [13].

Each layer is created in the first stage according to the defined profile and layer thickness. This can be achieved in several ways using different physical concepts. Cutting outlines from foil or sheet metal is the easiest way to achieve this.

In the second stage, each layer is connected to the previous layer, and the new layer then forms the top layer of the growing part. The physical model is built up layer by layer until the part is complete [14].

13.3.1 Creation of 3D Data Set

The first step of AM is the creation of a digital model. The necessary models are created as computer-aided design (CAD) files. The reverse engineering concept is also used to create digital files through the 3D scanning process. Converting CAD models to STL (stereolithography) files is a crucial step in additive manufacturing that distinguishes it from conventional manufacturing techniques. The slicer program is used to import the STL file after it is created. The software converts STL files to G-code after reading them. G-code is a language for numerical control programming (NC). In computer-aided manufacturing (CAM), it is used to control automated machine tools (such as CNC machines and 3D printers).

13.3.2 AM Process and Material Selection

After modeling, select the appropriate additive manufacturing material according to the requirements of the part and then select the corresponding additive manufacturing process. Material-related process parameters for the construction process are obtained from the manufacturer, mainly from databases or determined in separate tests. Finally, they are stored in material files.

13.3.3 AM Front-End Data Handling and Build Process

AM or front-end software, often referred to as "rapid prototyping software," is primarily built into the standard machine software that comes with the machine. Alternatively, it uses third-party software.

13.3.4 Post-Processing

Before the construction process begins, the part is positioned in the construction space and machine-related parameters such as post-coating time are set. The build process then begins and continues automatically layer by layer until the part is complete. Subsequently, the part goes through a machine-specific cooling process (if necessary) before being removed from the machine.

13.4 METAL ADDITIVE MANUFACTURING TECHNIQUES

There are two laser powder additive manufacturing technologies: one is PBF and the other is DED. PBF is a three-axis manufacturing process with a laser head on a fixed axis, and the table/bed can move horizontally and vertically. On the other hand, DED has five axes with table movement that is horizontal and vertical and can be rotated through 360°. Additionally, the laser head can be tilted or held at the desired angle. Both techniques use laser energy to melt metal powders. In the case of the DED process, powder is fed coaxially along with the laser source [15].

13.5 FABRICATION OF FUNCTIONAL COMPONENTS THROUGH HYBRID MANUFACTURING

It is possible to fabricate functional components by (i) solely involving single additive manufacturing method, (ii) combining conventional manufacturing and AM, (iii) combining two different AM techniques. These methods can be categorized as hybrid process, hybrid additive, and hybrid material manufacturing [16].

13.5.1 Hybrid Process

In this section, a hybrid process is considered as a manufacturing process that combines the potential of at least one additive with the potential of non-additive process steps. Usually only non-additively manufactured machines—mainly milling machines—are upgraded to hybrid production systems with add-on modules (hardware) that can be additively manufactured [17]. (Figure 13.2).

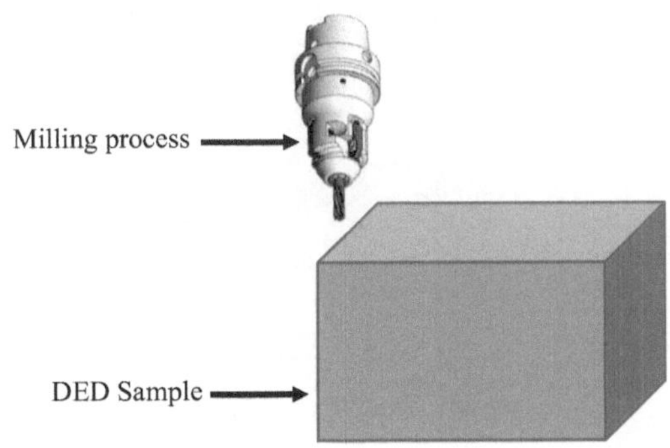

FIGURE 13.2 Hybrid process of additive manufacturing and milling.

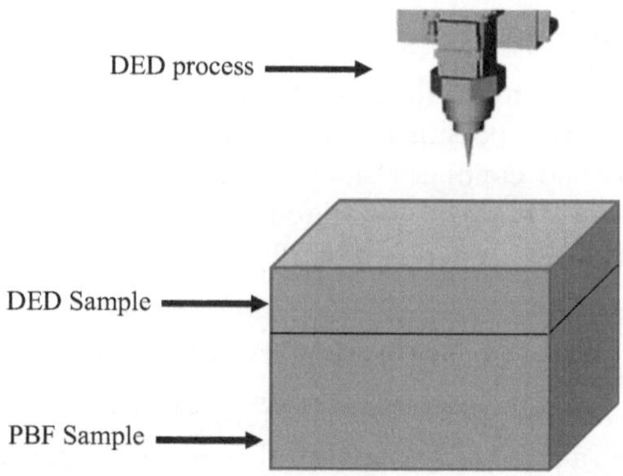

FIGURE 13.3 Hybrid manufacturing of PBF with DED.

13.5.2 Hybrid Additive Manufacturing

In this concept, two types of additive machines must perform a job. Metal additive manufacturing concepts such as PBF and DED are well suited for combined processes and jobs. Usually the substrate/subtractive products will be made using the PBF process and the DED process can be involved in rework/repair of defective PBF samples. Both processes use metal powder as the working material. Also, both machines can use the same powder or different powders under the desired working conditions [18] (Figure 13.3).

13.5.3 Hybrid Material

This concept involves one or more additive manufacturing methods. The only benefit is that two materials are used to make the sandwich structure. Materials of different properties are employed for fabricating functional components [19] (Figure 13.4).

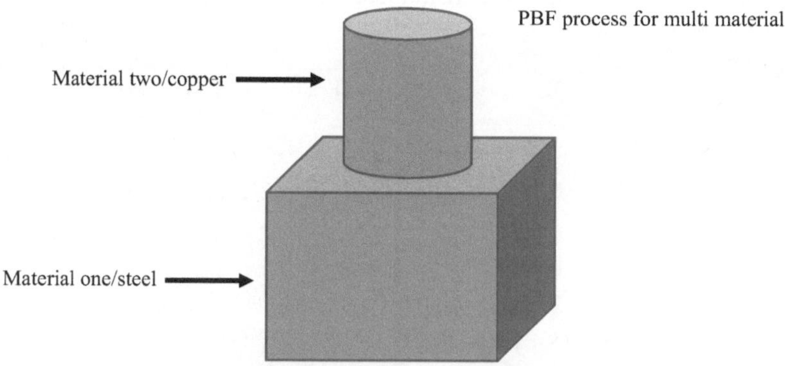

FIGURE 13.4 Hybrid manufacturing of steel with copper.

13.6 A CASE STUDY OF HYBRID TECHNOLOGIES USING ADDITIVE MANUFACTURING TECHNOLOGY

13.6.1 Machining of Additive Manufactured Parts

The aim of this concept is to propose a new idea of design for manufacturing concept; the additive manufacturing process is combined with machining process/machine. The DED process of additive technology has some limitations, like the edges and curves have not gotten an appropriate structure. A hybrid method of additive and machining is started a new way of 3D manufacturing and machining [20].

This DED process has used stainless-steel 316L material with the spherical shape of 50–150 μm with the average diameter of 85 μm. Table 13.1 shows the chemical composition of the powder. The DED component was fabricated by DMD 105-POM (USA) machine shown n Figure 13.5(a). The machine has a five-axis numerical controlled machine with the help of a computer-aided design (CAD) diagram [21].

Before finalizing the DED process parameters, Taguchi method of L9 optimization is tried to find out the best/suitable parameters. In the Taguchi method, the range followed a scanning speed of 300, 500, and 700 (mm/min); laser power of 400, 600, and 800 (W); powder feed rate of 2, 3, and 4 (g/min); and layer thickness of 200, 300, and 400 (μm). Based on the microhardness results, the DED processes are conducted. The maximum microhardness obtained by DED is 235 HV.

The following process parameters are used during manufacturing: laser power (600 W), hatch spacing (400 m), layer thickness (400 m), scanning speed (500 mm/min), and powder feed rate (4 g/min). Figure 13.5(b) depicts a micromachining machine that is controlled by computer numerical control programming, following

TABLE 13.1
Chemical Composition of Stainless-Steel 316L Powder for the DED Process

Element (% wt)	Ni	Mo	Mn	N	S	Cr	P	Fe	Si	C
DED	10.59	2.00	1.4	0.10	0.007	16.99	0.024	68.66	0.2	0.01

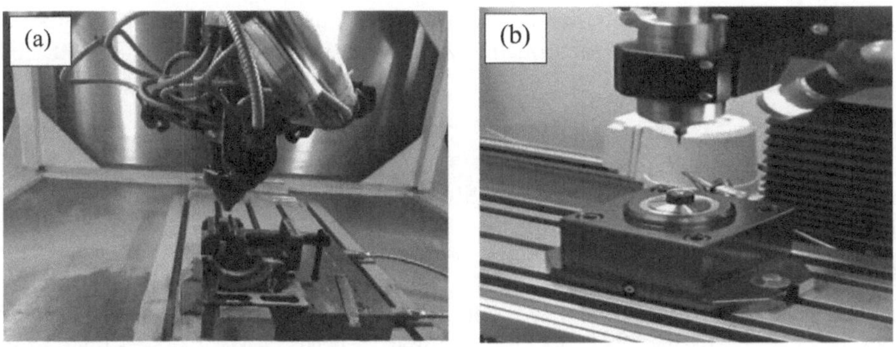

FIGURE 13.5 (a) During the DED technique, (b) during the micro-milling technique.

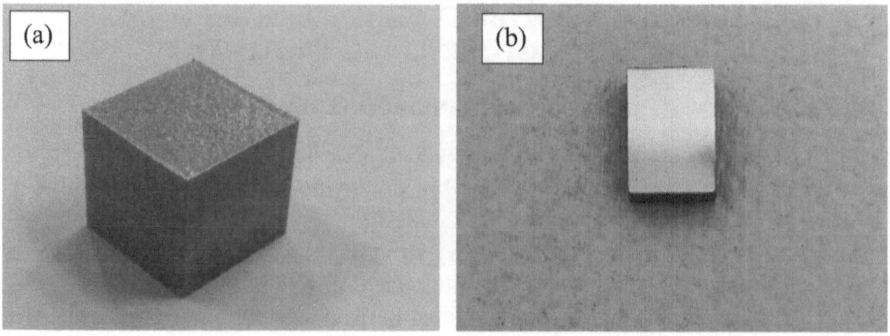

FIGURE 13.6 (a) Additively manufactured component, and (b) machined component.

the process parameters of feed rate (600 mm/min), cutting speed (500 mm/min), and cutting depth (4 mm).

Figure 13.6(a,b) shows the additively manufactured component and machined surface. In this, the hybrid process is successfully manufactured and machined.

13.6.2 The Hybrid Process of PBF and DED

This method of hybrid component is manufactured by PBF and DED processes of additive manufacturing. This method was recently used in DfAM (Design for Additive Manufacturing). The aim of this dual additive process is to repair/re-work the PDF component by the DED process [22].

The hybrid structure has used stainless-steel 316L metal powder for both PBF and DED processes. The chemical composition is shown in Table 13.2. The PBF uses the powder range between 5 μm to 50 μm and DED uses the powder range between 50 μm to 150 μm. Both powders are a spherical shape condition. Moreover, the average powder size of PBF and DED is 25 μm and 85 μm, respectively, using the PBF procedure with the M280-EOS machine. On the other hand, the DED process is done by a DMD 105-POM (USA) machine. Both machines are operated by a CNC process [23–25].

The PBF sample was fabricated with a size of 15 mm^3. Before the PBF final process parameters, the best/suitable parameters used the Taguchi method of L9 optimization. In the Taguchi method, the range were laser powers of 190, 200,

TABLE 13.2
Chemical Composition of Stainless Steel 316L Powder for the PBF and DED Process

Element (% wt)	Ni	Mo	Mn	N	S	Cr	P	Fe	Si	C
PBF	10.57	2.11	1.58	0.10	0.020	16.62	0.035	68.34	0.60	0.030
DED	10.59	2.00	1.4	0.10	0.007	16.99	0.024	68.66	0.2	0.01

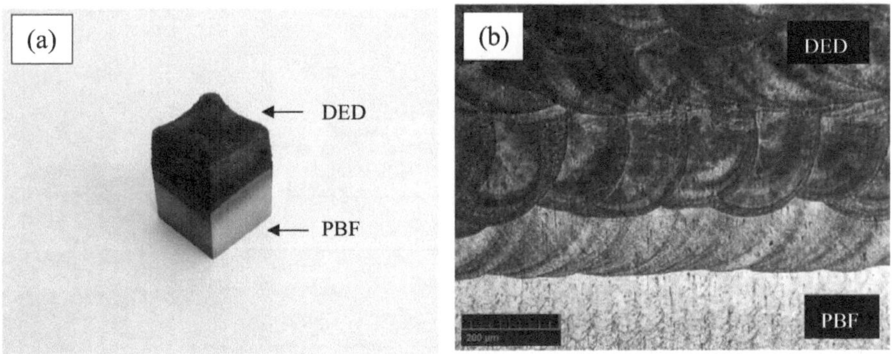

FIGURE 13.7 (a) Hybrid component by PBF and DED, and (b) microstructure image of hybrid component [1].

and 210 (W); scanning speeds of 800, 900, and 1,000 (mm/s); and layer thicknesses of 200, 300, and 400 (μm). Based on the microhardness results, the PBF processes were conducted. The maximum microhardness obtained by PBF was 260 HV. Manufacturing parameters include scanning speed (900 mm/s), hatching pitch (100 μm), laser power (200 W), and layer thickness (400 μm). Furthermore, the DED process was completed above the PBF component using the parameters of scanning speed (500 mm/min), laser power (600 W), hatch spacing (400 μm), layer thickness (400 μm), and powder feed rate (4 g/min). Figure 13.7 shows the hybrid component by PBF and DED, and the microstructure image of the hybrid component. The microstructure image conforming to the interface region and bonding of the two processes was good, and there were no pores and cracks found in the interface region.

The mechanical test of tensile test was conducted to the hybrid structure, using ASTM E8 standards. The tensile image is shown in Figure 13.8, and the results are shown in Table 13.3.

13.6.3 Hybrid Metals Manufactured by Powder Bed Fusion

In this concept, more than one material was used to fabricate the PBF component. The aim of this concept fabricated the hybrid sample using ALSi10Mg and AL7075 materials. Tables 13.4 and 13.5 show the chemical/elemental composition of the material. This PBF process used AlSi10Mg and Al7075 materials with the spherical shape of 25–70 μm with the average size of 30 μm. Based on the microhardness results, the PBF processes were conducted. The maximum microhardness obtained by AlSi10Mg was 125 HV, and Al7075 was 154HV.

The following parameters were used for manufacturing the base component of ALSi10Mg: the scanning speed (1,300 mm/s), hatching pitch (190 μm), laser power (370 W), and layer thickness (30 μm). On the other hand, the AL7075 component was fabricated using the following parameters: the hatching pitch (130 μm), layer thickness (40 μm), laser power (340 W), and scanning speed (1,300 mm/s). The hybrid structure is shown in Figure 13.9. The tensile results of AlSi10Mg and Al7075 are shown in Table 13.6.

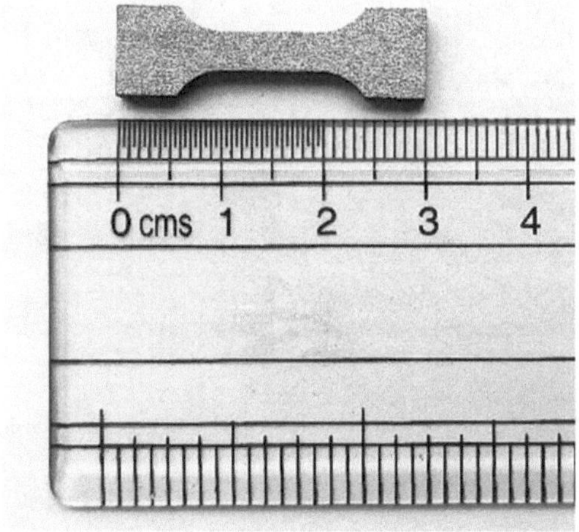

FIGURE 13.8 Tensile specimen scale dimensions [1].

TABLE 13.3
Tensile Results of PBF and DED (SS316L)

Process	Yield Strength [MPa]	Tensile Strength [MPa]	Elongation at Break [%]
PBF	484	652	48.3
DED	561	648	46.1

TABLE 13.4
Chemical Composition of ALSi10Mg Powder for the PBF Process

Element (% wt)	Al	Cu	Si	Mn	Fe	Zn	Mg	Ni	Ti	Pb	Sn
ALSi10Mg	Bal.	0.5	10	0.45	0.55	0.10	0.3	0.05	0.15	0.05	0.05

TABLE 13.5
Chemical Composition of AL7075 Powder for the PBF Process

Element (% wt)	Al	Si	Zn	Ti	Mg	Mn	Fe	Cr	Cu	Others Total	Others-Each
AL7075	Bal.	0.01	5.5	0.01	2.5	0.01	0.07	0.23	1.6	0.05	0.01

FIGURE 13.9 Hybrid structure of AlSi10Mg and Al7075.

TABLE 13.6
Tensile Results of AlSi10Mg and Al7075

PBF Process	Yield Strength [MPa]	Tensile Strength [MPa]	Elongation at Break [%]
AlSi10Mg	245	377	12.3
Al7075	103	227	11

13.7 CONCLUSION

The present study aimed at comparing the different hybrid process/structures. The microstructural and mechanical properties were conducted in SS316L, AlSi10Mg, and Al7075 specimens. The samples were prepared by laser additive methods technologies (PBF and DED). The following conclusions were drawn: (a) Maximum tensile strength and yield strength were obtained for SS316L, AlSi10Mg, and Al7075 specimens. (b) SS316L samples produced by the PBF process had high microhardness. (c) The AlSi10Mg had more elongation obtained compared with Al7075.

REFERENCES

[1] Kumaran, M. & Senthilkumar, V. (2021). Experimental characterization of stainless steel 316L alloy fabricated with combined powder bed fusion and directed energy deposition. *Welding in the World*, 65(7), 1373–1388.

[2] Kumaran, M. & Senthilkumar, V. (2021). Generative Design and Topology Optimization of Analysis and Repair Work of Industrial Robot Arm Manufactured Using Additive Manufacturing Technology. In IOP Conference Series: Materials Science and Engineering (Vol. 1012, No. 1, p. 012036). IOP Publishing.

[3] Martin, N., Hor, A., Copin, E., Lours, P., & Ratsifandrihana, L. (2022). Correlation between microstructure heterogeneity and multi-scale mechanical behavior of hybrid LPBF-DED Inconel 625. *Journal of Materials Processing Tech*, 303:117542.

[4] Cyr, E., Asgari, H., Shamsdini, S., Purdy, M., Hosseinkhani, K., & Mohammadi, M. (2018). Fracture behaviour of additively manufactured MS1-H13 hybrid hard steels. *Materials Letters*, 212:174–177.

[5] Müller, S. & Woizeschke, P. (2021). Feasibility of a laser powder bed fusion process for additive manufacturing of hybrid structures using aluminum-titanium powder-substrate pairings. *Additive Manufacturing*, 48:102377.

[6] Li, P., Gong, Y., Xu, Y., Qi, Y., Sun, Y., & Zhang, H. (2019). Inconel-steel functionally bimetal materials by hybrid directed energy deposition and thermal milling: Microstructure and mechanical properties. *Archives of Civil and Mechanical Engineering*, 19:820–831.

[7] Liu, Z., Ma, R., Guojian, Wang, W., & Liu, J. (2020). Laser additive manufacturing of bimetallic structure from Ti-6Al-4V to Ti-48Al-2Cr-2Nb via vanadium interlayer. *Materials Letters*, 263, 127210.

[8] Thompson, S. M., Bian, L., Shamsaei, N., & Yadollahi, A. (2015). An overview of direct laser deposition for additive manufacturing; Part I: Transport phenomena, modeling and diagnostics. *Additive Manufacturing*, 8, 36–62.

[9] Saboori, A., Aversa, A., Marchese, G., Biamino, S., Lombardi, M., & Fino, P. (2019). Application of directed energy deposition-based additive manufacturing in repair. *Applied Sciences*, 9(16), 3316.

[10] Dilberoglu, U. M., Gharehpapagh, B., Yaman, U., & Dolen, M. (2017). The role of additive manufacturing in the era of industry 4.0. *Procedia Manufacturing*, 11, 545–554.

[11] Kumaran, M., Senthilkumar, V., Panicker, C. J., & Shishir, R. (2021). Investigating the residual stress in additive manufacturing of combined process in powder bed fusion and directed energy deposition. *Materials Today: Proceedings*, 47, 4387–4390.

[12] Kumaran, M., Senthilkumar, V., Panicke, C. J., & Shishir, R. (2021). Investigating the residual stress in additive manufacturing of repair work by directed energy deposition process on SS316L hot rolled steel substrate. *Materials Today: Proceedings*, 47, 4475–4478.

[13] Häfele, T., Schneberger, J. H., Kaspar, J., Vielhaber, M., & Griebsch, J. (2019). Hybrid additive manufacturing – Process chain correlations and impacts. *Procedia CIRP*, 84, 328–334.

[14] Thompson, M. K., Stolfi, A., & Mischkot, M. (2016). Process chain modeling and selection in an additive manufacturing context. *CIRP Journal of Manufacturing Science and Technology*, 12, 25–34.

[15] Lewandowski, J. J., & Seifi, M. (2016). Metal additive manufacturing: a review of mechanical properties. *Annual Review of Materials Research*, 46, 151–186.

[16] Bambach, M., Sizova, I., Sydow, B., Hemes, S., & Meiners, F. (2020). Hybrid manufacturing of components from Ti-6Al-4V by metal forming and wire-arc additive manufacturing. *Journal of Materials Processing Technology*, 282, 116689.

[17] Yamazaki, T. (2016). Development of a hybrid multi-tasking machine tool: Integration of additive manufacturing technology with CNC machining. *Procedia CIRP*, 42, 81–86.

[18] Dilberoglu, U. M., Gharehpapagh, B., Yaman, U., & Dolen, M. (2021). Current trends and research opportunities in hybrid additive manufacturing. *The International Journal of Advanced Manufacturing Technology*, 113(3), 623–648.

[19] Sefene, E. M., Hailu, Y. M., & Tsegaw, A. A. (2022). Metal hybrid additive manufacturing: State-of-the-art. *Progress in Additive Manufacturing*, 7(4), 737–749.

[20] Kumaran, M., Senthilkumar, V., Panicker, C. T., & Shishir, R. (2022). Investigating the microhardness values of SS316L of hybrid additive manufacturing and micro-milling process. In *Recent Advances in Materials and Modern Manufacturing* (pp. 89–95). Springer, Singapore.
[21] Kumaran, M., Senthilkumar, V., Sathies, T., & Panicker, C. J. (2022). Effect of heat treatment on stainless steel 316L alloy sandwich structure fabricated using directed energy deposition and powder bed fusion. *Materials Letters*, 313, 131766.
[22] Kumaran, M., Senthilkumar, V., Justus Panicker, C. T., & Shishir, R. (2022). Concept design and analysis of multi-layer and multi-process piston of SS316L and AlSi10Mg by additive manufacturing. In *Recent Advances in Materials and Modern Manufacturing* (pp. 441–446). Springer, Singapore.
[23] Kumaran, M., & Senthilkumar, V. (2022). Influence of heat treatment on stainless steel 316L alloy manufactured by hybrid additive manufacturing using powder bed fusion and directed energy deposition. *Metals and Materials International*, 29, 1–18.
[24] Kumaran, M., Sathies, T., Balaji, N. S., Bharathiraja, G., Mohan, S., & Senthilkumar, V. (2022). Influence of heat treatment on stainless steel 316L alloy fabricated using directed energy deposition. *Materials Today: Proceedings*, 62, 5307–5310.
[25] Kumaran, M. (2022). Experimental Investigations on Directed Energy Deposition Based Repair of Stainless Steel 316L Alloy Substrate Manufactured through Hot Rolled Steel and Powder Bed Fusion Process. *Journal of Materials Engineering and Performance*, 1–12. 10.1007/s11665-022-07513-w

14 Smart Manufacturing Using 4D Printing

Dhanasekaran Arumugam
Center for NC Technologies, Department of Mechanical Engineering, Chennai Institute of Technology Madras, Chennai, Tamil Nadu, India

Christopher Stephen
Department of Mechanical Engineering, Vel Tech Rangarajan Dr. Sagunthala R&D Institute of Science and Technology, Chennai, Tamil Nadu, India

Arunpillai Viswanathan
Department of Mechanical Engineering, Chennai Institute of Technology Madras, Chennai, Tamil Nadu, India

Ajay John Paul
Doctoral Scholar, School of Mechanical Engineering, Kyungpook National University, Daegu, South Korea

Tanush kumaar
UG scholar, Department of Mechanical Engineering, Chennai Institute of Technology Madras, Chennai , Tamil Nadu, India

CONTENTS

DOI: 10.1201/9781003333760-14

14.1 INTRODUCTION

The 4th industrial revolution, also referred to as Industry 4.0, has increased the production line's automation and digitization while also introducing new digital technologies. In order for companies to remain competitive in the market, they must adopt manufacturing technologies that allow machine-to-machine and human-to-machine communications in a virtualized environment.

Modern digitalization methods, such combining the Internet with intelligent items, allow the interaction of components connected to manufacturing procedures. Examples of these technologies include artificial intelligence (AI) and smart additive manufacturing (SAM). A cyber-physical system, also known as the digital factory, smart factory, advanced manufacturing, smart company, online environment, smart company, or integrated industry, is created when the physical and digital worlds are combined.

A new idea called smart manufacturing aims to satisfy the manufacturing needs of the future. Distributed manufacturing systems with cyber capabilities that allow for more customization, quicker invention, and economically viable low production numbers are some of its distinguishing characteristics. Thus, the advancement of new communications technology and creative business models is hastening an industry transition that impacts current company processes and even the market's structure. Additive manufacturing (AM), also known as 3D printing, is one smart manufacturing technique that is quickly posing a serious threat to the manufacturing sector. The manufacturing industry has been focusing on AM, which has consistently been at the cutting edge of technological growth, in order to take advantage of its potential and hasten the digital transformation of the industry. AM improves businesses' capacity for digital user communication by speeding up on-site printing, mass production customization, and direct manufacturing while lowering waste, freight charges, and delivery time. It can also make supply chains less complex [1,2].

Manufacturing procedures are classified into four categories: I. Deformation: molding and casting, extruding, forging, powder metal processes, and sheet forming; II. Removal: carving, milling, drilling, turning; III. Joining: fastening, welding, and brazing; IV. Additive, this comprises of 7' 3D printing categories specified by ASTM. Unlike subtractive manufacturing methods, additive manufacturing is the process of combining materials to create items from three-dimensional model data, typically in layer-by-layer approaches [3].

With 3D printing, complex shapes can be produced with less material than with traditional production methods [4]. Industries that use 3D printing include those in the automotive, aerospace, aviation, medical, jewelery, health care, art, sculpting, fashion, architectural, food, and construction [5].

14.2 BRIEF HISTORY OF 3D PRINTING

Though the indication for 3D printing was first conceived in the 1970s, the first experiments only began in 1981. Between 1980–1981: Using the photo-hardening of polymeric polymers, Hideo Kodama of the Nagoya Municipal Industrial Research Institute developed and documented two of the first AM techniques. Sequences of cross-sections of the desired object are printed using the additive manufacturing technique known as stereolithography. After each cross-section is printed, a laser beam preferentially hardens a liquid ultraviolet light resin. It was invented in 1984 by Olivier de Witte (Cilas), Alain le Méhauté (CGE/Alcatel), and Jean-Claude André (CNRS). A patent submitted by Chuck Hull was approved in 1986. The method functioned by using ultraviolet laser light beams to harden resin that was contained in a vat cross section by cross section. Throughout most of AM, Hull was adopted. Still being used is indeed the stl file extension. Carl R. Deckard of the University of Texas at Austin invented the selective laser sintering technique in 1987. It usually functions by employing a strong laser beam to fuse powder particles preferentially along the cross sections of the desired form. The powder is typically preheated in the bed at a temperature just below the fusion point and can be made of plastic, metal, ceramic, or glass. R. F. Housholder submitted a patent application for a related method in 1979, but it was never used in practice. S. Scott Crump created and received a patent for the most widely used 3D-printing method called fused deposition modeling (FDM), in 1989–1990. It entails layer-by-layer deposition of fused material, most frequently plastic, in accordance with a .stl file. Inkjet printers were used to bind a bed of powder layer by layer in a method created by MIT in 1993. Sanders Prototype introduced a new method in 1993. It was founded on soluble supports for polymer jetting. The selective laser melting process was created in 1995 by the Fraunhofer Institute ILT in Aachen. Because metal alloys are used, the process produces precise and mechanically robust results and it can capable of working with complicated geometries. The Wake Forest Institute for Regenerative Medicine investigated bioprinting methods with success in 1999. The RepRap open-source project was founded in 2004 by Adrian Bowyer in an effort to democratize and spread additive manufacturing technology. In 2008 saw the Dutch launch of Shapeways. It presents itself as an online service that allows users to send 3D files to make things and have them delivered to a specified place. The widespread adoption of the technology in households will be substantially aided by a DIY kit for 3D printers created by Makerbot in 2009. Makerbot developed a DIY kit for 3D printers in 2009, which will aid in the technology's spread into more homes. In 2011, the Southampton University Laser Sintered Aircraft (SULSA), an unmanned aerial vehicle with a structure printed using a laser sintering machine with a resolution of 100 micrometres per layer, demonstrated the potential of 3D-printing technologies. Airbus Operation GmbH submitted a patent application in 2014 that describes how to 3D print a whole aeroplane construction. Additional advancement and approach combinations suggest a steady shift in the prospective utilization. Due to the increased output quality and the materials that can be used, AM, which was initially used to produce quick prototypes, particularly for engineering where the limited material availability and the lack of mechanical characteristics [6].

14.3 3D-PRINTING PROCESS

A three-dimensional model that is created using computer aided design software or one that is 3D skimmed from a real object is the first step in the printing process. The model is transformed into standard triangulation language (STL) format. According to the needed resolution, the STL format merely defines the surface of model using a system of triangles of various sizes. The triangular mesh more closely approximates the intended surface the smaller the triangles are, producing a surface (smoother).

Pre-processing refers to the process of setting up the appropriate mesh model. The second production stage is processing. Each succeeding layer of a 3D-printed object must be supported, either by the platform, the one before it, or additional parts. Once the model and supports are built in their proper and ideal orientation, the STL model is separated into layers using a plane parallel to the surface, known as the xOy plane. Then, layers are put in an Oz-direction one after another. Layer thickness is influenced by the printer, AM process, and quality standards. The printer is then given the sliced model. AM offers more effective material usage compared to traditional manufacturing processes. The model is removed from the printing platform when printing is finished, and postprocessing is used to improve the printed object even more. The as-built models are rinsed in a wash solution, most frequently isopropyl alcohol (IPA), to remove the liquid resin covering after photopolymerization takes place. These models are then either naturally exposed to sunshine or artificially cured with UV light to improve the mechanical qualities.

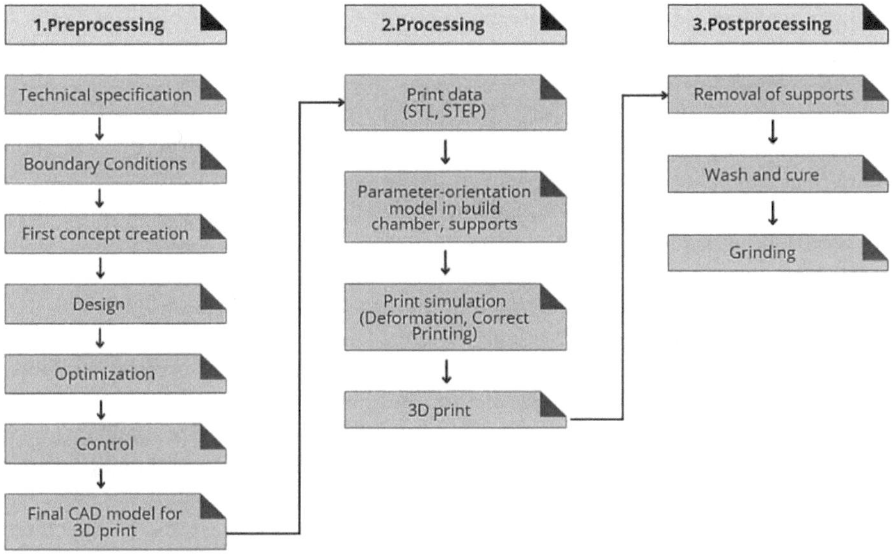

FIGURE.14.1 Steps used in additive manufacturing [7].

Aside from removing the support, other techniques include grinding, sealing, glueing, polishing, painting, varnishing, coating, sterilizing, inspecting, and measuring. Post-processing techniques specific to photopolymer parts exist. Notably, the temperature, rinse time, and curing time all have a considerable impact on the mechanical characteristics of the completed components [7]. Figure 14.1 shows the process used in additive manufacturing.

14.4 NEED FOR 4D PRINTING

Complex shapes and structures may now be created by manufacturers and researchers using 3D printing, which was previously thought to be unachievable using conventional production techniques. In the past three decades, printing technology has undergone constant development and has undergone significant change. Although complex, bio-inspired, multi-material designs can be created using 3D printing, this technology is not yet ready for application in large-scale manufacturing. It is not yet ready for widespread use due to its lengthy cycle time [8].

The barrier for 3D-printing technologies is also increased by the shortage of printing materials with excellent performance. Additionally, the creation of mostly static structures using one or more materials is the core emphasis of 3D printing. Traditional 3D printing uses commercial single or multiple filaments to create static structures; for the requirements of dynamic structures and the applications that they can be used for, such as soft grippers [9] and self-assembling space antennas [10], this is insufficient.

Similar to this, a novel strategy has been developed in response to the increasing need for flexible products in a variety of applications, such as self-folding packaging and adaptable wind turbines. In order to create a meta-material structure, researchers are currently exploring beyond conventional 3D printing, which creates structures from one material. Therefore, an inventive printing technology concept has arisen to modify their structure over time by utilizing the behavior of various materials. This concept is known as 4D printing.

Different materials that produce stacked structural reactions when activated by outside stimuli are combined to create the meta-material structure. Material anisotropy is created by the congruent printing of several materials and allows an item to modify its structure by elongating, bending, corrugating, and twisting along its axes. The quick development of smart materials and multi-material structures has given 4D-printing benefits over 3D printing in a number of recent areas. Four-dimensional printing modernizes the notion of change in the configuration over time, in contrast to three-dimensional printing, which depends on external inputs. Therefore, it is recommended to thoroughly pre-program 4D-printed structures using time-dependent deformations of products [11].

Smart material, which gives printed products more flexible, expandable, and malleable qualities to respond to particular stimuli, is the fundamental component of 4D-printing technology. It's been studied how 4D printing is being used more and more in the fields of aerospace, medicine, smart textiles, and responsive structures. In order to increase the potential applications of both 3D- and 4D-printing technologies, a new study topic branch from additive manufacturing called 4D printing is being investigated [12].

14.5 4D PRINTING

It has several definitions have been developed. The foremost is that 4D printing is just 3D printing that has been given more time. But the term that best represents 4D printing is as shown in Figure 14.2, where it is the gradual alteration of a 3D-printed structure's shape, characteristics, and usefulness over time as a result of exposure to heat, light, and pH. Different meta-material structures can be formed during 4D printing with modifications in the surroundings. The 4D-printed materials are ideal for producing toys, robots, lifters, micro-tubes, and lockers thanks to these properties.

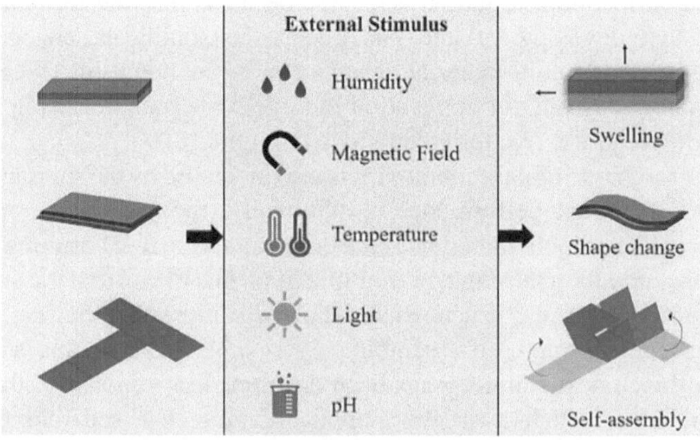

FIGURE 14.2 Stimuli and their responses in smart materials [18].

[13–16]. A 4D object can be created in the same way as any other 3D-printed object. The advanced materials used in 4D printing are different because they may be programmed to perform different functions when hot water, light, or heat is applied. Therefore, the 3D shape and behavior of a non-living entity can change over time [17].

14.6 FACTORS RESPONSIBLE FOR 4D PRINTING

The AM technique, the printing medium, the stimuli, the interaction mechanism, and modeling are the five factors that affect 4D printing [19]. First and foremost, think about the AM printing process. The AM technology enables the production of printing material from digital data provided by the computer without the need for intermediary equipment. Almost all of these methods can print a 4D material as long as the printing medium is available.

The second consideration is the printing medium, which is sandwiched layer by layer and must respond to stimuli. The terms "programmable materials" or "smart materials" are frequently used to describe these materials. The sort of smart materials will determine the stimuli employed, and how they react to those stimuli will determine how much they can adapt themselves.

The third component or feature will be the stimulus used during 4D printing. Stimuli might be biological, chemical, or physical. Physical stimuli such as UV light and other types of light are included. They also include moisture, temperature, magnetic, and electric energy. The use of oxidants and reductants, chemicals, pH, and other factors are examples of chemical stimuli. Biochemical stimuli include enzymes and glucose. When a stimulus is applied, the structure goes through physical or chemical changes, including stress relaxation, molecular mobility, and phase changes, which result in the structure deforming.

The interaction mechanism and mathematical modeling make up the fourth and fifth factors, respectively. Not all materials can undergo the appropriate metamorphosis when a stimulus is supplied to a substance that is intelligent. We must supply some engagement mechanism, such as mechanical loading or physical manipulation, in order to design a sequence of shape change. To determine how long the stimulus will continue to affect the smart material, mathematical modeling is necessary after providing the interaction mechanism. In Figure 14.3, every element influencing 4D printing is represented diagrammatically.

14.7 LAWS OF 4D PRINTING

The shape-morphing behaviors of almost all 4D structures are governed by three universal principles, despite the existence of a wide variety of materials and stimuli. In order to represent and forecast the fourth dimension, the laws shown in Figure 14.4 are used.

14.7.1 First Law

Almost all shape-morphing behaviors of multi-material 4D-printed structures are caused by a single fundamental event: the relative expansion between active and

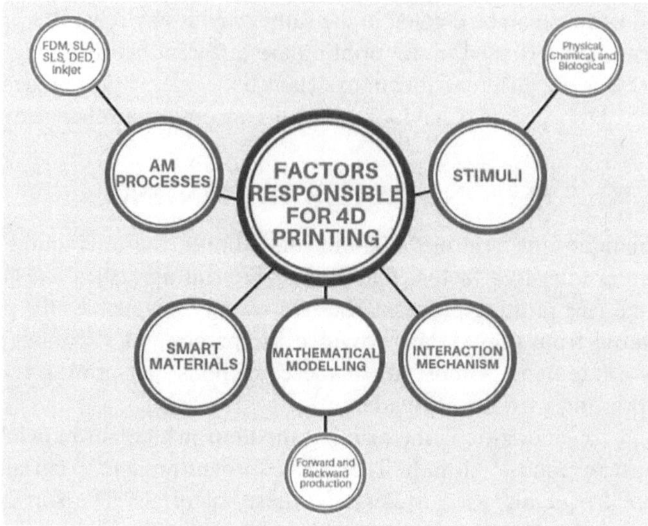

FIGURE 14.3 Factors in 4D printing [20].

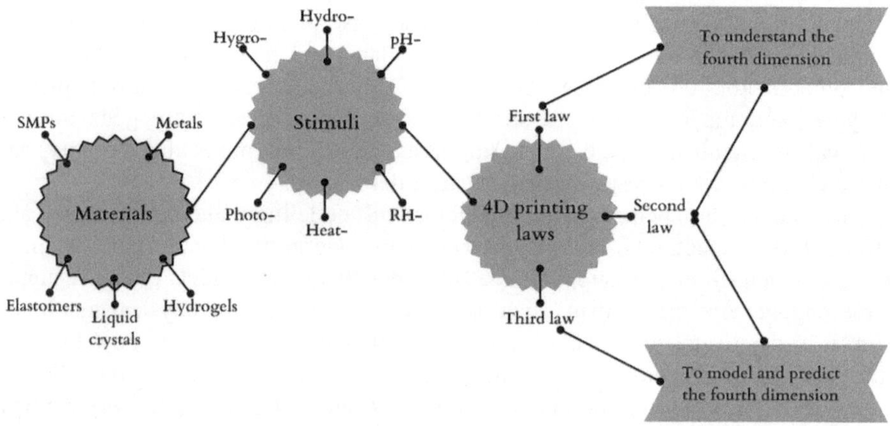

FIGURE 14.4 Laws of 4D printing [21].

passive materials. Almost all complicated shape-morphing behaviors associated with 4D printing, including twisting, coiling, and curling, are caused by this relative expansion. By creating diverse heterogeneous structures and encoding various sorts of anisotropy between active and passive materials, this relative expansion is made possible.

14.7.2 Second Law

Four types of physics—mass diffusion, thermal expansion, molecular transformation, and organic growth—affect the majority of the shape-morphing behaviors of multi-material 4D-printed objects. The following subsections discuss, quantify, and integrate these physics. Each one results in a greater proportion of active than passive materials and the consequent shape-changing behaviors in response to stimuli. Although it can be internal, the stimulus is usually external.

14.7.3 Third Law

According to the third rule of additive manufacturing, "almost all multi-material 4D printed structures demonstrate time-dependent shape-morphing behavior controlled by two 'types' of time constants." Depending on the 4D printing material and stimuli, these constants may be equal, enormous, or disappear in relation to one another. In addition, a mathematical equation for the fourth dimension was developed, which can be applied in upcoming hardware and software to simulate 4D structures.

14.8 TECHNIQUES USED IN 4D PRINTING

4D printing methods add time as the fourth dimension as opposed to producing static objects. The ability to respond to the environment, including temperature, light, water, pH, and other factors, allows a building that has been 3D printed via 4D printing to change over time. Although technology is still in its infancy, 4D printing has become an intriguing subset of manufacturing techniques that is catching the attention of both academics and business on a large scale. The basic idea is to employ 3D printing to control materials at the nano- and micro-scales in order to produce materials that can gradually change their structural features when used at the macro-scale. As shown below, nano- to macroscale 3D-printed objects can be made as smart devices, metamaterials, and origami using 4D printing for a range of practical uses in prototyping, aerospace, biomedical, etc. Table 14.1 lists the different AM procedures.

All of these 4D-printing techniques, are shown in Figure 14.5. Deformations can also develop over time in a single, inactive material by carefully regulating the printing conditions [22]. So long as the printing technology permits dynamic changes in the shape and functionality of the materials, we can create 4D structures. This section will concentrate on 4D-printed constructions that were created, utilizing various printing materials and processes.

14.8.1 Single Material

Structures printed using conventional 3D printers are made of a single material. Printing a single smart material in a 3D printer is the most basic method of 4D printing. Shape memory polymers (SMPs) and liquid crystal elastomers are the two single smart materials utilized for shape-alteration the most commonly (LCEs). Known as SMPs, this class of polymers may hold onto a transient shape while being subjected to heat or light and subsequently return to that shape. When stimulated externally, side-chain or main-chain LCEs with mesogens (liquid crystal molecules) may experience a significant constriction along the path of the mesogens. A single SMP or LCE can be printed using a one-way actuator, a two-way actuator with reversible motion, or both, depending on the situation.

TABLE 14.1
AM Process

Sl. No	Techniques	Media	Printing Mechanism	Materials
1	FDM	Solid	Solid materials are melted and printed in a layer-by-layer method	Thermoplastics (PLA, PET, PA, PA, ABS) and composites
2	SLS	Powder	Heat-induced sintering of powdered materials	Polymers and metals (glass, nylon, PS, aluminum, and titanium)
3	SLM		Sintering previously melted metallic powder	Metals, alloys, ceramics, composites
4	EBM-electron beam melting		Metallic powder that has already been melted is sintering	Metals, alloys, ceramics, composites
5	Binder jetting		Sintering powder materials upon heat	Polymers and polymeric, powders (PVA, starch, and cellulose derivatives)
6	SLA	Liquid	Using layers of liquid resin to cure and solidify	Polymers, ceramics, composites
7	Inkjet		Layer by layer, liquid solidifies	Tetrahydrofurfuryl methacrylate, Triethylene glycol dimethacrylate, wax
8	DLP		Layer by layer, liquid solidifies	Elastomers, metamaterials
9	DIW		Liquid layer by layer solidifying	Polymers, waxes, ceramics

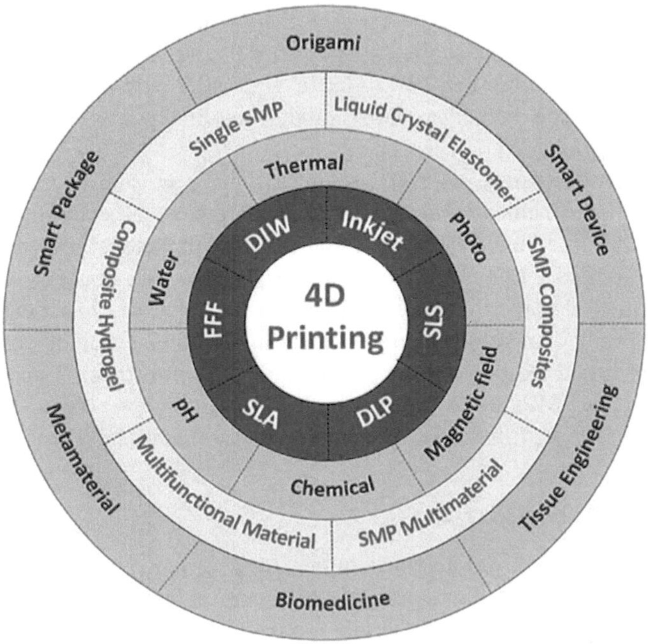

FIGURE 14.5 Various techniques used in 4D printing [23].

14.8.2 Multi-Materials

Multi-materials are a variety of various polymers with a particular geometric structure that have been 3D printed. Because printing these multi-materials is such a challenging process, there are currently only a few printers that can do it. Fiber-reinforced or bilayer SMP composites, multi-material SMPs, and desolvation-induced multi-materials are a few of the most often utilized multi-material systems for 4D printing. Composite hydrogels they are water and thermally responsive are another popular option. In the latter, shape modification is possible by making use of the volume reduction brought on by the desolvation of the final unreacted component.

Environmental stimuli can cause eigenstrains to form in a multimaterial structure; these eigenstrains depend on the volume fractions and relative positions of the various materials and can cause the structure's shape to alter. Only two of these ink-jet multi-material printers, the Poly Jet Connex series by Stratasy and the ProJet MJP series by 3D systems, are currently commercially available. These printers can only use the photopolymer resin-based inks manufactured by the firm. There is a different DIW printing method that can be used to print several materials. DIW was developed as a method for creating intricate ceramic structures using ceramic ink [24].

The ability of the ink to rapidly keep its structure after emerging from a nozzle makes DIW printing advantageous for applications involving the printing of various materials. A multi-material structure can be produced when numerous DIW printing nozzles comprised of various inks are employed. A relatively straightforward concept governs the multi-material structure 4D-printing technique. In order to print a structure, internal stress must be created in specific locations. The structure will experience more predictable shape changes after the stress is released. The internal stress required in this case can be produced via SMPs or the hydrogels' swelling. These restrictions can be removed by modifying the hydrogels with the aid of modern technologies [25].

SMPs are mostly used to print 4D structures. These are more durable and have a wide variety of mechanical properties, making them suitable for use in the construction of multi-materials-based 4D structures [26]. In contrast to hydrogels, SMPs may be programmed into a variety of temporary shapes and maintain that shape no matter how many deformations are applied. No matter how sophisticated they are, hydrogels can only change between two shapes. A further advantage of SMPs over SMAs is their broad temperature range [27] low cost and toxicity, and easy manufacturing [28]. The SMPs are recommended for 4D printing over other SMMs because of their quick reaction times, ease of printing, and dominant response strain [29].

14.8.3 Non-Active Materials

In the case of 3D- or 4D-printed structures, stress, whether intrinsic or induced in the smart materials, is the main cause of the transition from 2D to 3D and from 3D to 4D structures. Non-active materials don't experience this stress internally, though. The process of printing or programming generates these inherent tensions into the non-active materials. Due to its impact on the product's usage and durability, internal tension is frequently minimized during the application of non-active materials. However, if this internal tension is utilized constructively and under control, it can become a boon and be helpful in generating shape-changing geometries with inert materials. Thermoplastic polymers like acrylonitrile butadiene styrene and poly-lactic acid gain internal stress as a result of the rapid heating and cooling that occurs during FDM printing. The internal stress that has developed can be released by heating the polymer above its transition temperature, which causes the printed structure to take on new shapes.

Additionally, the internal stress can be released by heating the nozzle or platform and accelerating the fabrication process. By carefully managing the internal tension that develops in such materials, sophisticated shape-changing structures can be printed using non-active materials. Structures that are twisted and curved can be printed using the 4D printing of composites (4DPC) technique. The 4DPC differs from it in that it uses continuous fibre composite, which is used to make aircraft, turbine blades, and vehicles and has a high strength and stiffness.

4DPC-printed structures deform in terms of shape depending on the location, anisotropy, and stacking arrangement of the various composite material layers. Structures that are twisted and bent are the result of the laminate deforming after cooling and curing. Furthermore, the configuration of twisted and curved laminates was predicted using experimental data and the finite element method (FEM). A potential use for 4DPC is the printing of blades for hockey sticks and turbines, among other things.

The exceptional thermo-mechanical properties, dynamical behavior, and multiple uses of meta-materials, such as their zero Poisson's ratio, are well known. These are artificial materials that have been constructed using conventional 3D printing. FEM models were used to precisely mimic the behavior of the meta-structures during loading and unloading cycles, while experimental tests were used to demonstrate the reversible deformation behavior of the meta-structures. Different sectors of 4D structure applications have been made possible by the opportunity to experiment with varied stimuli, materials, composite materials, printers, and alterations.

14.9 MATERIALS USED IN 4D PRINTING [20]

Recent developments have improved the accuracy and adaptability of material arrangement in additive manufacturing, which is useful for 4D printing as well. They are frequently referred to as "smart materials" since the materials used in 4D printing have the ability to change their properties over time. Figure 14.6 shows different types of materials used in 3D- and 4D-printing methods.

These materials demonstrate behaviors such as self-assembly, self-healing, shape memory, and self-capability in response to an external input. Further, using materials that may change shape, 4D printing also uses materials that can change color when exposed to UV and visible light.

14.9.1 Moisture-Responsive Hydrogels

Due to the vast range of possible uses, materials that are sensitive to moisture or water have attracted a lot of interest. Because of their extraordinary ability to respond to moisture or water, these compounds are also known as hydrogels. When in contact with moisture, they actually belong to a class of cross-linked polymer chain 3D networks that can expand by up to 200% of their initial volume. Since they can be folded, bent, stretched, and expanded geometrically, hydrogels are also very capable of being printed to create a variety of structures.

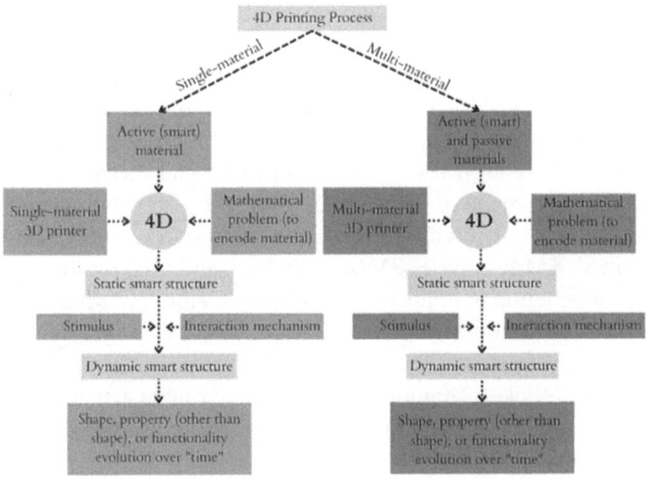

FIGURE 14.6 Types of materilas used in 3D- and 4D-printing methods [21].

These can be easily printed with direct ink and are very biocompatible. But their sluggish reversal reaction is the sole issue. The hydrogels must be engineered to add anisotropy to their swelling in order to get around this. Normally, hydrogels are immersed in liquid and left to absorb the liquid until they reach their saturation point. This approach has a drawback in that it limits the ability of hydrogels to regulate at the intermediate level. The aqueous medium's temperature can be managed to get around this.

14.9.2 Thermo-Responsive

These are the intelligent substances that react to stimuli that include thermal elements. Form change effect (SCE) and shape memory effect (SME) are the two main mechanisms driving the shape changes in these materials in response to thermo-stimuli (SME). SME occurs when external stimuli are used to restore the original shape of a distorted plastic material.

Materials with shape memory are examples of smart materials that exhibit the SME effect (SMMs). These materials fall under various categories, including shape-memory gels, shape-memory metals, shape-memory ceramics, shape-memory hybrids, and shape-memory polymers (SMGs). SMMs can also be classified into one-, two-, or three-way materials depending on how many times their shape is changed. One-way SMMs cannot recover the initial shape after deformation, however two-way and three-way SMMs can retrieve the original shape after deformation into a temporary shape utilizing an intermediary shape. The SMMs may display SCE in addition to SME depending on the environmental factors.

Due to their ease of printing, SMPs are the form of SMM that researchers utilize the most frequently. The SMPs can be distorted with the proper stimulation and then return to their original shape. The normal range of operating temperatures for SMPs is higher than their typical glass transition temperature (Tg). When cooled, they have a transitory shape that is not affected by external loads. They are configured to operate above Tg and in particular temperature and mechanical circumstances. They maintain their original shape when the temperature is raised back up to Tg. This is because polymer chains below Tg have low internal energies and can't move around freely, which makes the material rigid and glassy. But when Tg is exceeded, the polymer chains are given energy, which makes them move. The substance consequently takes on a rubbery texture and is flexible. The SMPs, such as thermoset and thermadapt SMPs, can be customized to take advantage of their unique properties for printing applications.

The SMPs have two or three intermediate states, and a stable intermediate state may be maintained. SMA, another SMM often utilized in 4D printing, has the ability to change shape in reaction to temperature variations. Above a typical temperature for SMAs called the transition finishing temperature, they have high yield strength and a Young's modulus of elasticity that makes them extremely elastic. The most widely used SMA is nitinol (Nickel-Titanium), which has favorable actuation characteristics, high strain and strength, stable cyclic features that make it more biocompatible, and good SME attributes.

Even though these Cu-based SMAs are less ductile than other SMAs, they have been used in a number of research projects. Using SLM printing technique, another class of SMAs based on iron (Fe) have also been investigated. These materials have pseudo-elastic stresses and are inexpensive. SMAs have mostly been researched in relation to their prospective applications in biomedicine, such as orthopaedic, orthodontic, and surgical treatments, as well as surgery, physiotherapy, etc.

14.9.3 Photo-Responsive

For the deformation of intelligent materials, light also serves as a subliminal stimulus. A smart material, often referred to as a photo-responsive material, heats up when a section of it is exposed to light because the exposed area absorbs the light. Heat also stimulates the deformation of smart materials, as was already mentioned, causing the photoresponsive material to change shape. Because light doesn't directly cause a change like heat or moisture, it acts as an indirect stimulus.

14.9.4 Electro-Responsive

Electricity is a similar indirect stimulation to light in that it has been demonstrated to have a heating impact on the materials it passes through. So-called electro-responsive materials are those that deform as a result of their responsive behavior to the electric current.

14.9.5 Magneto-Responsive

Another indirect stimulus that might cause smart materials to deform is a magnetic field or magnetic energy. Magneto-responsive materials are those that can deform in reaction to magnetic energy, enabling the printing of four-dimensional structures. Magneto-responsive materials have the potential to be used to print metals and polymers. The only restriction is that for the magnetic field to be altered, the print size must be extremely small and light.

14.9.6 Piezoelectric Responsive

When subjected to mechanical stress, piezoelectric materials have the ability to generate a charge. Piezoelectricity is the process of creating electric charges as a result of a mechanical force. There is a lot of promise for 4D-printed structures made of magneto-responsive materials in the printing of metals and polymers; the only limitation is that the print size needs to be small and light in order to be impacted by the magnetic field. Piezoelectric materials are also suitable for 4D printing since they may distort when subjected to mechanical forces. The phenomenon is straightforward: when stress is applied, charges are generated, and these charges eventually cause changes to the structure.

14.9.7 pH Responsive

These are intelligent materials that can adjust their shape and volume in response to pH levels and values. These materials can be used for 4D-printing technology since the shape distortion occurs in response to pH values. Since polyelectrolytes feature an ionizable side group, they can accept or give protons when the pH value changes, making them pH-responsive polymers that have been employed in 4D printing.

Electrostatic repulsion causes the polymer chain to stretch when a proton is released, deforming the structure. When a proton is accepted, the structure returns to its original shape. Polyelectrolytes have polyanions or polyacids as the ionizable side chains and polycations or polybases as the functional group.

The side chains stretch at higher pH levels, releasing the proton, then receive it at lower pH values. At higher pH levels, the functional group, however, takes the proton and releases it at lower pH values. pH-responsive materials have applications in soft robotics, biocatalysts, actuators, valves, medication delivery, and colloid stabilization. Additionally, one or two stimuli may be responsive to the smart materials. SMPs, for instance, may respond to changes occurs in electricity, light and heat, while composite materials respond to a variety of stimuli.

14.10 PROPERTIES OF MATERIALS USED IN 4D PRINTING

After its introduction, 4D printing has been touted as having a variety of applications in several technological and scientific sectors. This technique has significantly decreased the volume needed for storage in printed structures and has made a variety of changes possible for printed devices and structures [30]. A 3D printer can create complicated structures directly, or it can create simple 3D structures that are subsequently put together to form structures which are complex in nature [31]. Due to 4D-printing technology, printing is now dynamic, allowing for simple selection of various techniques and materials depending on the task at hand. 3D printing, in comparison, was static and employed the same substance for a variety of uses. Structures created with 4D printing have the capacity to self-assemble and self-heal for a variety of uses. The three characteristics of the 4D-printed structures are self-assembly, self-adaptability, and self-repair, which increased their usage in various applications.

14.10.1 Self-Assembly

Self-assembly is the capacity of a material to join together to build distinct 3D-printed structures before evolving into a challenging 4D structure. When employed in adverse working conditions like deep space applications, these 4D materials can be quite useful. Transporting such 4D-printed goods to space stations is an option as its components can self-assemble. This property can be applied to build satellites and antennae [19]. Using 4D-printed buildings allows for the removal of mistakes or flaws in the creation. It is possible to insert such structures into the faulty area, where these materials can repair themselves using their self-assembly property. However, it should be highlighted that not every application will benefit from the 4D-printed materials' ability to self-assemble. When such an operation is necessary, this characteristic will prove useful [32].

14.10.2 Self-Adaptability

The constructions created via 4D printing may adjust to the surrounding circumstances and weather. For instance, in certain applications, 4D-printed material assemble quickly while it takes more time to construct others. This could be referred to as the material's self-capability. Some substances undergo form and structural changes as a result of cooling and heating. This could be interpreted as their inherent ability to adapt to the circumstances. The ability of 4D-printed materials to be self-adaptive can be highly useful for building medical tissues and equipment.

14.10.3 Self-Repair

The ability of 4D-printed materials to self-correct mistakes and flaws makes them particularly valuable in a variety of applications. Self-repair is the common term for this characteristic. Self-repairing 4D-printed materials can be used where self-healing, recycling, etc. are necessary [33]. Already, self-healing inks have been developed, but applications need to find ways to use these materials [34].

14.11 APPLICATIONS OF 4D PRINTING [20]

Despite the fact that the 4D-printing technique is still in developing stage, it has already garnered a lot of interest from several academic and technological fields. 4D-printed materials are used in military vehicles that can change their shape to adapt in diverse conditions. A military uniform created using 4D printing could alter its shape when it comes into contact with a sharp object. Printed things are capable of adapting to changing environmental conditions while taking user requirements into account. As a result of their smaller volume, 4D-printed objects occupy less space and labor during shipping and assembly. 4D-printed objects can also be used to meet new requirements due to their independency. They have a huge potential for usage in space-related projects.

14.11.1 Medical

This is one of the industries that is growing as a result of 4D printing, where we the replicas of various organs can be printed to study their structure and behavior to find effective treatments for illnesses which them. Furthermore, the 4D-printed nano-biomedical devices, like biosensors, bioactuators, and biorobots, can help to monitor the physical changes in cells and tissues. In addition to performing surgery, the biorobots can deliver medicines and treatments.

By utilizing their self-deformation, 4D printing can be particularly effective in addressing abnormalities in the field of orthopedics. The generated orthopedic models can also alter their characteristics in response to applied stimuli. The orthopedic components made using 4D printing can be used in complex surgeries. The field of orthopedics has a tremendous amount of potential for 4D printing to create the parts used for surgeries [35].

Using smart materials, which are flexible and self-compatible and can be genetically more suited to the patient, it will be able in the future to print organs like the heart, liver, and kidney [36]. The skin transplant can be created via 4D printing and will match the patient's natural skin tone as well as to treat skin burns [37]. The stents that are created using the 4D-printing technique can be inserted into the body through small incisions to provide the desired results.

A nano 4D-printed device can also be introduced into the body to carry out the required activity. Since the structure or functionality of complex materials produced by 4D printing can change over time and provide the cells with the required stress, it has huge to be employed as a scaffold. The 4D-bioprinted materials offer precise control to carry and deliver drugs, biomolecules, and cells in a programmable way by self-folding, self-unfolding, swelling, and deswelling.

Dynamic materials are designed to react to specific environmental stimuli, such as temperature changes, changes in magnetic fields, changes in humidity, pH, and ions in the surrounding environment.

Due to these characteristics, scientists can study variations in the micro-environment by detecting changes in the dynamic materials' structure. For instance, a poly diacrylate-based liquid-responsive hydrogel sheet with reversible mobility can be produced. The film displays diverse degrees of color, spontaneous deformation, and movement in the presence of various humidity levels. Further, many properties of polymeric dynamic materials allow each of these activities to be engaged by a particular external stimulus [18].

A new branch of tissue engineering known as bioprinting was developed as a result of the scarcity of transplantable organs and the limitations of conventional tissue engineering. The process of creating biological components such as cells, tissues, drugs, etc. through a material transfer technique referred to as bioprinting. This technology is used to perform biological operations. Bioprinting emerged as a result of increased organ transplants.

Bioprinting provides a variety of benefits over conventional tissue engineering, including the ability to build enormous tissue-engineered objects and produce tissues with a high density. With the advancements in bioprinting, the use of biocompatible materials for 3D printing can change over time [38]. The change of a 3D structure's physical, chemical, and biological makeup as well as its traits and shape are referred as evolution. As a result of biological developments, 4D bioprinting has been extended to incorporate the functional maturation and transformation of 3D cells or tissues over time [39].

4D bioprinting using a biomaterial that can change its shape is used to mature the 3D-printed objects. The capability of 4D bioprinting to construct complex structures is the primary basis for its categorization in the field of tissue engineering. The highest resolution of all bioprinting techniques is found in 4D bioprinting, which enables the inclusion of more information and details into the tissue. In 4D bioprinting, material is removed from a cartridge using a laser pulse, which is then applied to the substrate layer by layer.

Materials used for 4D bioprinting should respond to pressure, pH, temperature, moisture, light, electric and magnetic energy, and other parameters. It is possible to develop biological structures that can alter their shape in response to inputs.

The biological applications of 4D bioprinting, such as medication administration, tissue production, and proper organ regeneration and transplantation, have proved to have significant potential [40].

Due to the development of biomaterials, 4D bioprinting is also becoming a more active area of research. The material soybean epoxidized acrylate (hMSCs) is used to print scaffolds to support multipotent human bone marrow mesenchymal stem cells [41].

14.11.2 Soft Robotics

For the development of robots in classical robotics, materials with high hardness such as metals, hard plastics, and ceramics can be employed [42]. These robots are made for certain environments and uses, and they are not resistant to any environmental factors. They are unable to achieve significant deformations and are unable to carry out flexible duty [43]. To get around these restrictions, a branch known as soft robotics was created, which can create robots that are flexible like humans, can change how rigid they are, and can adapt to their surroundings. Soft and intelligent materials like electroactive polymers are needed for soft robotics (EAPs). Due to these materials, a smooth contact with delicate objects might emerge, offering traditional robots a better tolerance towards destructive pressures. Soft robotics is an area that benefits most from the flexibility, deformability, and environmental adaptability of 4D-printed structures. It is possible to create the actuators for soft robots utilizing 4D-printed structures.

Dielectric elastomer actuators (DEAs) were traditionally made by hand, but Rossiter et al. used 3D-printing techniques to make them. This eliminated the issue of time and labor consumption. The DEAs are part of 4D EAP smart materials that react to electrical stimulation [44]. As the DEA receives electrical energy, it transforms it into mechanical energy, which causes the actuator to deform and cause the robot to move.

The ability to embed parts into unique components during 3D printing and then use external stimuli to have the parts self-assemble after the product is manufactured is another way to apply 4D printing to soft robotics [45]. Soft actuators (SA) are inexpensive, simple to make, and environmentally benign, and they can be beneficial in soft robotics as well.

The SMPs have also been used as actuators (shape change), but robotics hasn't used them extensively because of their poor reaction. But as 4D-printing technology advances, it is now possible to create intricate structures containing SMPs that may be activated by a variety of triggers. In particular circumstances, these can serve as actuators [46].

Despite significant advancements and progress in the field of 4D-printed soft robotics, modeling and complete control of these robots remain challenging tasks. This is because these robots have nonlinear dynamics, temporally variable physical and chemical characteristics, and material stiffness [47]. 4D printing has enormous promise for use in soft robotics, but more materials of this type will need to be created in the future to enable improved actuation and control. Another issue that needs to be resolved in the future is the actuation performance and robustness of 4D-printed robotics.

14.11.3 Self-Evolving Structures

Self-evolving structures that can regain their previous shape upon contact with water can be created using 4D printing. The group created numerous intricate structures that displayed deformations including folding and stretching. The structures' mechanical degeneration was brought on by the constant wetting and drying of the structures. However, these issues can soon be resolved by investigating novel multi-materials with self-evolving properties. Additionally, other causes, like as temperature and UV exposure, can be investigated for such materials in addition to moisture.

There are several uses for these self-evolving structures. For instance, these buildings can be easily assembled in deep water, which is challenging to accomplish with other structures. This may also be helpful for fixing things in deep water. The study of deep-water ecosystems and plants can be advanced by instruments with such printed parts that can be assembled there. The diving suit used by divers in deep water can be improved with the help of such self-evolving 4D-printed constructions.

14.11.4 Origami

The art of origami involves folding a piece of paper into a three-dimensional object. One of origami's primary characteristics is size reduction, which has received a lot of attention in the modern world because many different fields call for the compacting of enormous objects into small spaces. Origami is beneficial for constructing things like cartons, shopping bags, car airbags, and electronics (photovoltaic solar cells), all of which need to be compact and have characteristics that can change shape. Due to the high cost of traditional origami, the concept of active origami was created. Designing an origami structure that can fold on its own is known as active origami, and for these smart materials like SMPs, light-activated polymers, etc. are needed. Due to its ease of use and low cost of production, 4D printing can be helpful in the development of active origami structures [48],

Thus, the potential for active origami using multi-materials like PACs can be explored. The PACs' hinges were printed on flat polymer sheets, creating active origami structures that could change shape to take the form of a cube, pyramid, or aeroplane. Utilizing various intelligent materials as well as cutting-edge computational tools and methods, active origami can be improved even further to produce more flexible 4D-printed active origami structures. Additionally, consecutive folding allows for the printing of more intricate active origami structures, with none of the components interfering with one another as they fold.

14.11.5 Aerospace

The production of components for space missions is crucial to the aerospace sector. The components must be produced for less money and have a long life span. The 4D-printed constructions can survive harsh environmental conditions and are made at a reasonable cost. Additionally, these structures are adaptable to their environment. Thus, the aerospace sector is discovering possible uses for 4D printing.

The possibility of using such materials in future space missions has been suggested by the International Space Station (ISS3D)'s printing of ABS structures in microgravity. Additionally, they may be produced inside the spacecraft and have a short printing time, which lessens the need for components from Earth. The utilization of self-sustaining materials is one of the mission's main components. The 4D structures can influence the course of such missions since they have the capacity to self-assemble in accordance with the circumstances.

The shape memory alloy nitinol, which is composed of nickel and titanium, is extremely well-liked in the space industry and has been created utilizing 4D-printing methods. The satellites, self-sustaining tools for space exploration, can be made in part from 4D materials. The light weight of structures created with 4D printing reduces a component's mass by 80%. PEEK, a thermoplastic that can withstand high temperatures and stresses with ease, has been printed using the SLS technology and is utilized to make parts for satellites and spacecraft.

14.11.6 Sensors and Flexible Electronics

The creation of sensors can benefit from 4D printing, due to the 4D-printed structures' responsiveness to factors including pH, humidity, temperature, stress, and strain. For certain stimuli, they can therefore serve as sensors. Additionally, 4D printing is economical and produces lightweight, sensitive, accurate, and highly responsive structures. As a result, 4D printing will develop and modernize sensors. SMAs can be used to assess temperature, strain, and detect wear and damage inside a structure in addition to acting as actuators.

Intelligent vehicles and aircraft can be produced using ultrasonic additive manufacturing (UAM), which mixes metals and smart materials to print intelligent structures. UAM is a sort of 3D printing that may be used to manufacture 4D structures and intelligent materials. UAM-printed structures have less mass, which is advantageous for the transportation sector.

Construction mistakes can be fixed using 4D printing since the parts may be shipped to the defective area and self-assemble there with the application of the right stimulus. Self-healing structures can be created by utilizing the self-repairing capabilities of 4D-printed structures. The durability and dependability of the material systems will be improved by the self-healing materials. [49]. Because 4D buildings have the potential to repair themselves, they can be reused, which reduces material consumption and protects the environment from waste.

4D printing can be used in the textile sector to create textiles that would adapt to changing environmental circumstances and change as a result, improving comfort and ventilation [50]. The color and texture of the surface of 4D-printed textiles can also alter in response to environmental changes. Materials like PLA, ABS, and nylon could be printed on the exteriors of fabrics such as cotton, polypropylene, polywood, and polyester [51]. In addition, 4D printing can improve the quality of household appliances, making them more cosy and moisture- and heat-resistant. In response to conditions and surroundings.

An airplane's 4D-printed wings can adjust to various air conditions and offer better lift and safety. There is evidence that leaf springs made of composite materials manufactured via 4D printing are equivalent to conventional springs [52]. The wind turbine blades were manufactured using 4D printing without the use of typical electromechanical devices like actuators and sensors. These blades' 4D printing increased system and energy production control as well. The same research team also unveiled a smart photovoltaic system that can deform in accordance with the availability of sunshine and offers an effective solar energy production system [53].

14.12 OTHER APPLICATIONS OF 4D PRINTING [54]

The proposed 4D-printing technique is a more ideal option in the current competitive structure of manufacturing and its sub-domains due to the amazing overall precision, perfection, and quality solutions to the complicated industrial difficulties. The fundamental idea behind 4D is that manufacturing strategies are used to model and build structures made of one specific material or mixtures of active and passive materials that travel to a specific stimulus on a pre-programmed basis. These manufacturing strategies are used in conjunction with CAD tools. Reactive outcomes have recently occurred, such color changes, self-healing, or biological behavior. For material scientists and engineers, the rising collection of materials for form memory offers a considerably greater range of stimulants, transition states, and material characteristics.

However, fusing designs with other materials or combining them into composites may result in even better design potential. A smart 3D object created using this printing procedure that uses smart materials may experience form changes as a result of exposure to water, heat, light, electricity, or magnetic fields. In 4D printing, the printed object form is meticulously pre-programmed in conjunction with the intended shape-shift. As a result of its computational folding, one of the key advantages of 4D printing is that it can create objects that are larger than printers. Items too large for a printer can be streamlined into secondary forms for 3D printing when printed 4D things change size, shrink, and grow. The field of materials can undergo a huge transformation thanks to 4D printing.

Applications for medical 4D printing assist in creating a tangible 3D model by connecting layers of smart material using computer-controlled CAD data. A new generation of printing equipment for medical research is actively being developed by numerous enterprises and research organizations. The hard materials used in 4D printing are part of an intelligent structure that can change shape in response to external elements including sunlight, water, and other environmental influences.

Applying 4D printing to items that we use on a daily basis, at least for the majority of us, is the most beneficial use of this technology. These include clothing and footwear that adapt to environmental changes in order to better serve our changing demands and respond to our evolving needs. Shelters, equipment, and devices can be printed flat or infinitely small and transported to disaster zones.

Researchers are currently printing biocompatible, automatically adaptable materials that can be introduced into the human body and used as directed. Ultimately, this module can be put together without outside interference that could change its form and functionality. Without requiring sensors or electrical devices, 4D-printing technology might realize these smart curtains by focusing instead on changing heat levels during the course of the day to change their structure. With 4D printing, fashion could see a similar transformation. One idea is that changes in temperature and activity might affect clothing. By altering their design, shoes made using 4D printing can offer greater comfort and amortization during running.

14.13 ROLE OF 4D PRINTING IN THE FIELD OF MANUFACTURING

The major aspects or broadly categorized qualities well related with the 4D-printing applications include process development, materials for smart purposes, flexibility in shapes and dimensions, and better programming. These primary characteristics are further divided into a number of sub-characteristics, including the extensive 3D process, the use of hydrogel composites and biomaterials, support for structural alterations, and instruction in thermomechanical facts and multi-material printing features [55,56].

In addition to being more useful, 4D printing is also a better way to make things. It is a method that can also reduce the labor expenses associated with manufacturing. 3D printers allow for mass production of items that are exclusively used in environmental settings. The medical equipment, nanotechnology, and biomaterials industries are those where 4D printing is most likely to be used, according to the industry.

Other businesses that were early users of 3D printing include the automotive, aerospace, and defense sectors. After 3D printing, this innovation would become the subsequent game-changing development technology. To change their form over time, researchers are experimenting with converting 3D printing into 4D printing. One of the key advantages is that almost any type may be printed physically, allowing product designers to focus on quality rather than the shape created by the new mainstream [57–59].

An improvement to 3D printing is 4D printing, which enables 3D-printed items to change their shape and material properties in response to pre-programmed stimuli such immersion in water, exposure to heat, electricity, UV light, or other energy sources. Materials are coded during production with instructions that cause the material to react and adapt to environmental conditions, making it intelligent.

Intelligent materials and nature are used in 4D printing to create shapes and functional objects. Shape memory polymer and hydrogel tinctures are two of the main active polymers used in 4D printing. A hydrogel is a colloid gel that is made up of an interconnected network of hydrophilic polymer chains with water serving as the dispersion medium. Strongly absorbent polymer networks might be natural or manufactured. Functional, multicore linkages between network chains and hydrophilic functional groups connected to the polymeric backbone allow hydrogels to absorb water. Numerous natural and artificial materials can be made using the hydrogels concept. When incorporated into a non-swelling polymer or filament, hydrogel expands and causes mismatch stresses between the two materials when dissolved in a solvent. However, 4D hydrogel printing has a slower speed of actuation and less stiffness [60–62].

Instead of printed hydrogels, 4D printing with smart materials has a relatively high output speed and a somewhat steep end structure. If the form memory polymer's temperature rises above the point at which it changes shape, the distorted structure reverts to its original state. The photoisomerization stimulus is utilized to stimulate the 4D-printed component. Additional activation techniques for 4D printing include the use of wetness, heat and tension combinations, and heat from the light source [63–66].

By using fewer parts for motion output, this function lowers the weight of 3D-printed components on board. The development does away with the requirement for onboard CPUs, cameras, power storage engines, etc. utilizing 4D printing to advance manufacturing

A few of the most prominent 4D additive manufacturing techniques used for the completion of common and complex jobs are liquid solidification, material extrusion, material jetting, and powder solidification. These suggested 4D-printing technologies use stereolithography, direct laser printing, hydrogel extrusion, fused deposition modeling, shape memory alloys, selective laser melting, inkjet printing, and others [67,68].

Several 4D-printed polymer bilayer actuators can change shape in response to light. The photo-active layer for bilayer actuators is printed on many supporting layers using a newly synthesized linear azobenzene polymer. Bilayer actuators can create complex machinery, including unmanned aircraft, artificial muscles, and platforms for the distribution of biological drugs. A hearing aid consists of a top end and a second bottom earpiece, with the top end of the earpiece being designed to fit into the user's ear canal and the second bottom earpiece being designed to fit along the first curve of the ear channel. In this, a soft-tuning member makes up at least a portion of the adaptable earpiece on the first-end earpiece [69–71].

In several industries, such as aerospace, transportation, consumer goods, and medical equipment, 4D manufacturing is quickly replacing conventional manufacturing. Despite the widespread use of 3D printing in several business sectors, there is still opportunity for technological advancement due to the increasingly sophisticated materials. Prototypes have been employed in the automotive, medical, and aviation industries while research and development are still ongoing. In order to layer-by-layer repair a substance, 4D-printing technology makes use of stereo principles, which regulate how UV light is employed throughout the printing process.

The shape of a four-dimensional device will change over time, even if it is printed as a three-dimensional form. When heated by hot water or heat, smart materials used in 4D printing, including programmable materials, exhibit a variety of functionalities. Over time, a deceased entity's form and behavior are impacted by this factor [72–74].

In order to create 4D-printed functional gadgets, practical goods like mobile devices are first manufactured along with a form change. These functional features, such as optical or conductivity qualities, may also occur during a shape-shift phase in actual 4D printing. For 3D-printed constructions, the time-sensitive mechanical features, such as tissue maturation, degradability, self-healing, and color change, are also referred to as 4D printers [75,76].

Bi-layer SMP composites and thermal composite hydrogel are utilized in smart material photo polymeric materials for 4D printing. If these materials are subjected to sunlight, fire, dampness, water, etc., their properties alter. The 4D printers could be useful in creating implants and smart medical devices. There is no need for techniques to model and create actuators because it is challenging for researchers to construct more complex structures. Additionally, the 4D-printing techniques themselves lack the accuracy needed to incorporate designs from nanoscale chemistry to macroscale mechanics in the material composition and work [77,78].

Many industries might benefit from the new technology. It has applications in a variety of industries, including aircraft, engineering, autos, manufacturing, defense, and medical. Wide-ranging frameworks are built using 4D printing, and they change throughout time from one type to another. The 4D printer will be used by the developers to schedule different material qualities. It comprises geometry, different water-absorbing traits, and other material-specific details. The conversion of non-electronic materials is also possible using this embedded programming technique. There are numerous ways to implement 4D printing. Real modifications are necessary in the chemistry of materials having modifiable characteristics. Design ideas are now taking shape and utilizing these new materials [79,80].

This is the result of thriving cooperation in a market environment that is rapidly changing. In this fast-changing environment, it is projected that the rising need for technical innovation will enhance 4D-printing technology. As an addition to 3D printing, 4D printing offers improved uniformity, dependability, and performance. Early forms of 4D printing include adaptable structures that can change when subjected to pressure, gravity, air, or water [81,82].

Structures printed using conventional 3D printers are made of a single material. In theory, 4D printing is a production technique that can create clothing that alters far more than just form. With 4D printing, the designers may program the items to change color or adopt an entirely new theme. Additionally, 4D-printed clothing may be sufficiently self-programmed to protect wearers against changes in form and other features under adverse conditions [83–85].

The development of better conducting polymers and electronic equipment to build organic thin-film transistors on plastic foils would be aided by the use of 4D-printing technology. One of the main potential benefits of 4D printing, whether it is used in business or not, is that parts can self-heal. If the actual measurements or qualities of the object change unexpectedly, intelligent materials should be programmed to adapt.

Despite the fact that 4D printing has only just become popular, the supply chain, which includes quicker shipment and reduced storage space, clearly benefits. 4D printing reduces transportation and processing expenses to a significant extent, opening up a wide range of applications for the supply chain [86–88].

A few of the most prominent 4D additive manufacturing techniques used for the completion of common and complex jobs are liquid solidification, material extrusion, material jetting, and powder solidification. These suggested 4D-printing technologies use stereolithography, direct laser printing, hydrogel extrusion, fused deposition modelling, shape memory alloys, selective laser melting, inkjet printing, and others [68,89].

14.14 CHALLENGES [90]

Innovative new technologies, like 4D printing, are motivating and propel the manufacturing sector forward. However, as this analysis has made clear throughout, more basic research and development in 4D printing is needed. This technology presents considerable difficulties because not everyone can easily access it. For instance, biocompatibility is a fundamental criterion in the medical industry, but the majority of the materials reported for 4D printing is not.

So, the fundamental issue here is a lack of immediately applicable practical application. It has been shown that 4D-printed materials adapt to several stimuli, including temperature, water, pH, and light. However, it is still difficult to understand and anticipate how the material will behave when exposed to multiple stimuli at once. The full recovery or reversibility of the printed material to its original shape or configuration once a stimulus is withdrawn could also be difficult, so this issue needs to be further researched.

It is challenging to compare typical isotropic materials with 3D-printed constructions because of their anisotropic nature. When contemplating 4D-printed materials, where the material evolves over time, the stakes are considerably higher. This idea raises the question of whether the materials' qualities are stable throughout time or if they change over time. There has to be further research because this has not been addressed.

Comparing 3D- and 4D-printed materials in terms of applications is still difficult. For instance, a variety of materials are manufactured in 3D for high-temperature applications. As of now, the majority of materials created by 4D-printing technologies are polymer based and have few high-temperature applications. Printable materials with the proper viscosities and low glass transition temperatures are the best materials to use in 4D printing. The need for viable substitutes for SMPs as 4D-printing materials depends on having these two things.

Additionally, by developing mechanical qualities that are more compatible with 4D printing, smart effects can be activated in traditional 3D formations. The slow deterioration of the smart structure following a number of cycles is one issue that 4D-printed systems encounter. This field has to advance to offer new materials and fabrication techniques to overcome the deterioration problem. Additionally, machine learning is a resource that might be applied by integrating it into 4D-printing algorithms to create a printing process that is intelligent and self-sufficient.

14.15 FUTURE SCOPE

There is a lot of room for growth in the relatively new and intriguing study area of 4D printing. Due to the fact that 4D technology is a development of 3D printing, there are inherent difficulties with resolution, material restrictions, mechanical qualities, and the viability of creating complicated shapes. The FDM, SLA, selective laser melting, and selective laser sintering were the foundational technologies for 4D printing. A few tens of micrometers are the typical resolution of commercialized 3D poly-jet printers.

The production of tissue, scaffolds, or micro- and nano-actuators requires high resolution for accurate 3D form fabrication and moving. Because the nozzle size influenced how rough the surface of the printed object was, increasing resolution also resulted in better surface quality. Through the nozzle of a 3D printer, polymeric materials for 4D printing were injected. Printed products exhibited anisotropic mechanical properties because of the layer-by-layer production process, which was influenced by adhesion between the polymer strands. Adhesion and interlayer behavior would be a barrier to the adoption of functional polymeric materials because the majority of PACs are multi-material systems.

Another barrier to commercialization was the 4D-printed devices' long-lasting active motion. An important problem was found to be the mechanical deterioration that took place during thermo-mechanical cycling in printed multi-material systems. Chemical deterioration occurred as well throughout the curing procedure. It was discovered that using ultraviolet rays to cure photopolymers damaged the polymer and degraded its characteristics. The network structure was greatly impacted by the thermal energy as well.

The constraint on polymer or materials for 4D printing was brought on by this material deterioration. In the biomedical industry, improvements in biocompatible materials would be problematic because the SLA technique, which can achieve high resolution, cannot use biocompatible materials that have received FDA approval. These biocompatible materials can be printed using FDM, SLS, and inkjet processes without undergoing any chemical alterations.

The range of 4D-printing applications can also be expanded in terms of a variety of external stimuli thanks to a variety of printable materials. Additionally, the glass transition temperature of things produced by 4D printing was too high for human body. In 4D-printed items, the glass transition temperature is a factor that impacts active motion since permanent shape is regained above the glass transition temperature. Co-polymerizing to alter the length of crosslink polymers could be used to adjust the glass transition temperature. Materials for 4D printing should be printable and have the right viscosity and low glass transition temperature.

Wood fibers and botanical cellulose, which exhibit the above-described smart behavior, make excellent candidate materials for 4D printing in place of SMPs. The intelligent impacts of 3D buildings, however, can be sparked by acquiring a fundamental understanding of the mechanical characteristics of materials appropriate for 4D printing. Therefore, for 4D printing, it's crucial to understand the structure and mechanical qualities. Simulators like ABAQUS and COMSOL have been used to do simulations. The behavior of the stress-strain curve with regard to temperature, deformation, and bending behavior in 1D or 2D comprised the majority of the reported simulation results. Experimental findings were used to determine characteristics like thermo-mechanical properties based on the traditional lamina theory or SMPs' behavior model. Due to the intricacy of 3D models, many simulations were performed using simpler models that had rigid and actuating elements that were described in 1D or 2D.

A 3D model is required for self-collision and optimal design, even though actuation of a 4D-printed device might be anticipated with a 2D model. The deformation behavior model of SMPs was used to examine the active motion caused by SME. However, the deformation mismatching, swelling effect, or optimized structure were to blame for the active motion of 4D-printed items. The challenge of creating a theoretical model for PAC would be left for later effort.

The additive manufacturing innovation known as 4D printing has significant potential for biomedical uses as well. Using 4D printing, it is possible to create patient-focused items including artificial organs, vascular, and dental implants. A medical staple or micro-gripper with a self-tightening mechanism that produces defined stress in response to input was also developed [91]. By overcoming PAC's poor mechanical strength and durability, 4D printing will be expanded to create artificial muscles with peak stress levels between 40 and 80 MPa. Besides, the sectors of sensors and electrical devices may also benefit from the use of 4D-printing technology.

14.16 CONCLUSION

The advancements and uses of 4D printing were covered in this chapter. By incorporating time, it improves on 3D printing that has already been developed. Depending on the stimuli they are exposed to, 4D structures can morph over time. These buildings can put themselves together, fix themselves, and adapt to changes on their own. Applications for 4D printing can be found in the fields of medicine, soft robotics, self-healing goods, active origami, etc. Organ printing may become a reality thanks to 4D bioprinting, which has the potential to physically re-create the biological development process. Future discoveries of more intelligent materials and new, more effective printing techniques will greatly expand the potential for 4D printing. It is possible to develop 4D printing, which is currently primarily used for structural modifications, to achieve multifunctional applications. Additionally, 4D printing has the potential to be used in a variety of other industries in the future. In conclusion, 4D printing will advance as a new discipline of controlled smart structure printing and develop in a variety of different applications.

REFERENCES

[1] N. Araújo, V. Pacheco, and L. Costa, Smart additive manufacturing: The path to the digital value chain, *Technologies*, 9 (2021), 88. 10.3390/technologies9040088

[2] T. Phillips, J. Allison, C. Seepersad, and J. Beaman, Smart manufacturing in additive manufacturing, *Smart Manufacturing*, 2020 Elsevier Inc., 10.1016/B978-0-12-82002 8-5.00007-2

[3] F. Aqlan and C. de Vries, Mariea Sargent, Andrew Valentine, using 3D printing to teach design and manufacturing concepts, Proceedings of the 2020 IISE Annual Conference L. Cromarty, R. Shirwaiker, P. Wang, eds.

[4] https://3dprinting.com/what-is-3d-printing/- accessed on 01-08-2022:11.57 am.

[5] https://manufactur3dmag.com/3d-printing-applications/-accessed on 01-08-2022:12. 05 pm.

[6] I. Paoletti and L. Ceccon, The evolution of 3D printing in AEC: From experimental to consolidated techniques, 10.5772/intechopen.79668

[7] M. Pagac, J. Hajnys, Q.-P. Ma, L. Jancar, J. Jansa, P. Stefek and J. Mesicek, A review of vat photopolymerization technology: Materials, applications, challenges, and future trends of 3D printing, *Polymers*. 13 (2021), 598. 10.3390/polym13040598

[8] W. Gao, Y. Zhang, D. Ramanujan, K. Ramani, Y. Chen, C. B. Williams, C. C. L. Wang, Y. C. Shin, S. Zhang, and P. D. Zavattieri, Comput. *Aided Des.* 69 (2015), 65.
[9] J. C. Breger, C. Yoon, R. Xiao, H. R. Kwag, M. O. Wang, J. P. Fisher, … and D. H. Gracias, Self-folding thermo-magnetically responsive soft microgrippers, *ACS Appl. Mater. & Interfaces.* 7 (5) (2015), 3398–3405.
[10] F. Momeni, X. Liu, and J. Ni F., A review of 4D printing, *Mater. & Des.* 122 (2017), 42–79.
[11] E. Hawkes, B. An, N. M. Benbernou, H. Tanaka, S. Kim, E. D. Demaine, … and R. J. Wood, Programmable matter by folding, *Proceedings of the National Academy of Sciences.* 107 (28) (2010), 12441–12445.
[12] S. E. Bakarich, R.Gorkin III, M. I. H. Panhuis, and G. M. Spinks, 4D Printing with mechanically robust, thermally actuating hydrogels, Macromol. Rapid Commun. 36 (12) (2015), 1211–1217.
[13] S. Tibbits, C. McKnelly, C. Olguin, D. Dikovsky, and S. Hirsch, in: Proc. Of the 34th Annual Conference of the Association for Computer Aided Design in Architecture vol. 539, 2014.
[14] B. An, and D. Rus, in: Presented at IEEE International Conference on Robotics and Automation, 2012. USA.
[15] Y. Liu, W. Zhang, F. Zhang, X. Lan, J. Leng, S. Liu, X. Jia, C. Cotton, B. Sun, B. Gu, and T.-W. Chou, Compos. *B. Eng..* 153 (2018), 233.
[16] K. K. Shigetomi, H. Onoe, and S. Takeuchi, Han, *PLoS ONE* 12 (2012), 1.
[17] https://www.sculpteo.com/en/3d-learning-hub/best-articles-about-3d-printing/4d-printing-technology/-accessed on 22-07-2022:7.37 am.
[18] W. Zhou, Z. Qiao, E. N. Zare, J. Huang, X. Zheng, X. Sun, M. Shao, H. Wang, X. Wang, D. Chen, J. Zheng, S. Fang, Y. M. Li, X. Zhang, L. Yang, P. Makvandi, and A. Wu, 4D-Printed dynamic materials in biomedical applications: Chemistry, challenges, and their future perspectives in the clinical sector, *J. Med. Chem.* 2020 (63) (2020), 8003–8024, 10.1021/acs.jmedchem.9b02115
[19] F. Momeni, S. M. M. Hassani, N. X. Liu, and J. Ni, Mater. *Des.* 12 (2017), 42.
[20] A. Ahmed, S. Arya, V. Gupta, H. Furukawa, A. Khosla, 4D printing: Fundamentals, materials, applications and challenges, *Polymer.* 228 (2021) (2021), 123926, 10.1016/j.polymer.2021.123926
[21] F. Momeni, and J. Ni, Laws of 4D printing, *Engineering.* 6 (2020) (2020), 1035–1055, 10.1016/j.eng.2020.01.015
[22] Z. Zhao, J. Wu, X. Mu, H. Chen, H. J. Qi, and D. Fang, Macromol. *Rapid Commun.* 13 (2017).
[23] X. Kuang, D. J. Roach, J. Wu, C. M. Hamel, Z. Ding, T. Wang, M. L. Dunn, and H. J. Qi, Adv. Funct. *Mater.* 29 (2018), 1805290.
[24] J. A. Lewis, J. E. Smay, J. Stuecker, J. Cesarano, and J. Am, Ceram. *Soc.* 12 (2010), 3599.
[25] Q. Zhao, J. Sun, Q. Ling, and Q. Zhou, Langmuir 5 (2009) 3249] [J.P. Gong, Y. Katsuyama, T. Kurokawa, Y. Osada, *Adv. Mater.* 14 (2003), 1155.
[26] M. C. Mulakkal, R. S. Trask, V. P. Ting, and A. M. Seddon, *Mater. Des.* 160 (2018), 108.
[27] S. Erkeçoglu, A. D. Sezer, and S. Bucak, Smart drug deliv. *Syst.* 1 (2016).
[28] M. D. Monzon, R. Paz, E. Pei, F. Ortega, L. A. Suarez, Z. Ortega, M. E. Aleman, T. Plucinski, and N. Clow, Int. J. Adv. Manuf. *Technol.* 89 (2017), 1827.
[29] J. M. Pearce, *Science* 6100 (2012), 1303.
[30] T. Campbell, S. Tibbits, and B. Garret, Sci. *Am.* 311 (2014), 60.
[31] Y. Zhou, W. M. Huang, S. F. Kang, X. L. Wu, H. B. Lu, J. Fu, H. Cui, and J. Mech. Sci. *Technol.* 10 (2015), 4281.
[32] S. Tibbits, in: Proc. 31st Annual Conference of ACADIA, 2011. Banff.

[33] T. Raviv, and L. Striukova, Sci. *Rep*. 4 (2014), 7422.
[34] S. Bauer, S. B. Gogonea, I. Graz, M. Kaltenbrunner, C. Keplinger, and R. Schwodiauer, Adv. *Mater*. 26 (2014), 149.
[35] B. Gao, Q. Yang, X. Zhao, G. Jin, Y. Ma, and F. Xu, *Trends Biotechnol*. 34 (2016), 746.
[36] J. Gosnell, T. Pietila, B. P. Samuel, H. K. N. Kurup, M. P. Haw, and J. J. Vettukattil, J. Digit. *Imag*. 6 (2016), 665.
[37] P. He, J. Zhao, J. Zhang, B. Li, Z. Gou, M. Gou, and X. Li, *Burns Trauma*. 5 (2018), 1
[38] Y. C. Li, Y. S. Zhang, A. Akpek, S. R. Shin, and A. Khademhosseini, *Biofabrication*. 9 (2016), 012001.
[39] N. J. Castro, C. Meinert, P. Levett, and D. W. Hutmacher, Curr. Opin. *Biomed. Eng*. 2 (2017), 67.
[40] D. A. Zopf, A. G. Mitsak, C. L. Flanagan, M. Wheeler, G. E. Green, S. Hollister, Otolaryngol. *Head Neck Surg*. 152 (2015), 57.
[41] S. Miao, W. Zhu, N. J. Castro, M. Nowicki, X. Zhou, H. Cui, J. P. Fisher, and L. G. Zhang, *Sci. Rep*. 6 (2016), 27226.
[42] C. Yang, X. Tian, T. Liu, Y. Cao, and D. Li, *Rapid Prototyp. J*. 1 (2017), 209–215.
[43] S. H. Ahn, K. T. Lee, H. J. Kim, R. Wu, J. S. Kim, and S. H. Song, *Int. J. Precis. Eng. Manuf.* 4 (2012), 631.
[44] A. O'Halloran, F. O'Malley, and P. McHugh, *J. Appl. Phys*. 7 (2008), 071101.
[45] L. Justin, M. Amelia, and B. Christopher, in: Presented at 22nd Annual International Solid Freeform Fabrication Symposium, 2011. Austin, TX.
[46] M. López-Valdeolivas, D. Liu, D. J. Broer, and C. Sánchez-Somolinos, *Macromol. Rapid Commun*. 5 (2018), 1700710.
[47] C. Della Santina, R. K. Katzschmann, A. Biechi, D. Rus, in: Presented at IEEE International Conference on Soft Robotics, 2018, pp. 46–53. RoboSoft.
[48] Q. Ge, C. K. Dunn, H. J. Qi, and M. L. Dunn, *Active origami by 4D printing, Smart Mater. Struct*. 9 (2014), 094007.
[49] M. Röttger, T. Domenech, R. van der Weegen, A. Breuillac, R. Nicolaÿ, and L. Leibler, *Science* 356 (2017), 62.
[50] Y. S. Zhang, K. Yue, J. Aleman, K. Mollazadeh-Moghaddam, S. M. Bakht, J. Yang, W. Jia, V. Dell'Erba, P. Assawes, S. R. Shin, M. R. Dokmeci, R. Oklu, and A. Khademhosseini, *Ann. Biomed. Eng*. 1 (2017), 148.
[51] P. Cordier, F. Tournilhac, C. Soulie-Ziakovic, and L. Leibler, *Nature*. 451 (2008), 977.
[52] S. Ramesh, C. Usha, N. K. Naulakha, C. R. Adithyakumar, and M. Reddy, *Mater. Sci. Eng*. 1 (2018), 012123.
[53] S. V. Huo, *Compos. Struct*. 210 (2019), 869.
[54] A. Haleem, M. Javaid, R. P. Singh, and R. Suman, Significant roles of 4D printing using smart materials in the field of manufacturing, *Advanced Industrial and Engineering Polymer Research*.
[55] S. K. Leist, and J. Zhou, Current status of 4D printing technology and the potential of light-reactive smart materials as 4D printable materials, *Virtual Phys. Prototyp*. 11 (4) (2016), 249–262.
[56] S. K. Sinha, Additive manufacturing (AM) of medical devices and scaffolds for tissue engineering based on 3D and 4D printing, in: *3D and 4D Printing of Polymer Nanocomposite Materials*, Elsevier, 2020, pp. 119–160.
[57] K. Deshmukh, M. T. Houkan, M. A. AlMaadeed, and K. K. Sadasivuni, Introduction to 3D and 4D printing technology: State of the art and recent trends, 3D and 4D rint. Polym. *Nanocomp. Mater*. (2020), 1–24.
[58] M. Javaid, and A. Haleem, Exploring smart material applications for COVID-19 pandemic using 4D printing technology, *J. Indust. Integrat. Manag*. 5 (4) (2020), 481–494.

[59] Y. S. Lui, W. T. Sow, L. P. Tan, Y. Wu, Y. Lai, and H. Li, 4D Printing and stimuli responsive materials in biomedical aspects, *Acta Biomater.* 92 (2019), 19–36.
[60] C. Lin, L. Liu, Y. Liu, and J. Leng, 4D printing of bioinspired absorbable left atrial appendage occluders: A proof-of-concept study, *ACS Appl. Mater. Interfaces.* 13 (11) (2021), 12668–12678.
[61] G. Sossou, F. Demoly, H. Belkebir, H. J. Qi, S. Gomes, and G. Montavon, Design for 4D printing: A voxel-based modeling and simulation of smart materials, *Mater. Des.* 175 (2019), 107798.
[62] A. Nishiguchi, H. Zhang, S. Schweizerhof, M. F. Schulte, A. Mourran, and M. Mooller, 4D printing of a light-driven soft actuator with programmed printing density, *ACS Appl. Mater. Interfaces* 12 (10) (2020), 12176–12185.
[63] J. Lee, H. C. Kim, J. W. Choi, and I. H. Lee, A review on 3D printed smart devices for 4D printing, *Int. J. Prec. Eng. Manuf. Green Technol.* 4 (3) (2017), 373–383.
[64] F. B. Coulter, and A. Ianakiev, 4D printing inflatable silicone structures, 3D Print. *Addit. Manuf.* 2 (3) (2015) 140–144.
[65] M. H. Ali, A. Abilgaziyev, and D. Adair, 4D Printing: A critical review of current developments, and future prospects, *Int. J. Adv. Manuf. Technol.* 105 (1) (2019), 701–717.
[66] Y. Y. C. Choong, S. Maleksaeedi, H. Eng, J. Wei, and P. C. Su, 4D Printing of high-performance shape memory polymer using stereolithography, *Mater. Des.* 126 (2017), 219–225.
[67] A. Haleem, and M. Javaid, Expected role of four-dimensional (4D) CT and four dimensional (4D) MRI for the manufacturing of smart orthopaedics implants using 4D printing, *J. Clin. Orthopaed. Trauma.* 10 (2019), S234–S235.
[68] Y. Jiang, J. Leng, and J. Zhang, A high-efficiency way to improve the shape memory property of 4D-printed polyurethane/polylactide composite by forming in situ microfibers during extrusion-based additive manufacturing, *Add. Manuf.* 38 (2021), 101718.
[69] S. V. Hoa, Factors affecting the properties of composites made by 4D printing (moldless composites manufacturing), *Adv. Manuf. Polym. Compos. Sci.* 3 (3) (2017), 101–109.
[70] S. Y. Hann, H. Cui, M. Nowicki, and L. G. Zhang, 4D Printing soft robotics for biomedical applications, *Add. Manuf.* 36 (2020), 101567.
[71] M. P. Caputo, A. E. Berkowitz, A. Armstrong, P. Müllner, and C. V. Solomon, 4D Printing of net shape parts made from Ni-Mn-Ga magnetic shape-memory alloys, *Add. Manuf.* 21 (2018), 579–588.
[72] D. Grinberg, S. Siddique, M. Q. Le, R. Liang, J. F. Capsal, and P. J. Cottinet, 4D Printing based piezoelectric composite for medical applications, *J. Polym. Sci. B Polym. Phys.* 57 (2) (2019), 109–115.
[73] H. Ding, X. Zhang, Y. Liu, and S. Ramakrishna, Review of mechanisms and deformation behaviors in 4D printing, *Int. J. Adv. Manuf. Technol.* 105 (11) (2019), 4633–4649.
[74] Y. Zhang, Q. Wang, S. Yi, Z. Lin, C. Wang, Z. Chen, and L. Jiang, 4D Printing of magnetoactive soft materials for on-demand magnetic actuation transformation, *ACS Appl. Mater. Interfaces.* 13 (3) (2021), 4174–4184.
[75] A. Haleem, M. Javaid, and R. Vaishya, 4D printing and its applications in orthopaedics, *J. Clin. Orthopaed. Trauma.* 9 (3) (2018), 275.
[76] S. Chung, S. E. Song, and Y. T. Cho, Effective software solutions for 4D printing: A review and proposal, *Int. J. Prec. Eng. Manuf. Green Technol.* 4 (3) (2017), 359–371.
[77] M. Zarek, N. Mansour, S. Shapira, and D. Cohn, 4D printing of shape memory based personalised endoluminal medical devices, *Macromol. Rapid Commun.* 38 (2) (2017), 1600628.

[78] S. Saska, L. Pilatti, A. Blay, and J. A. Shibli, Bioresorbable polymers: advanced materials and 4D printing for tissue engineering, *Polymers*. 13 (4) (2021), 563.

[79] Q. Ge, C. K. Dunn, H. J. Qi, and M. L. Dunn, Active origami by 4D printing, *Smart Mater. Struct.* 23 (9) (2014), 094007.

[80] S. Valvez, P. N. Reis, L. Susmel, and F. Berto, Fused filament fabrication-4D-printed shape memory polymers: A review, *Polymers*. 13 (5) (2021), 701.

[81] R. Ashima, A. Haleem, S. Bahl, M. Javaid, S. K. Mahla, and S. Singh, Automation and manufacturing of smart materials in additive manufacturing technologies using internet of things towards the adoption of industry 4.0, *Mater. Today: Proc.* (2021).

[82] K. Ahmed, M. N. I. Shiblee, A. Khosla, L. Nagahara, T. Thundat, and H. Furukawa, Recent progresses in 4D printing of gel materials, *J. Electrochem. Soc.* 167 (3) (2020), 037563.

[83] A. Chadha, M. I. Haq, A. Raina, R. R. Singh, N. B. Penumarti, and M. S. Bishnoi, Effect of fused deposition modelling process parameters on mechanical properties of 3D printed parts, *World J. Eng.* 16 (4) (2019), 550–559.

[84] J. Wang, Z. Wang, Z. Song, L. Ren, Q. Liu, and L. Ren, Biomimetic shapeecolor double-responsive 4D printing, *Adv. Mater. Technol.* 4 (9) (2019), 1900293.

[85] R. Aziz, M. I. Haq, and A. Raina, Effect of surface texturing on friction behavior of 3D printed polylactic acid (PLA), *Polym. Test.* 85 (2020), 106434.

[86] P. R. Reddy, and P. A. Devi, Review on the advancements of additive manufacturing-4D and 5D printing, *Int. J. Mech. Prod. Eng. Res. Dev.* 8 (4) (2018), 397–402.

[87] Y. Zhang, L. Huang, H. Song, C. Ni, J. Wu, Q. Zhao, and T. Xie, 4D printing of a digital shape memory polymer with tunable high performance, *ACS Appl. Mater. Interfaces*. 11 (35) (2019), 32408–32413.

[88] H. Hassanin, A. Abena, M. A. Elsayed, and K. Essa, 4D printing of NiTi auxetic structure with improved ballistic performance, *Micromachines*. 11 (8) (2020), 745.

[89] A. Haleem, and M. Javaid, expected role of four-dimensional (4D) CT and four-dimensional (4D) MRI for the manufacturing of smart orthopaedics implants using 4D printing, *J. Clin. Orthopaed. Trauma.* 10 (2019), S234–S235.

[90] A. Hassan, Alshahrani, Review of 4D printing materials and reinforced composites: Behaviours, applications and challenges, *J. Sci.: Adv. Mater. Devices*. 6 (2021), 167–185.

[91] D.-G. Shin, T.-H. Kim, and D.-E. Kim, Review of 4D printing materials and their properties, *International Journal of Precision Engineering and Manufacturing-Green Technology*, Vol. 4, No. 3, pp. 349–357.

15 Developments in 4D Printing and Associated Smart Materials

Ganesh P. Borikar, Ashutosh Patil, and Snehal B. Kolekar
School of Mechanical Engineering MIT World Peace University Pune, India

CONTENTS

DOI: 10.1201/9781003333760-15

15.1 INTRODUCTION

To create a static structure utilizing digital data in 3D coordinates, 3D printing performs a major role, while 4D printing introduces a new technique that is capable of changing the printed configuration with time when external stimuli are applied. Because these structures are 4D printed and capable of changing shape and function, smart materials and smart design are required for 4D printing rather than 3D printing. This indicated that the structures that will be 4D printed need to be meticulously planned by considering the expected time-dependent changes of the object. In the view of researchers at the Massachusetts Institute of Technology (MIT), 4D printing refers to the production of 3D-printed structures that are adaptable and programmable as they evolve [1]. Intelligent materials can detect stimuli in their surroundings and respond appropriately. Thus, one of the definitions of smart materials is to provide a powerful and intelligent response to a product that could potentially provide a wide range of advanced features that would otherwise not be possible. Three prerequisites must be met for 4D printing. There is a condition of only using stimuli-responsive composites as materials, which are combinations or blends of multiple materials with different properties stacked layer by layer. The second condition is to animate the object through a stimulus like humidity, water, heating, wind, cooling, magnetic energy, gravity, or even ultraviolet (UV) light. The final condition is the time of stimulation, and the ultimate result is the changed state of an object. Certain achievements in this field of research include scientists from the University of Wollongong's involvement in Australia's development of hydrogel. Shape-memory polymer fibers have also been investigated for their incorporation into composite materials. And the use of materials that react when they come into contact with water, such as water-absorbing materials that can change shape or respond when immersed in water. Researchers have successfully demonstrated 4D-printing capabilities in a variety of ways using various materials.

15.2 LITERATURE REVIEW

- In this paper, smart materials and 4D-printing technology are discussed in relation to their most recent advances and uses. It has been suggested that intelligent materials can be produced by blending functional and intelligent materials, and these types of materials are better suited to 4D printing and summarizes the authors' research on shape memory polymers (SMP) and the arm-type additive layered manufacturing (ALM) technique with thermosetting polyurethane [2].
- When used with stimuli-responsive materials, pre-programmed and printed reactions to external stimuli are possible. The author explores novel approaches that have been identified and starts to think about their future influence [3].

- Researchers demonstrated that hardwood grain patterns could be printed, leading to single-curve and double-curve surfaces. Furthermore, thanks to printing actuators with different infill heights, the various parts of the wooden actuator can react in a regulated passive manner. Instead of using the wood's typical natural properties, the study pushes the limits of how printed wood responds to humidity. It introduces a collection of regulated printed hygroscopic features [4].
- This study highlights the existing limitations in current additive manufacturing (AM) technologies and offers numerous possibilities for improved utilisation and production. In this article, a comprehensive overview of additive manufacturing of shape memory polymer composites (SMPC) is presented, along with a list of promising research areas for future research [5].
- The study discusses materials, methods, computer-aided design (CAD), and applications, as well as future work suggestions. In this article, 4D printing and functionally graded additive manufacturing (FGAM) are analyzed together, with FGAM characterized as just one additive manufacturing process that involves material gradational blending [6].
- The case studies demonstrate a specific subset of additive manufacturing in which adaptable, biomimetic composites may be designed to reconfigure or have integrated features or functions that modify themselves in response to external inputs [7].
- Model-based studies demonstrate the created façade's strong operability and formal reversibility. Based on the findings, the author highlighted the potential value of this module when it comes to designing and building adaptable buildings. The design process, simulation, and manufacture of 4D-printed and parametrically movable façades are all covered in this work for the first time [8].
- The customized 4D-printed rings can adapt to changes in environmental circumstances such as load and temperature due to a one-way shape memory effect (OWSME) and a two-way shape memory effect (TWSME). The first NiTi diamonds made by additive manufacturing have been published, presenting revolutionary solutions in the jewelery industry [9].
- Users of fused-deposition modeling (FDM) 3D printing will benefit from this additional knowledge by researching 4D printing more effectively. In this review, the main findings are the materials and composites that are used in 4D printing with fused deposition modeling (FDM), along with some limitations [10].
- The current research focuses on the varied capacities of smart materials (SM), which are employed as raw materials in 4D printing. Depending on the application, 4D-printed parts are capable of altering shape and self-assembling to accomplish a function. After that, the generated object can return to its "memorized" shape when prompted by the stimulus [11].

15.3 4D PRINTING

3D-printing technology uses digital data in the form of 3D coordinates to create static structures, whereas 4D printing introduces a new technique capable of changing the printed configuration over time when exposed to external stimulus. The process of printing structures by 4D printing as well as 3D printing is mostly similar to the design of products using 3D modeling tools such as computer-aided design to print using a 3D printer. It only takes smart materials and smart design to distinguish 3D and 4D printing. When 4D printed, structures are exposed to an external stimulus; their shape changes over time. 4D-printed structures must be meticulously pre-programmed to accommodate the expected time-dependent object deformations [12]. 4D-printing technology has also been shown to be more accurate than physically stretching or stressing the material as it returns to its original form. Building sophisticated structures with more durable materials, which the old manufacturing system cannot easily do, is a significant step forward. The concept of 4D printing has emerged, providing an inventive new solution to expanding printing technologies. Intelligent materials differ from traditional 3D-printing materials in terms of thermomechanical characteristics and other material qualities, as well as shape-changing attributes. However, using 3D-printing technology, printable items are characterized by rigidity [13].

15.3.1 4D Printing Using Polyurethane (PU) and Polyurethane Composites

Polymer architectures with PU are recommended for SMP. Changing the chemical composition of PU is a mechanism of the shape memory effect (SME). Glassy thermoplastic polyurethane (TPU) (40% weight) and semi-crystalline polyurethane (PU) can be combined to create a dependable SMP (60% wt.). The foundation of the building contributes to the tower's glassy look as a hard segment or a sharp point of SMP, and semi-crystalline as a transition segment. There are many more PU configurations available. It is based on a thermoset material and has been evaluated using non-FDM 4D printing. Like PLA, blended composites such as TPU exist; however, unlike PLA, they have been demonstrated to possess form memory [10].

15.3.2 4D Printing Using Polylactic Acid (PLA) and Polylactic Acid Composites

Lactide is produced by digesting carbohydrate sugars to produce PLA, a polymer. PLA is a biodegradable and renewable substance, thermoplastic with biocompatible and nontoxic properties. These features pique the reader's curiosity. The general public is curious about the polymer's possible uses in the realms of medicine, environmental protection, and some other things. There are a number of circumstances in which environmental inertness is advantageous. Another well-liked material for FDM printing is PLA. A melting point of 150–180°C is rather low. In PLA, the form of the material can be remembered. Polymer physical relations were used to demonstrate shape memory capabilities. The printed sample's architecture is maintained by the physical connections of the PLA polymer chain. When a substance reaches the glass transition temperature (Tg), which is between 60 and 70°C, it becomes very elastic, and the interactions between these entanglements can return to their previous configuration after being temporarily stretched and deformed. PLA is commonly thought of as a fragile material due to its application in 3D printing. As a result, research has been done to improve these mechanical qualities by printing with PLA composites. PLA/clay/wood nanocomposites, graphene, and PLA/wood biocomposites are examples of composites that have been investigated. Therefore, studies on 4D PLA printing using FDM may be divided into three groups: studies on pure PLA, studies on mixed PLA composites, and studies on PLA layer composites and preforms [10].

15.4 SMART MATERIAL

15.4.1 Shape Memory Materials (SMM)

When suitably stimulated, shape-changing materials may change shape and then revert to their original shape once the stimulus is withdrawn. A shape memory material can be divided into one-way and two-way types, and it is necessary to program them. Furthermore, a three-way shape memory material has been established. The triple SMM consists of one permanent and two temporary forms. The SMP can transform from a transient to a memorized form in response to external stimuli. To program shape memory, SMPs use less energy, and in terms of recovery, they are more durable. Based on comparisons between synthetic polymers and the extracellular matrix (ECM) proteins found in nature, biodegradable and biocompatible materials have already demonstrated great promise in tissue engineering. Immunological responses to synthetic polymers employed in biomedical applications were either minimal or non-existent. Mitsubishi Heavy Industries (MHI) developed a polyurethane SMP that can be processed more easily and across a wider temperature range. Composite materials made of long fibers combined into polymeric resins are required for 4D printing. By varying the gel's location and concentration of crosslinking, 4D printing can use different hydrogel-based components with chemical and thermal properties. By integrating diverse functional chemical elements, smart multi-materials can be produced. A functional group can be activated via both permanent and ephemeral forms. Graphene nanocomposite-based SMP systems can achieve a higher degree of deformation while increasing the composite's strength. Hydrogels are stimulus-sensitive materials that have been used to create artificial muscles that can function similarly to real muscles. The potential is to experience enormous deformations and huge displacements and to transform chemical energy into mechanical energy. Hydrogels, on the other hand, may react to a variety of stimuli. Among shape memory materials (SMMs) that exhibit the SME effect are shape memory gels (SMGs) and shape memory alloys. In addition, SMMs are broken down into three groups based on the number of form alterations: one-way, two-way, and three-way materials. After deformation, one-way SMMs should not permit recovering the original shape; however, through the use of an intermediate shape, two- and three-way SMMs enable recovery of the original shape following deformation into a new shape [14].

15.4.2 Stimulus-Responsive Single SMP

SMP requires programming. SMPs initially deform at temperatures above their transition temperature. In shapeshifting, the SMP heats above its Tt, and then, due to entropic elasticity, it recovers to its previous shape achieved during the recovery phase. Intricate designs could be printed on this type of biodegradable and biocompatible SMP for flexible electronics or medicinal devices. After printing the temporary shape for the SM components, conductive carbon nanotube ink was programmed into the shape to create a smart circuit. We printed sunflowers with photoresponsive SM petals in a temporary shape (closed shape) at low temperatures, then slowly illuminated them until they returned to their original shape at high illumination levels. Synthetic polymers used in biomedical applications elicited little or no immune response when compared to the natural extracellular matrix (ECM) proteins. Composite materials made of long fibers mixed with polymeric resins are required for 4D printing. 4D printing uses chemically and thermally reactive hydrogel-based components by altering their location and cross-linking density. By integrating diverse functional chemical elements, smart multi-materials can be produced. To activate functional groupings, permanent as well as temporary shapes can be used [15].

With an obvious nematic–isotropic transition temperature (TNI), liquid crystal elastomers (LCEs) thermally activated by the main chain have recently been employed to print LCEs. When heated and cooled along the print direction, 40% reversible contraction was shown by the produced LCE filament. The eyestrain in the hinge was caused by the hydrogel expansion, which also caused a bending distortion. To produce components with reversible shape-changing properties, hydrogel was combined with SMPs and 3D printed using a Connex Objet260 3D printer. The elastomer column and sandwich construction turned the hydrogel's triaxial swelling into bending, which is an in-plane force, which, when heated again, would return to its former shape. A hydrogel composite created by 4D-printing material by Gladman et al. could be printed for biomimetic 4D printing by combining innovative hydrogel materials with geometric designs into a programmable bilayer architecture. Additionally, the possibility of using digital light processing (DLP) to make composite hydrogels capable of changing form has been explored.

SMPs are a kind of polymer that, in reaction to external stimuli like heat or light, may keep a transient form and return to its initial shape. The SMP must reach the programming stage and the recovering stage in order to engage in shapeshifting behaviors. The SMP is crushed at a temperature above a range (Tt) while in the programming step; after cooling the SMP below Tt, it is finally unloaded. The SMP has the deformed shape coded (or fixed). Shapeshifting happens during the recovery phase when the SMP is heated beyond Tt, and then entropic elasticity causes it to revert to its previous form. It should be emphasized that a magnetic field may be used to regulate thermally induced SMP from a distance. The hysteresis effect, joule heating, and photothermal effect can all be used to heat thermally induced SMP [16].

15.4.3 Shape Memory Alloys

Shape memory alloys (SMAs) have been around for a long time; SMAs were first reported by a Swedish metallurgist in 1932 when they discovered the gold and cadmium alloy exhibited a rubberlike characteristic; this is now known as pseudo-elasticity. SMAs are nothing but a metallic alloy that goes through a phase change from solid to solid. From the higher temperature phase to the lower temperature phase, or austenite to martensite, phase change is connected with the key mechanism of shape memory alloys due to the crystalline structure change of alloys that results in macro-scale effects. By utilizing selective laser melting (SLM) and selective laser sintering (SLS), the additive fabrication of Fe-Mn-Al-Ni, Cu-Al-Ni, Ni-Ti, and Ti-Ni-Cu alloys is described in many publications on shape memory alloys. 3D printing is typically very good for obtaining good strength qualities at high loading of metallic structures; thus, SLS and SLM techniques are commonly associated with SMEs. In a linear wire stain, the induced temperature was changed when a current was given to SMA, and then the wire stain was translated into significant displacements, simulating worm muscle tetanus.

To create a 4D-printed magnetic shape memory alloy of Ni-Mn-Ga powder bed binder, jetting was used. Parts with high porosities that may be controlled by varying packing densities and particle size distribution are generally produced by binder jetting; a 70.43% porosity was found for the SMA by Caputo et al. They investigated how different porosity and density alloys affected mechanical and shape memory properties. Andani et al. discovered that, compared with porous alloys, more dense alloys have a higher transformation temperature (10–14°C) and a higher modulus of elasticity. Recoverable strain by 5% and a partial recovery of the superelastic response are revealed by subjecting SMA to cyclic compression testing. It is demonstrated that while retaining shape memory effects, an 86% decrease in the elastic modulus of the porous alloys indicates that these objects could be incorporated for biomedical implants like bone constructions because they are lightweight and 4D printable.

The shape memory impact is a sort of smart material that may directly turn heat energy into mechanical activity (SME). SMAs' transition between two unique crystalline phases, such as the small temperature stage (martensite) and the large temperature stage (austenite), results in the formation of SMEs. The small temperature stage (martensite) and the large temperature stage (austenite) results in the formation of SME. SMA damage during the martensitic phase may cause a reverse phase shift if the alloy is heated above a certain temperature. Martensite will transform into austenite in the crystalline structure, returning the alloy to its original form [17].

It is utilized to achieve 4D printing via non-uniform swelling-induced Eigen stresses by adjusting the crosslinking density as well as the placement of responsive gels. A bilayer hinge composite hydrogel is employed for direct-ink-writing (DIW) printing and exhibits reversible shape distortion in response to both temperature changes and hydration. Several classes of materials have been investigated for use in morphing structures. SMPs are distinguished from hydrogels by their programmability. Bilayer structures can be utilized to generate multifunctional components by combining LCE and inactive or active materials. Shapeshifting for 4D printing with larger and more reversible complexity is made possible by using LCE for direct printing of multi-materials. The fabrication of digital materials with different properties (or digital SMPs) by 3D printing allows for a multitude of subsequent transformations and shapeshifting. Kuang et al. have created a high-strain, repeated self-healing, and complex topology semi-interpenetrating polymer network elastomer by using UV-assisted DIW printing. The linear polycaprolactone (PCL) chain diffused and re-entangled in fracture interfaces as a result of heating, resulting in recurrent crack healing. Furthermore, the high-strain shape memory effect may aid in massive crack healing.

As a result of chemically separating fiber-like sections in the hydrogel composite sheets, Wu et al. have created hydrogel composite sheets that contract and stretch differently in response to external stimuli. This was done through the use of a multistep curing procedure. To create a composite hydrogel containing fiber-like regions (photoinitiator), N, N′-methylenebisacrylamide (crosslinker), and 2–2′-azobis (2-methylpropionamidine) were mixed in an aqueous solution of 2-acrylamide-2-methylpropane sulfonic acid (AMPS), N, N′-methylenebisacrylamide (crosslinker), and 2-azobis (2-methylpropionamidine). It was possible to modify the stripe characteristics by mixing PNIPAm and PNIPAm/PAMPS striping. This was due to the dehydration of PNIPAm produced by heating or ionic strength, as well as the water content of PAMPS [16].

15.4.4 Metamaterials

Victor Veselago initially suggested metamaterials in 1968 and designed materials that can achieve qualities not often present in natural materials, giving rise to the term “smart materials.” To create curved metamaterials that have unique electromagnetic properties and functionality, 3D printing is used. As individual unit cells, metamaterials are carefully designed to function in a way that’s not inherent in the material. This results in complicated and convoluted unit cells, requiring high resolutions to achieve such minute details. In addition, some metamaterial properties only work at the nm scale, and as most additive manufacturing technologies operate between 200 and 50 m, this can be a challenge for most additive manufacturing technologies. AM can manufacture intricate designs that traditional manufacturing techniques cannot; the development of complex metamaterials in the future is highly realistic due to advances in AM technologies that enhance resolution. However, with recent advances in micro-stereolithography, resolutions of up to 40 nm are now possible. Using Ion et al.’s door handle as an example, when torque is applied, the polymer’s inherent properties would fracture; however, careful arrangement of the unit cell ensures that the door handle can function without the use of hinges. The responsive metamaterials with vibration-canceling capabilities could find use in fast-adapting helmets, smart wearables, and soft robotics [18].

Kuang et al. created active metamaterials by combining DIW with shape-memory composites, while Roach et al. created shape-shifting metamaterials by mixing thermally inert materials with active LCE. Chen et al. in a recent study propose a photo masking-based customized DIW method for printing multi-material lattice structures that follow controlled buckling and fatigue fracture development paths. By inserting hard magnetic particles into a soft matrix phase, researchers have developed magnetoactive metamaterials that can shapeshift quickly and repeatably. They have also made breakthroughs in water and thermo-responsive materials. Ze et al. and Ma et al. were the first to do so in this respect, extending the responsiveness of active materials to multiple-responsive magnetic SMPs (both magnetic and thermal stimuli) [18].

15.5 SHAPE CHANGE MECHANISM

The concept of using a material's change in shape is not new to 4D printing. This property is used in a number of different shape programming techniques. Certain methodologies make it simple to convert 2D polymeric sheets to 3D sheets. For such treatments, polymeric sheets are excellent because they are very flexible, cost-effective, and lightweight. A real-world application of shape-changing from 2D to 3D can be found in this method's portability and affordability. Water absorption has been used to turn 2D food ingredients into 3D shapes, which has been researched for edibles.

15.5.1 Active Origami and Self-Folding Techniques

In order to create self-actuating structures that can fold and unfold like origami paper techniques in the real world, active origami uses specialized smart materials. By constructing structures that can fold and occupy smaller places, engineering problems involving large volumes of storage space may be resolved using this technique. Active origami structures were created using polymeric sheet hinges paired with printed active composites (PAC). These connected sheets would take the desired shape by folding to create intricate shapes like pyramids and 3D boxes when external stimuli are applied to them. Origami tubes can also be used to construct 3D structures, robotics, actuators, and space deployments using active origami techniques.

For the form-programming of thin polymeric sheets, light-activated self-folding is used. This approach uses the same folding and cutting processes as origami and kirigami to achieve the required shape. When polymeric sheets are exposed to near-infrared light, they will fold to form origami structures or, when smeared in black ink, pop up to form kirigami creations. Figure 15.1 shows many geometries and combinations generated through light-activated self-folding. Shape-changing structures are

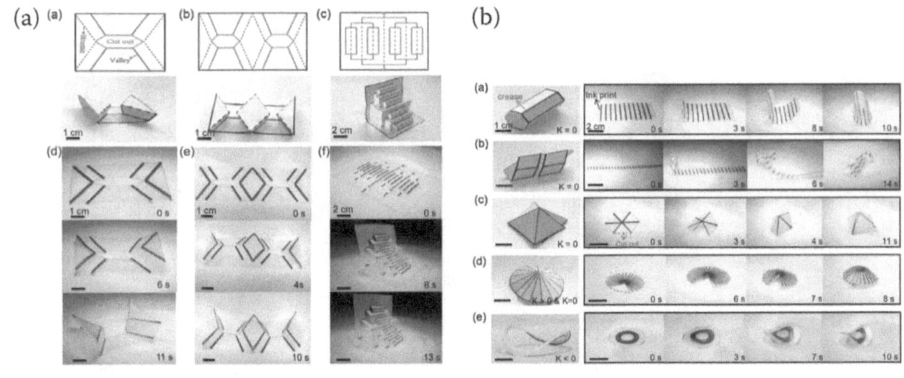

FIGURE 15.1 (a) Kirigami techniques, (b) origami techniques [19].

often created using self-folding techniques. When heated, SMP-based materials that self-fold are intended to create intricate helical hinge structures.

During self-folding, a collision index is created to forecast the collision. If these collisions were controlled, the SMP would be free to follow the right folding sequence. Self-folding techniques were used to build hydrogel-based microgrippers for surgical and soft robot applications.

15.5.2 Stimuli-Based Actuation

The 4D-printed product's distinctive shape modification is a result of the stimulus's influence. These 4D-printed goods include a variety of smart materials that are activated by various inputs.

15.5.2.1 Temperature-Induced Actuation

In response to thermal changes, shape-changing polymers can change their structure. The temperature at which SMPs become momentarily deformed is known as the "glass transition temperature" (Tg). When reheated over TG, however, with the aid of a recovery force, they are able to return to their prior state. During the printing process, SMPs are subjected to a certain amount of internal stress. When exposed to heat stimuli, these tensions are released. A microstructure transformation occurs as a result of this process, and the polymeric substance might change shape. As illustrated in Figure 15.2, improvements in the research of thermally sensitive SMPs have led to the development of a variety of new and enhanced shape memory cycles. To achieve the two-way shape memory effect (2W-2SM), an additional programming step has been added to the traditional one-way shape memory

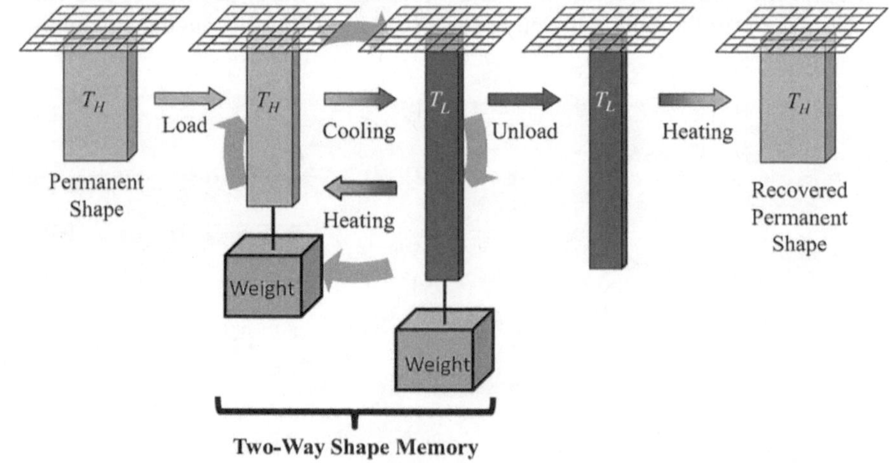

FIGURE 15.2 One-way and two-way shape memory effect [21].

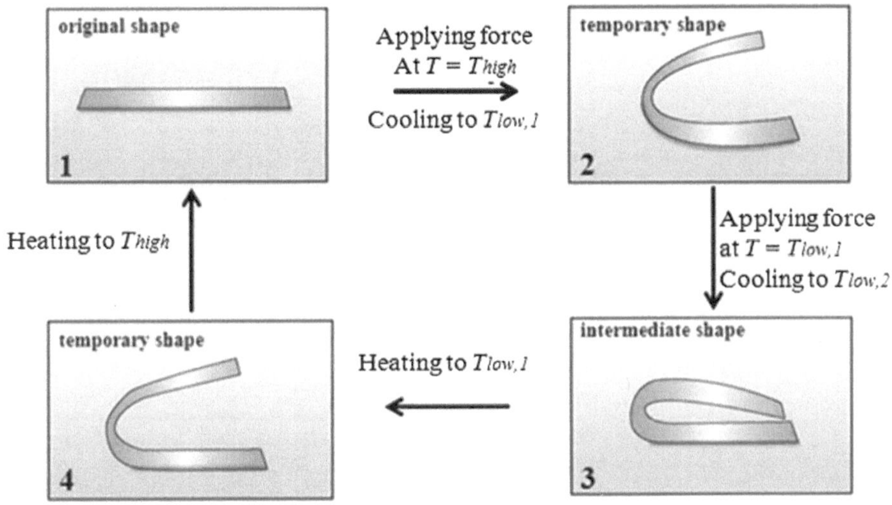

FIGURE 15.3 The triple shape memory effect [22].

effect (1W-1SM). This 2W-2SM is intriguing since it allows for numerous memory form cycles. We highlight one such cycle of different memory forms [20] Figure 15.3.

Since then, several contraptions have been created that make use of these effects. The 2W-2SM has been used to create a polymer actuator that can work without any external load. The 2W-2SM protocol-compliant thermo-responsive polymer laminate development is well under way. In the biomedical field, cross-linked 2W-2SM polymers are commonly utilized to make artificial muscles. Research has been conducted to induce a certain shape-changing sequence in SMPs in response to heat stimulation. Glass transition temperatures are written on samples of epoxy-based SMPs of various compositions. This made it possible for these samples to revert to their previous form by adhering to a predetermined process. Self-interlocking and helical structures were then created using this technology [23].

15.5.2.2 Moisture or Solvent-Induced Actuation

The plasticizing effect is responsible for the form change in polyurethane SMPs caused by water or solvent. As the solution's ambient temperature decreases, the polymer softens and has a lower transition temperature. The low transition temperature is triggered by the form change mechanism, and the shape recovery process follows. SMPs are composed of a mixture of hydrophilic and hydrophobic polymer groups that, when exposed to water, have a shape-changing effect. To create several sets of characteristics, a hydrophilic polyurethane block copolymer was combined in various ratios with hydrophobic polyhedral oligomeric silsesquioxane (POSS) for studying the characteristics of one such group. As the POSS component grew, the material's melting point and crystallinity also increased. A considerable form change has been seen in several composite materials when magnetically activated in an appropriate solvent, as seen in Figure 15.4. These extremely flexible materials are ideal for creating biocompatible structures.

FIGURE 15.4 Magnetically actuated material exhibiting shape change insolvent [24].

15.5.2.3 Magnetically Induced Actuation

When exposed to alternating magnetic fields, iron oxide (Fe_3O_4) and other magnetic nanoparticles in shape-memory nanocomposites show shape-changing characteristics. Due to their ability to change shape when exposed to alternating magnetic fields, Fe_3O_4-based poly (lactic acid)-based SMPs were used to manufacture the scaffold using 4D printing. These scaffolds might be employed for biomedical stents because they expand in response to alternating magnetic fields. Likewise, the SMP system is made up of cross-linked poly(-caprolactone) and Fe_3O_4 nanoparticles that can be used as stents and drug delivery systems. These SMPs have a higher form recovery ratio because they are composed of cross-linked polymer topologies. Because of their super magnetic properties, the Fe_3O_4 particles change shape when exposed to electromagnetic fields. In response to the external magnetic field, these particles transform the energy, which then begins to modify their shape.

Using electromagnetic induction, the SMP can generate the thermal energy required to achieve shape change. Medical devices that change shape when operated by alternating magnetic fields also employ magnetically operated SMPs, such as nickel-zinc ferrite-loaded devices [24].

15.6 APPLICATIONS

15.6.1 Mechanical Actuators

15.6.1.1 Thermo-Responsive Smart Gripper

They created a multi-material, structured smart gripper using the 4D-printing process. They created shape-memory structures by printing them with photo-curable methacrylate copolymer resin on an additive manufacturing system based on projection

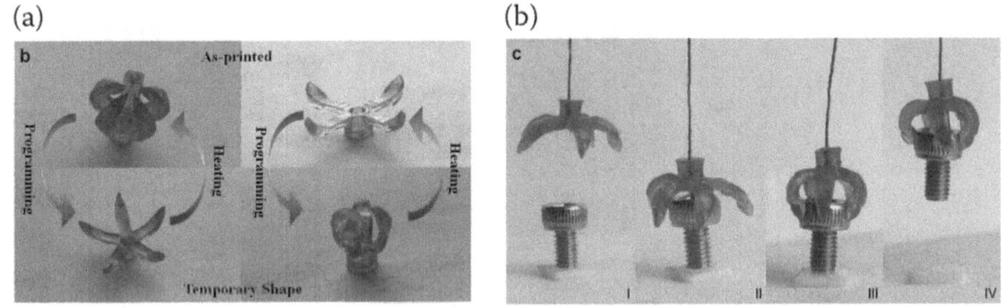

FIGURE 15.5 (a) The transformation of a printed structure into a temporary shape, (b) lifting the bolt with a stepwise operated smart gripper [25].

micro-stereolithography (PSL). In drug delivery systems, printed multi-material structures (Figure 15.5) are actuated by heat and can be used as micrograppers [14].

15.6.1.2 Magnetically Activated Smart Key-Lock Connectors

They created a smart lock key connector using the 4D-printing process. A printing material for 3D printing is made up of reactive diluents, modified alumina platelets, polyurethane acrylate (PUA) oligomers, rheology modifiers, and texturable actuators. The low magnetic field activates these material composites. The internal cuboid's reduced concave shape is altered into a swelling convex shape by administering a stimulus, and it becomes entangled with the outer cuboid. This actuator is used mostly for attaching, namely, to fasten the two tubes (Figure 15.6.b), mechanical fasteners, as well as in the human body to create physical connections between muscles [14].

15.6.1.3 Adaptive Metamaterials

Using FlashForge's FDM machine, beams are created out of thermo-responsive, functionally graded polyurethane metamaterial. Printed beams become active when their glass transition temperature is exceeded by thermal conditions. Self-folding

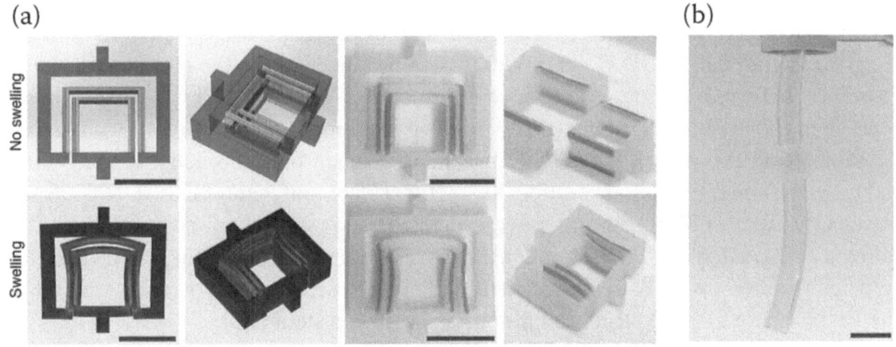

FIGURE 15.6 (a) (top) Before the form alteration, sketches, and real printed key-lock objects (a) (bottom) Finite element calculations and actual photographs of the key-lock structure following the form change, (b) two tubes are fastened together using a cuboid [24].

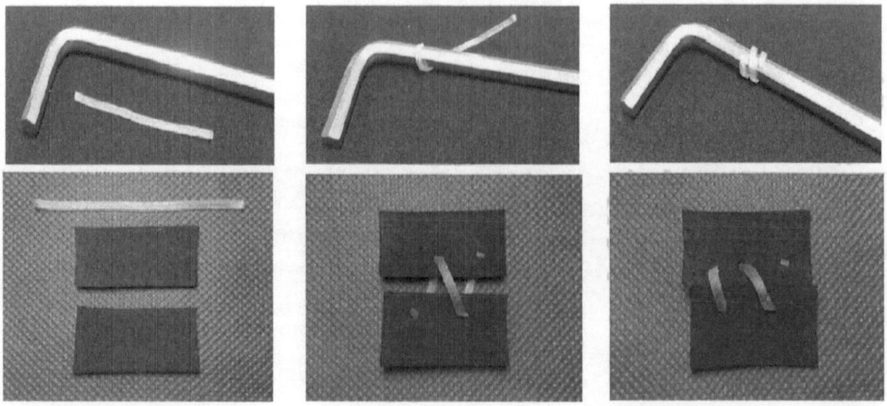

FIGURE 15.7 4D-printed self-conforming devices (on the left). (a) Straightened beam and hex key, (b) conforming to a radius of 6.8 mm, and (c) configuration following full shape recovery. (Right) Self-tightening fiber manufactured in 4D. (d) Straightened beam and soft polymeric substrate, (e) creating a loose knot, (f) heating causes fiber shrinkage [27].

beams and self-coiling springs are used as examples in the medical field to demonstrate their application. Figure 15.7 illustrates the series representation of a beam self-coiling [26].

15.6.2 Bio-Medical Applications

15.6.2.1 Tracheal Stent

They investigated how various illnesses can alter the human respiratory system's tracheobronchial tract architecture. Artificial stents were created to keep the trachea open. However, it frequently develops problems, resulting in a shortened stent life. With stent migration, there was also a severe issue. To address this issue, a C-shaped flap of the trachea is being constructed, which is a fully comfortable gadget that can be put inside the trachea. Using Body Parts3D software, the digital model of the trachea is obtained, which requires data from an MRI scan of the respiratory system in humans. From the digital data, a 3D CAD model is then produced using the software package. The STL file format is then used to save the created CAD model. This STL file is used as an input for stereolithography (SLA) printing machine software. Figure 15.8 depicts the fabrication process. Because the prosthetic trachea must be activated at human body temperature (37°C), a composite methacrylate polycaprolactone (PCL) precursor was used for SLA printing polymeric material. A 98% shape recovery ratio was attained for printed materials containing 88% methacrylate. Furthermore, it is anticipated that the specially designed stent will prevent stent migration and stent failure due to the trachea's true structural alignment with the stent [23].

15.6.2.2 Adaptive Scaffold

Using soyabean oil epoxidase acrylate, they created a biocompatible scaffold for use as a printing medium. This resin is in high demand for bioprinting with

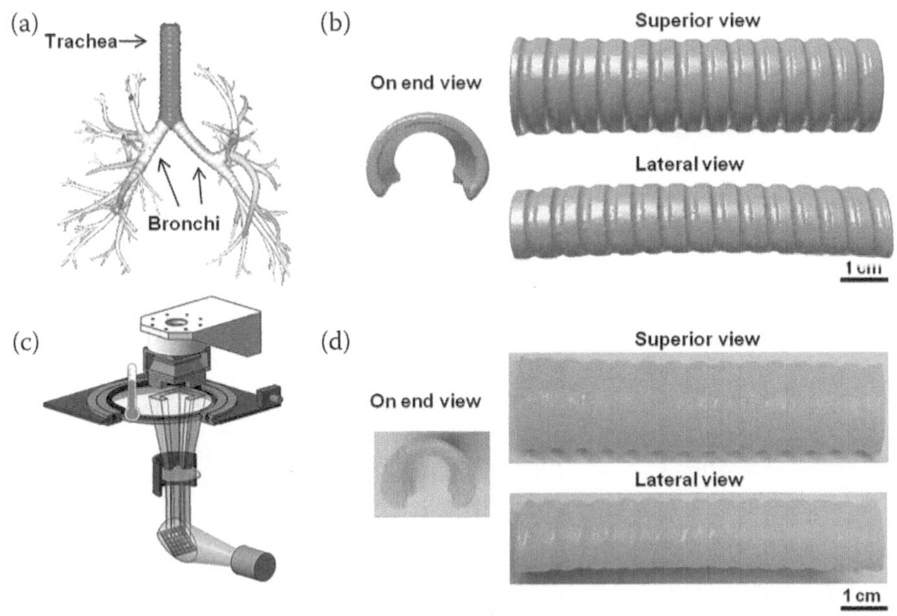

FIGURE 15.8 Depicts the airway stent fabrication technique. a) An MRI scan yielded a digital representation of a middle-aged male's tracheobronchial tree. b) The structure's CAD model. c) The digital models are produced on a commercial SLA printer using a customized heating bath that is heated above the melting point of the resin. d) The printed tracheal stents have a fixed form [28].

3D stereolithography. At human body temperature (37°C), shape recovery is achieved, which is a significant benefit for 4D bioprinting applications. These factors can promote the development of multipotent mesenchymal stem cells (HMSCs) in human bone marrow (Figure 15.9).

15.6.3 Aerospace and Aeronautic

The field of aircraft may benefit from 4D printing. For instance, it's envisioned that 4D printing would be utilized to produce aircraft and drone wings with altered shapes for improved performance. The space industry may employ 4D printing to produce lightweight, space-recoverable structures like solar panels. Self-deployable reflectors are commonly used in spacecraft. By using 4D-printed materials with adjustable volume and size, this is made feasible.

15.6.4 Building and Construction

Material developers are focusing on developing self-healing, self-organizing, and reproducible materials in addition to traditional structural ones. This program might potentially be used for self-cleaning, which is particularly useful in tall buildings where cleaning is difficult. Understanding the diverse responses of these materials to various stimuli requires further fundamental research.

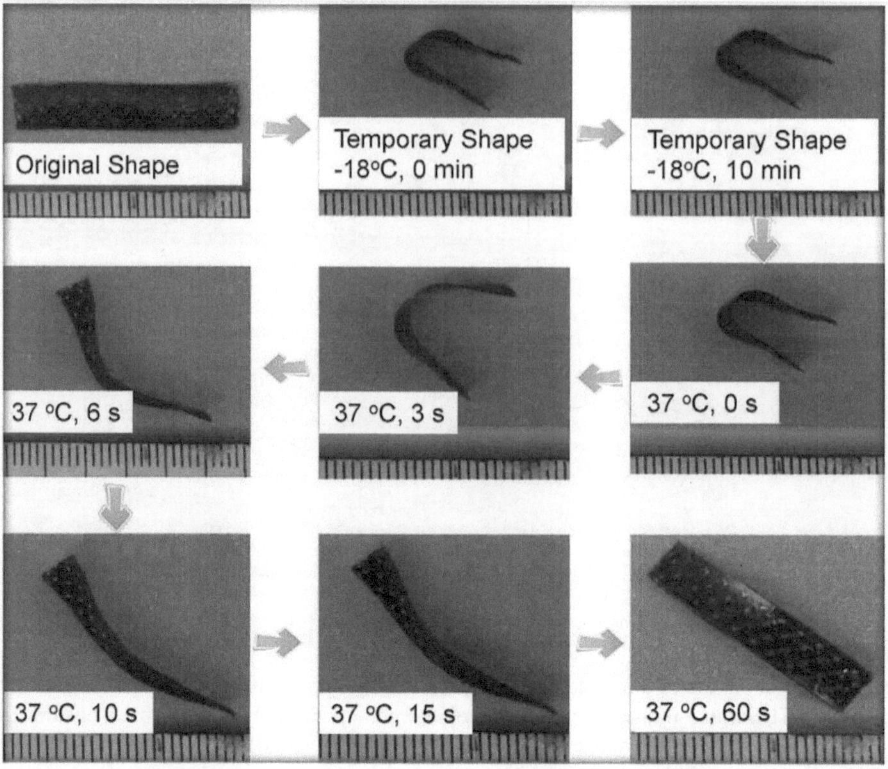

FIGURE 15.9 Shows the behavior of a scaffold changing from a temporary shape at −18°C to a permanent unfolded structure at 37°C in a time-lapse of 60 seconds [29].

15.7 CHALLENGES AND FUTURE SCOPE

Despite being a relatively new technology, 4D printing can solve a variety of problems in the real world. However, there are various problems in this field that must be solved. Problems such as support structures, especially for internal structures that are difficult to access, printing multiple material groups at the same time (such as polymers and metals), a lack of reasonably priced print-ready materials, and extended print times cannot be addressed by current 3D printers. The technology that we are using must be significantly upgraded, or new technology that overcomes the current challenges needs to be developed. Five-axis 3D printing is becoming more common as a means of solving problems that arise with the technology. Another concern is the limitation of 4D-printed objects' mechanical qualities due to desired modifications to their form or properties. Structures like the progressive folding of a locker require certain polymer ratios. Smart printed materials have evolved in recent years, exhibiting a wide range of mechanical properties; this research field is crucial for the advancement of 4D printing. Accurate or delayed actuation, a lack of resources, and limited control over intermediate stages of deformation are further problems with 4D printing. Future research should focus on more effective methods for applying stimuli, such as ways to better manage moisture absorption in hydrogels or how to provide heat in thermo-responsive SMPs. These developments may potentially result in improved actuation accuracy. Furthermore, the fabrication of sophisticated structural modifications can lead to undesirable property changes. Greater flexibility and applicability will be made possible by comprehending the effects of structural pattern size and transformation mechanics, proving that 4D printing has a promising future ahead of it.

REFERENCES

[1] S. Ramesh, S. Kiran Reddy, C. Usha, N. K. Naulakha, C. R. Adithyakumar, and M. Lohith Kumar Reddy, "Advancements in the research of 4D printing-A review," in *IOP Conference Series: Materials Science and Engineering*, vol. 376, no. 1, Jun. 2018, doi: 10.1088/1757-899X/376/1/012123

[2] X. Li, J. Shang, and Z. Wang, "Intelligent materials: A review of applications in 4D printing," *Assem Autom*, vol. 37, no. 2, pp. 170–185, 2017, doi: 10.1108/AA-11-2015-093

[3] E. Pei, "4D printing: Dawn of an emerging technology cycle," *Assem Autom*, vol. 34, no. 4. Emerald Group Holdings Ltd., pp. 310–314, Sep. 09, 2014. doi: 10.1108/AA-07-2014-062

[4] R. El-Dabaa and I. Salem, "4D printing of wooden actuators: encoding FDM wooden filaments for architectural responsive skins," *Open House Int*, vol. 46, no. 3, pp. 376–390, Jan. 2021, doi: 10.1108/OHI-02-2021-0028

[5] I. T. Garces and C. Ayranci, "Advances in additive manufacturing of shape memory polymer composites," *Rapid Prototyp J*, vol. 27, no. 2, pp. 379–398, Jan. 2021, doi: 10.1108/RPJ-07-2020-0174

[6] E. Pei, G. H. Loh, D. Harrison, H. de Amorim Almeida, M. D. M. Verona, and R. Paz, "A study of 4D printing and functionally graded additive manufacturing," *Assem Autom*, vol. 37, no. 2, pp. 147–153, 2017, doi: 10.1108/AA-01-2017-012

[7] E. Pei, "4D printing - Revolution or fad?," *Assem Autom*, vol. 34, no. 2, pp. 123–127, 2014, doi: 10.1108/AA-02-2014-014

[8] H. Yi, "4D-printed parametric façade in architecture: prototyping a self-shaping skin using programmable two-way shape memory composite (TWSMC)," *Eng. Constr. Archit. Manag.*, vol. ahead-of-print, no. ahead-of-print, Jan. 2021, doi: 10.1108/ECAM-05-2021-0428

[9] A. Nespoli, N. Bennato, E. Bassani, and F. Passaretti, "Use of 4D-printing and shape memory alloys to fabricate customized metal jewels with functional properties," *Rapid Prototyp J*, vol. ahead-of-print, no. ahead-of-print, Jan. 2021, doi: 10.1108/RPJ-06-2021-0156

[10] J. Carrell, G. Gruss, and E. Gomez, "Four-dimensional printing using fused-deposition modeling: a review," *Rapid Prototyping J*, vol. 26, no. 5. Emerald Group Holdings Ltd., pp. 855–869, May 19, 2020. doi: 10.1108/RPJ-12-2018-0305

[11] S. S. Mohol and V. Sharma, "Functional applications of 4D printing: a review," *Rapid Prototyp J*, vol. 27, no. 8, pp. 1501–1522, Jan. 2021, doi: 10.1108/RPJ-10-2020-0240

[12] J. Choi, O. C. Kwon, W. Jo, H. J. Lee, and M. W. Moon, "4D printing technology: A review," *3D Printing and Additive Manufacturing*, vol. 2, no. 4. Mary Ann Liebert Inc., pp. 159–167, Dec. 01, 2015. doi: 10.1089/3dp.2015.0039

[13] A. Haleem, M. Javaid, R. P. Singh, and R. Suman, "Significant roles of 4D printing using smart materials in the field of manufacturing," *Advanced Industrial and Engineering Polymer Research*, vol. 4, no. 4. KeAi Communications Co., pp. 301–311, Oct. 01, 2021. doi: 10.1016/j.aiepr.2021.05.001

[14] A. Ahmed, S. Arya, V. Gupta, H. Furukawa, and A. Khosla, "4D printing: Fundamentals, materials, applications and challenges," *Polymer*, vol. 228. Elsevier Ltd, Jul. 16, 2021. doi: 10.1016/j.polymer.2021.123926

[15] H. A. Alshahrani, "Review of 4D printing materials and reinforced composites: Behaviors, applications and challenges," *J Sci: Adv Mater Devices*, vol. 6, no. 2. Elsevier B.V., pp. 167–185, Jun. 01, 2021. doi: 10.1016/j.jsamd.2021.03.006

[16] X. Kuang *et al.*, "REVIEW www.afm-journal.de Advances in 4D Printing: Materials and Applications," 2018, doi: 10.1002/adfm.afdm201805290

[17] Z. X. Khoo *et al.*, "3D printing of smart materials: A review on recent progresses in 4D printing," *Virtual Phys Prototyp*, vol. 10, no. 3, pp. 103–122, Jul. 2015, doi: 10.1080/17452759.2015.1097054

[18] K. R. Ryan, M. P. Down, and C. E. Banks, "Future of additive manufacturing: Overview of 4D and 3D printed smart and advanced materials and their applications," *Chem Eng J*, vol. 403. Elsevier B.V., Jan. 01, 2021. doi: 10.1016/j.cej.2020.126162

[19] Q. Zhang *et al.*, "Origami and kirigami inspired self-folding for programming three-dimensional shape shifting of polymer sheets with light," *Extreme Mech Lett*, vol. 11, pp. 111–120, Feb. 2017, doi: 10.1016/j.eml.2016.08.004

[20] Z. Zhang, K. G. Demir, and G. X. Gu, "Developments in 4D-printing: a review on current smart materials, technologies, and applications," *Int J Smart Nano Mater*, vol. 10, no. 3, pp. 205–224, Jul. 2019, doi: 10.1080/19475411.2019.1591541

[21] K. K. Westbrook *et al.*, "Two-way reversible shape memory effects in a free-standing polymer composite," *Smart Mater Struct*, vol. 20, no. 6, Jun. 2011, doi: 10.1088/0964-1726/20/6/065010

[22] S. Erkeçoglu, A. D. Sezer, and S. Bucak, "Smart Delivery Systems with Shape Memory and Self-Folding Polymers," in *Smart Drug Deliv Syst*, InTech, 2016. doi: 10.5772/62199

[23] S. Joshi *et al.*, "4D printing of materials for the future: Opportunities and challenges," *Appl Mater Today*, vol. 18, Mar. 2020, doi: 10.1016/j.apmt.2019.100490

[24] D. Kokkinis, M. Schaffner, and A. R. Studart, "Multimaterial magnetically assisted 3D printing of composite materials," *Nat Commun*, vol. 6, Oct. 2015, doi: 10.1038/ncomms9643

[25] Q. Ge, A. H. Sakhaei, H. Lee, C. K. Dunn, N. X. Fang, and M. L. Dunn, "Multimaterial 4D Printing with Tailorable Shape Memory Polymers," *Sci Rep*, vol. 6, Aug. 2016, doi: 10.1038/srep31110

[26] A. Rayate and P. K. Jain, "A Review on 4D Printing Material Composites and Their Applications," 2018. [Online]. Available: www.sciencedirect.comwww.materialstoday.com/proceedings2214-7853

[27] M. Bodaghi, A. R. Damanpack, and W. H. Liao, "Adaptive metamaterials by functionally graded 4D printing," *Mater Des*, vol. 135, pp. 26–36, Dec. 2017, doi: 10.1016/j.matdes.2017.08.069

[28] M. Zarek, N. Mansour, S. Shapira, and D. Cohn, "4D Printing of Shape Memory-Based Personalized Endoluminal Medical Devices," *Macromol Rapid Commun*, vol. 38, no. 2, Jan. 2017, doi: 10.1002/marc.201600628

[29] S. Miao *et al.*, "4D printing smart biomedical scaffolds with novel soybean oil epoxidized acrylate," *Sci Rep*, vol. 6, Jun. 2016, doi: 10.1038/srep27226

16 Role of Smart Manufacturing Systems in Improving Electric Vehicle Production

Akash Rai and Gunjan Yadav
Swarnim Startup and Innovation University, Gandhinagar, India

CONTENTS

16.1 INTRODUCTION

Smart manufacturing is a mixture of both digital and physical technology that drives the production and management of machines and tools efficiently [1]. Smart manufacturing is a part of industrial revolution 4.0 that includes the use of manufacturing machinery with the help of various computing terminologies like cloud computing, artificial intelligence, machine learning, and data analytics. The electric vehicle is the future of automobiles because it is four times more efficient than gasoline cars and eco-friendly for the environment [2]. Many countries and their governments with private players are working hard to encourage demand for

DOI: 10.1201/9781003333760-16

electric cars. Many goals have been set by counties for EV adaptation in the future. Now the question comes of how to cope with demand and supply [3]. And the answer is smart manufacturing systems. Mega factories and giga factories in the modern world use intelligent robotics systems to produce cars at a high volume while wasting little in the way of resources like time, materials, and energy [4]. These highly specialized robots can do a variety of tasks, including welding, painting, and reassembly. For automatic driving, safety, and security of EVs, data analytics and artificial intelligence have been applied in high-end electric vehicle cars.

16.2 LITERATURE WORK ON SMART MANUFACTURING APPLICATIONS IN THE ELECTRIC VEHICLE DOMAIN

From past research, it had been observed that smart manufacturing systems had been adopted by industries for increasing productivity and decreasing cost but only a few types of research have been done on electric vehicle smart manufacturing [5,6]. In Table 16.1, it has been observed that battery management in electric vehicles and automated guided vehicles is mainly focused on smart manufacturing.

16.3 RESEARCH METHODOLOGY ADOPTED FOR THE STUDY

This study is going to be a mixture of ISM (Interpretative Structural Modeling) and a MICMAC analysis matrix this helps to identify and interpret the key enablers in smart electric vehicle manufacturing systems. ISM approach helps to enable the linkage among different variables [11]. It helps to make a hierarchical structure for existing variables [3]. In Figure 16.1, the process tree of interpretative structural modelling structure is defined [12–14].

TABLE 16.1
Literature Review on Smart Manufacturing Applications in the EV Domain

Reference	Year	Area Focused
[7]	2022	Role of Big Data in Smart Factories Taking EV Batteries as Their Part of Study
[2]	2022	Use of Machine Learning Algorithm for Enhancing Vehicle Control
[4]	2020	Flexible Manufacturing System for Autonomous Vehicles
[8]	2021	Pricing Strategies for Electric Vehicle Charging
[9]	2021	An Algorithm Had Been Developed for Surveillance with the Use of AI in Autonomous Vehicles
[10]	2015	Focused on Vehicle Performance Measures Using MATLAB and Advisor Software

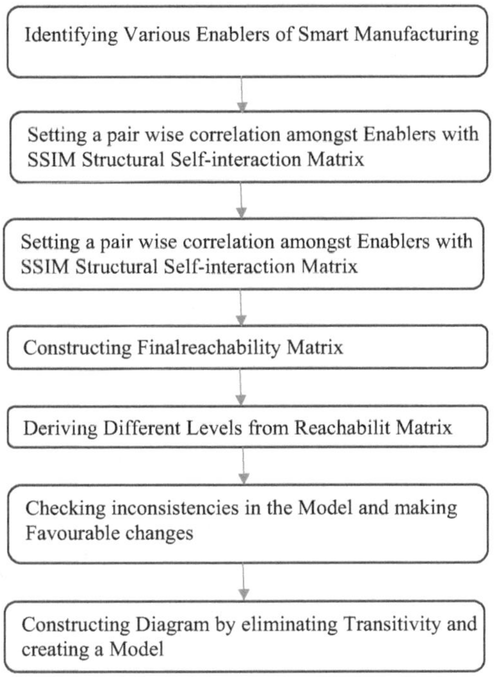

FIGURE 16.1 ISM process tree.

16.4 IDENTIFICATION OF ENABLERS IN SMART MANUFACTURING SYSTEMS IN THE EV DOMAIN

In this analysis, 24 enablers have been identified with the help of previous research. These enablers have been further included in the ISM model for analysis of the enablers. In Table 16.2, a descriptive analysis had been done.

TABLE 16.2
Enabler Identification

Sr. No	Name of EV Enablers	Description
1	Top level management	Top-level management helps to form policies and goals of the organization and this branch is the final decision maker.
2	Government policy structure	These are the external factors that influence the working behavior of a company i.e., these are the set of rules made by the government to encourage the organizational ecosystem.
3	Technological development in EV	Technology includes invention and innovation in the existing ecosystem. Technological advancement is required due to high completion amongst the company.

(Continued)

TABLE 16.2 (Continued)
Enabler Identification

Sr. No	Name of EV Enablers	Description
4	Financial policy	These policies are rules and regulations defined to control accounting and financial practices.
5	Strategic planning framework	Strategies are the variable inputs that are required for running a business consistently. These are a set of plans and procedures that help in decision making.
6	Customer demands	Demand is playing a crucial role in the existence of any product or service in the industry. Customer demand forecasting becomes very important in any industry.
7	Battery technology development	On EVs battery is the most expensive part and the battery influence the driving range of the vehicle.
8	Charging infrastructure for EV	Infrastructure includes charging stations and battery swapping stations for easy adaptability.
9	Availability of smart manufacturing	For fast, convenient, and low-cost manufacturing smart manufacturing systems have to be adopted. It includes the use of IT. In other words, a collaboration of physical and digital assets in the industries.
10	Employee motivation	Employees are the frontline of any business. Reward and recognition programs encourage employees to increase creativity and commitment toward work.
11	Customer feedback	This is a response after using any product or service from customers i.e., rating and suggestions after using the product or service.
12	Battery management system	Battery management system regulates and monitors the battery performance, including efficiency and battery safety management in an electric vehicle.
13	Space availability fast charging technology	Fast charging includes high power and voltage management system that required more space for the power grid at charging stations.
14	AI-assisted robotics for parts manufacturing	AI-assisted robotics is an example of smart manufacturing technology in which programmed assisted robots assemble the vehicle parts.
15	Reward and reorganization	Reward and recognition mean recognition of work done by the employees and providing them financial or non-financial benefits.
16	Effective communication and coordination	Communication and coordination are the two powerful tools in the organizational hierarchy. Better communication increased better coordination among the employees.

TABLE 16.2 (Continued)
Enabler Identification

Sr. No	Name of EV Enablers	Description
17	Use of AI-based chatbots for customer queries, feedback, and suggestions	Due to the high customer queries and feedback, the AI bot helps to manage the general queries of customers i.e., virtual assistant agents and other self-assisted chat boxes on the company's websites.
18	Combination of existing gasoline infrastructure with electric charging	Due to the limited space, a new EV charging Infrastructure is a challenging task, hence updating the existing gasoline stations can help to cut off the space requirement.
19	Automated and programmed manufactured environment	The automated environment consists of programmed machines and AI robots that integrally work to achieve specific goals in the manufacturing systems.
20	Co-ordination amongst employees	Employees are the core of any organization and coordination is the key to increased efficiency, better coordination amongst the employees, and increases the productivity of work.
21	Enhance customer satisfaction	A fast and convenient manufacturing process can help to fulfill huge demand for products by the customers the other with the use of AI effective feedback can be recorded.
22	Optimum utilization of available resources	Need and demand is unlimited and resources are limited, hence the proper allocation and utilization of resources help to manage the upcoming demands of the customers.
23	Better coordination among employees	Better coordination among the employees automatically increases the productivity of the firm.
24	Improve electric vehicle manufacturing using smart manufacturing	With the help of all of the above enablers, the manufacturing system of EVs can be Improved.

16.5 VAXO RELATIONSHIP IDENTIFICATION

The electric vehicle enablers taken in the study were compared with each other using the rule of V, A, X, and O, while their relationship among them, i and j, were used in Table 16.3.

1. V denotes that i will help to attain j.
2. A denotes that j will help to attain i.
3. X denotes that i and j will help to attain each other.
4. O denotes that i and j have no relation.

TABLE 16.3
VAXO Analysis

	Name of EV Enablers	24	23	22	21	20	19	18	17	16	15	14	13	12	11	10	9	8	7	6	5	4	3	2
E1	Top-level management	V	V	V	V	V	V	V	V	V	V	V	V	V	V	V	V	V	V	V	V	V	V	A
E2	Government policy structure	V	V	V	V	V	V	V	V	V	V	V	V	V	V	V	V	V	V	V	V	A	A	
E3	Technological development in EV	V	V	V	V	V	V	V	V	V	V	V	V	V	V	V	V	V	V	V	V	A		
E4	Financial policy	V	V	V	V	V	V	V	V	V	V	V	V	V	V	V	V	V	V	O	V			
E5	Strategic planning framework	V	V	V	V	V	V	V	V	V	V	V	V	V	V	V	V	V	O	V				
E6	Customer demands	V	V	V	V	V	V	V	V	V	V	V	V	V	V	V	V	A	A					
E7	Battery technology development	V	V	V	V	V	V	V	V	V	V	V	V	V	V	V	V	A						
E8	Charging infrastructure for EV	V	V	V	V	V	V	V	V	V	V	V	V	V	V	V	A							
E9	Availability of smart manufacturing	V	V	V	V	V	V	V	V	V	V	V	V	V	V	A								
E10	Employee motivation	V	V	V	V	V	V	V	V	V	V	V	V	V	O									
E11	Customer feedback	V	V	V	V	V	V	V	V	V	V	V	V	O										
E12	Battery management system	V	V	V	V	V	V	V	V	V	V	V	A											
E13	Space availability fast charging technology	V	V	V	V	V	V	V	V	V	V	O												
E14	AI-assisted robotics for parts manufacturing	V	V	V	V	V	V	V	V	V	V													
E15	Reward and recognition	V	V	V	V	V	V	V	V	V														

E16	Effective communication and coordination	V	V	V	V	V	V	A	V
E17	Use of AI-based chatbots for customer queries, feedback, and suggestions	V	V	V	V	V	A	V	
E18	Combination of existing gasoline infrastructure with electric charging	V	V	V	V	A	V		
E19	Automated and programmed manufactured environment	V	V	V	V	V			
E20	Coordination amongst employees	V	V	A	V				
E21	Enhance customer satisfaction	V	A	V					
E22	Optimum utilization of available resources	V	V						
E23	Better coordination among team	V							
E24	Improve electric vehicle manufacturing using smart manufacturing								

In Table 16.3, VAXO analysis had been done with the help of the ISM model enablers in the field of electric vehicles. Relationships amongst the enablers are explained below.

1. The relationship between enabler 1 and enabler 5 is indicated by the letter V in Figure 16.3, which shows how enabler 1 helps to realize enabler 5 strategic framework objectives.
2. The relationship between enabler 5 and enabler 6 is indicated by the letter A. In Figure 16.3, strategic structure enablers help to obtain customer demand.
3. In Figure 16.3, the availability of smart manufacturing and the availability of a charging infrastructure for EVs are interrelated; hence, this link has been marked by the letter X.
4. In Figure 16.3, the variables for the battery management system and customer feedback do not appear to be related in any way, as indicated by enablers 11 and 10.

16.6 FINAL REACHABILITY MATRIX

The reachability matrix denotes the impact of one variable on another.analysis had been done in Table 16.4. The Final Reachability matrix is developed with the help of the VAXO analysis table.

The analysis had been defined as follows:

If E2 influences the 2nd variable, this relation is denoted by 1.
If the 2nd variable influences the E2 variable, it is denoted by 0.
If E2 and 2nd are both influenced by each other, it is denoted by 1.
If E2 and the 2nd variable do not have any relation, it had been denoted by 0.

16.7 MICMAC ANALYSIS

A graph is created as part of the MICMAC analysis to classify components according to their driving and dependent forces [15–17]. For the study's findings and conclusions, MICMAC analysis is utilized to categorize the components and validate the interpretive structural model factors [18–20].

In Figure 16.2, MICMAC analysis has been done. The outcomes include four factors that drive the structure that consists of top-level management, government policy structure, technological development in the EV, and financial policy. These four are the driving forces for enhancing smart manufacturing in the EV domain.

16.7.1 Cluster I–Autonomous of EV Enablers

Autonomous or excluded enablers are the enablers that belong to this cluster. The bottom-left area of the graph contains these enablers, which take into account weak driving and weak reliance power. The characteristics displayed by these

TABLE 16.4
Final Reachability Matrix

EV Enablers	1	2	3	4	5	6	7	8	9	10	11	12	13	14	15	16	17	18	19	20	21	22	23	24
E1	0	1	1	1	1	1	1	1	1	1	1	1	1	1	1	1	1	1	1	1	1	1	1	1
E2	0	0	0	0	1	1	1	1	1	1	1	1	1	1	1	1	1	1	1	1	1	1	1	1
E3	0	0	0	0	1	1	1	1	1	1	1	1	1	1	1	1	1	1	1	1	1	1	1	1
E4	0	0	0	0	1	0	1	1	1	1	1	1	1	1	1	1	1	1	1	1	1	1	1	1
E5	0	0	0	0	1	1	0	1	1	1	1	1	1	1	1	1	1	1	1	1	1	1	1	1
E6	0	0	0	0	0	0	0	0	1	1	1	1	1	1	1	1	1	1	1	1	1	1	1	1
E7	0	0	0	0	0	0	0	1	1	1	1	1	1	1	1	1	1	1	1	1	1	1	1	1
E8	0	0	0	0	0	0	0	0	0	1	1	1	1	1	1	1	1	1	1	1	1	1	1	1
E9	0	0	0	0	0	0	0	0	0	0	1	1	1	1	1	1	1	1	1	1	1	1	1	1
E10	0	0	0	0	0	0	0	0	0	0	0	1	1	1	1	1	1	1	1	1	1	1	1	1
E11	0	0	0	0	0	0	0	0	0	0	0	0	1	1	1	1	1	1	1	1	1	1	1	1
E12	0	0	0	0	0	0	0	0	0	0	0	0	0	1	1	1	1	1	1	1	1	1	1	1
E13	0	0	0	0	0	0	0	0	0	0	0	0	0	0	1	1	1	1	1	1	1	1	1	1
E14	0	0	0	0	0	0	0	0	0	0	0	0	0	0	1	1	1	1	1	1	1	1	1	1
E15	0	0	0	0	0	0	0	0	0	0	0	0	0	0	0	1	1	1	1	1	1	1	1	1
E16	0	0	0	0	0	0	0	0	0	0	0	0	0	0	1	1	1	0	1	1	1	1	1	1
E17	0	0	0	0	0	0	0	0	0	0	0	0	0	0	0	1	0	1	0	1	1	1	1	1
E18	0	0	0	0	0	0	0	0	0	0	0	0	0	0	0	0	1	0	1	0	1	1	1	1
E19	0	0	0	0	0	0	0	0	0	0	0	0	0	0	0	0	0	0	0	1	1	1	1	1
E20	0	0	0	0	0	0	0	0	0	0	0	0	0	0	0	0	0	0	1	0	1	0	1	1
E21	0	0	0	0	0	0	0	0	0	0	0	0	0	0	0	0	0	0	0	1	0	1	0	1
E22	0	0	0	0	0	0	0	0	0	0	0	0	0	0	0	0	0	0	0	0	0	0	1	1
E23	0	0	0	0	0	0	0	0	0	0	0	0	0	0	0	0	0	0	0	0	0	0	0	1
E24	0	0	0	0	0	0	0	0	0	0	0	0	0	0	0	0	0	0	0	0	0	0	0	0

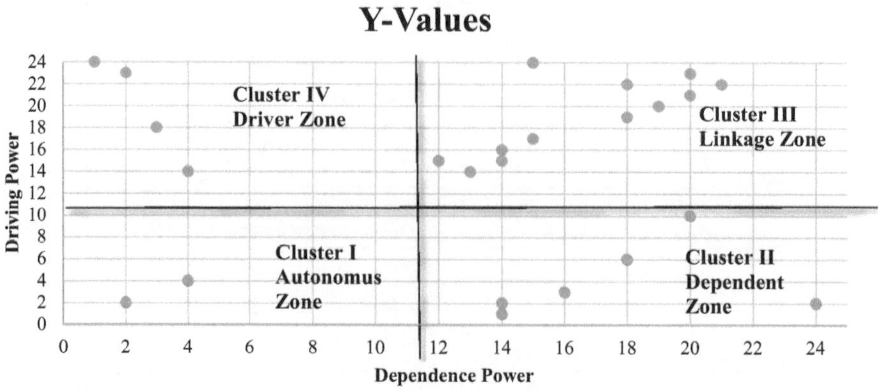

FIGURE 16.2 Cluster diagrams of EV enablers.

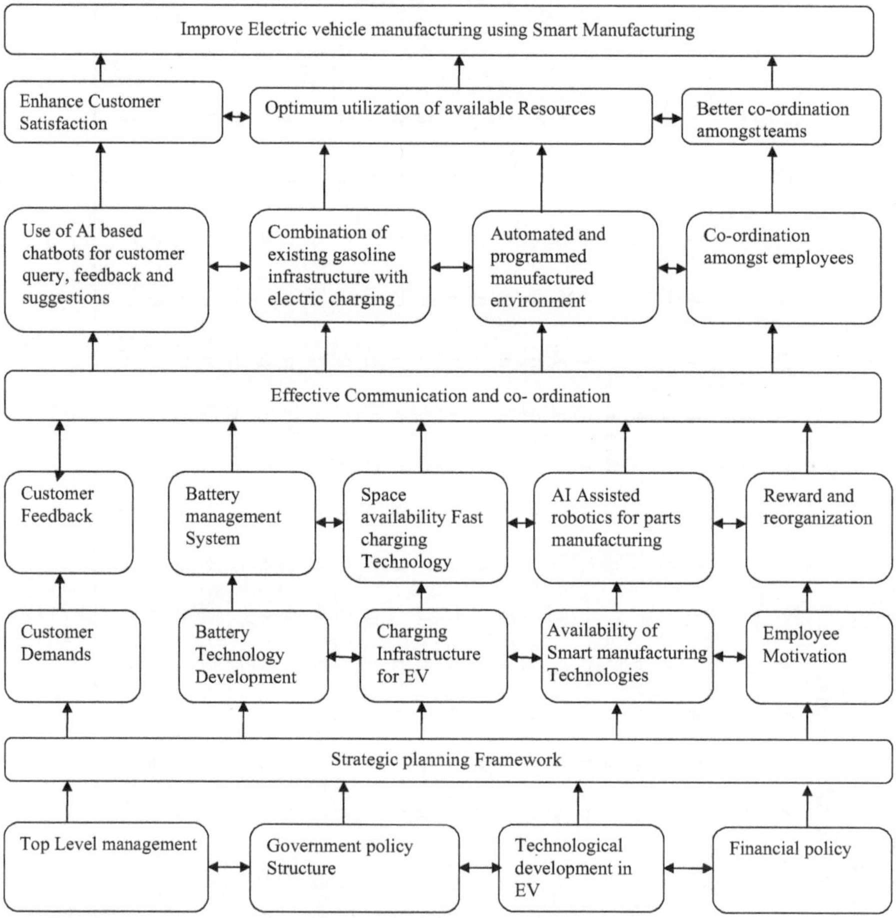

FIGURE 16.3 ISM model of EV enablers.

enablers are inconsistent with the overall system. In Figure 16.2, two autonomous zones have been defined as space availability, fast charging technology, and combination of existing gasoline infrastructure with electric charging have been identified in this zone.

16.7.2 Cluster II–Dependence Zone of EV Enablers

These enablers are located at the bottom-right corner of the graph and consider poor drive and strong reliance power. These enablers display output parameter characteristics across the system. The six EV enablers identified include improve electric vehicle manufacturing using smart manufacturing, enhance customer satisfaction, optimum utilization of available resources, better coordination amongst teams, automated and programmed manufactured environment, and use of AI-based chatbots for customer queries, feedback, and suggestions.

16.7.3 Cluster III–Linkage of EV Enablers

These enablers are located in the unstable top-right region of the graph and show the characteristics of strong driving and strong reliance power. Small changes to these enablers quickly change the behavior of other enablers. In this cluster, 12 enablers have been identified i.e., battery management systems, space availability fast charging technology, availability of smart manufacturing, battery technology development, AI-assisted robotics for parts manufacturing, availability of smart manufacturing, reward and reorganization, employee motivation, customer demands, customer feedback, coordination amongst employees, and automated and programmed manufactured environment.

16.7.4 Cluster IV–Driver Zone of EV Enablers

The enablers found in this cluster have a high driving power and a very low interdependence. These enablers are located in the top-left corner of the graph and function as initiators in the implementing process. They have the characteristics of strong driving and weak reliance power. These enablers function as input in the implementation process and aid in the achievement of other enablers. Four facilitators fall within this cluster in the current investigation that includes top-level management, government policy structure, technological development in EV, and financial policy.

16.8 ISM Model Development with EV Enablers

The ISM methodology begins with the identification of factors that are pertinent to the issue or problem [3]. The choice of a contextually appropriate subordinate relation is followed for process execution [21–24]. A structural self-interaction matrix (SSIM) is created based on a pair-wise comparison of variables once the contextual relationship has been determined [13,14]. In Figure 16.3, the relationship amongst the variables have been showcased.

16.9 CONCLUSION AND FUTURE RECOMMENDATIONS

At the present time, smart manufacturing is the key component of the automobile industry. To fulfill e-mobility demand in the economy, the adaptation of smart manufacturing is required. To deliver the expected quality in optimum time, it is required to adopt smart manufacturing units. In this study 24, electric vehicle enablers have been taken into consideration for this study. It is required to identify the relationships among these enablers to make the EV manufacturing process efficiently. For this, the ISM model was adopted. This model was based on a hierarchical-based approach where relationships among different variables have been analyzed. Further VAXO analysis and MICMAC have been applied and analysis was carried out. Relationships among the enablers are considered to be more specific and key-driven factors in ISM analysis. Further MICMAC has been applied to identify the significant outcomes.

The findings of the study suggested that out of 24 enablers, 4 are identified as the driving forces that have held the overall base structure of ISM that includes top-level management, government policy structure, technological development in EV, and financial policy; these four drive the whole manufacturing cycle in the electric vehicle production process. Technological development was the initial phase in an electric vehicle in terms of efficient battery manufacturing until smart charging infrastructure was developed, on the other hand, top-level management is the decision makes related to various production processes. Government policies are important to promote EVs to common people by providing incentives to manufacturers as well as buyers. Due to the high owning cost of EVs and for charging infrastructural development, financial policies are important factors to be taken into consideration.

In the future, there should be more factors that can be included to enhance the output by including more enablers and drivers so the results may be improved. On the other hand, further improvement can be identified by using grey wolf optimization and the convolution neural network model that is based on AI and ML approaches for decision making.

REFERENCES

[1] Khan, U., & Haleem, A. (2015). Improving to the smart organization: An integrated ISM and fuzzy-MICMAC modelling of barriers. *Journal of Manufacturing Technology Management.*

[2] Bitsch, G., & Schweitzer, F. (2022). Selection of optimal machine learning algorithm for autonomous guided vehicle's control in a smart manufacturing environment. *Procedia CIRP*, *107*, 1409–1414.

[3] Luthra, S., Kumar, V., Kumar, S., & Haleem, A. (2011). Barriers to implement green supply chain management in automobile industry using interpretive structural modeling technique: An Indian perspective. *Journal of Industrial Engineering and Management (JIEM)*, *4*(2), 231–257.

[4] Cronin, M. C., Awasthi, M. A., Conway, M. A., O'Riordan, D., & Walsh, J. (2020). Design and development of a material handling system for an autonomous intelligent vehicle for flexible manufacturing. *Procedia Manufacturing*, *51*, 493–500.

[5] Poduval, P. S., & Pramod, V. R. (2015). Interpretive structural modeling (ISM) and its application in analyzing factors inhibiting implementation of total productive maintenance (TPM). *International Journal of Quality & Reliability Management*, *32*(3), 308–331.

[6] Prasanna, M., & Vinodh, S. (2013). Lean Six Sigma in SMEs: An exploration through literature review. *Journal of Engineering, Design, and Technology*, *11*(3), 224–250.

[7] Kahveci, S., Alkan, B., Ahmad, M. H., Ahmad, B., & Harrison, R. (2022). An end-to-end big data analytics platform for IoT-enabled smart factories: A case study of battery module assembly system for electric vehicles. *Journal of Manufacturing Systems*, *63*, 214–223.

[8] Lin, J., Xiao, B., Zhang, H., Yang, X., & Zhao, P. (2021). A novel underfill-SOC based charging pricing for electric vehicles in smart grid. *Sustainable Energy, Grids and Networks*, *28*, 100533.

[9] Nguyen, M. T., Truong, L. H., Tran, T. T., &Chien, C. F. (2020). Artificial intelligence-based data processing algorithm for video surveillance to empower industry 3.5. *Computers & Industrial Engineering*, *148*, 106671.

[10] Kumar, A. G., Anmol, M., & Akhil, V. S. (2015). A strategy to enhance electric vehicle penetration level in India. *Procedia Technology*, *21*, 552–559.

[11] Yadav, G., & Desai, T. N. (2017). Analyzing lean six sigma enablers: A hybrid ISM-fuzzy MICMAC approach. *The TQM Journal*.

[12] Wu, W. S., Yang, C. F., Chang, J. C., Château, P. A., & Chang, Y. C. (2015). Risk assessment by integrating interpretive structural modeling and Bayesian network, case of offshore pipeline project. *Reliability Engineering & System Safety*, *142*, 515–524.

[13] Yadav, G., Mangla, S. K., Luthra, S., & Rai, D. P. (2019). Developing a sustainable smart city framework for developing economies: An Indian context. *Sustainable Cities and Society*, *47*, 101462.

[14] Pansare, R., Yadav, G., & Nagare, M. R. (2022). Development of a structural framework to improve reconfigurable manufacturing system adoption in the manufacturing industry. *International Journal of Computer Integrated Manufacturing*, 1–32. 10.1080/0951192X.2022.2090604

[15] Samantra, C., Datta, S., Mahapatra, S. S., & Debata, B. R. (2016). Interpretive structural modeling of critical risk factors in software engineering project. *Benchmarking: An International Journal*.

[16] Tripathy, S., Sahu, S., & Ray, P. K. (2013). Interpretive structural modelling for critical success factors of R&D performance in Indian manufacturing firms. *Journal of Modelling in Management*.

[17] Valmohammadi, C., & Dashti, S. (2016). Using interpretive structural modeling and fuzzy analytical process to identify and prioritize the interactive barriers of e-commerce implementation. *Information & Management*, *53*(2), 157–168.

[18] Yadav, N., & Sagar, M. (2015). Modeling strategic performance management of automobile manufacturing enterprises: An Indian context. *Journal of Modelling in Management*.

[19] Zargun, S., & Al-Ashaab, A. (2014). Critical success factors for lean manufacturing: A systematic literature review an international comparison between developing and developed countries. *Advanced Materials Research*, *845*, 668–681.

[20] Pansare, R., & Yadav, G. (2022). Repurposing production operations during COVID-19 pandemic by integrating Industry 4.0 and reconfigurable manufacturing practices: an emerging economy perspective. *Operations Management Research*, 1–20.

[21] Psomas, E. L. (2013). The effectiveness of the ISO 9001 quality management system in service companies. *Total Quality Management & Business Excellence*, *24*(7–8), 769–781.

[22] Psychogios, A. G., Atanasovski, J., &Tsironis, L. K. (2012). Lean Six Sigma in a service context: A multi-factor application approach in the telecommunications industry. *International Journal of Quality & Reliability Management*.

[23] Rajaprasad, S. V. S., & Chalapathi, P. V. (2015). Factors influencing implementation of OHSAS 18001 in Indian construction organizations: An interpretive structural modeling approach. *Safety and health at work*, *6*(3), 200–205.

[24] Sahu, A. R., Shrivastava, R. R., & Shrivastava, R. L. (2013). Critical success factors for sustainable improvement in technical education excellence: A literature review. *The TQM journal*.

17 Safety Management with Application of Internet of Things, Artificial Intelligence, and Machine Learning for Industry 4.0 Environment

Sandeep Chhillar and Pankaj Sharma
Department of Mechanical Engineering, JECRC University, Jaipur

Ranbir Singh
Department of Mechanical Engineering, BML Munjal University, Haryana

CONTENTS

17.1 INTRODUCTION

Industrialization has experienced several mutations since its evaluation. Steam power replaced electricity in machinery to reduce human efforts and was termed the first industrial revolution. First industrial revolution was focused on the concept of power distribution from a central location. Machines then become rigid and light with faster operation times in the second industrial revolution. Electronics-based

DOI: 10.1201/9781003333760-17

automation with a focus on performance optimization started the third industrial revolution. Ergonomics, safety, and flexibility are strengths of automation and are featured with optimization and improvement capabilities.

I4.0 is one of the primary outcomes of strategic initiatives by Germany's government in 2011 [1]. It is the trend name of automation with data over internet and security in production systems. The term I4.0 (I4.0) is generated with the introduction of a new industrial revolution. I4.0 is a new definition to industrial engineering. Artificial intelligence (AI) and cloud and Internet of Things (CoT and IoT) are the main features of I4.0. I4.0 aims to develop and transform machinery into adaptable and responsive modes. Germany conceived that the I4.0 vision needed to be spread worldwide. It aims to achieve an increase in productivity with reduction in production cost and lead times by real-time inter- and intra-communication, big data analysis, machine learning, monitoring and control in real time, machine-to-machine interactions, and machines with decision-making capabilities.

Safety norms and system also need to be updated philosophically, strategically, and technologically as the manufacturing and production philosophies; strategies and technologies get updated. More than a hundred recent industrial accidents prove that the lives of men, women, and children are always at risk due to location of industries and warehouses near residential areas and the hazardous working conditions. There is an urgent need of upgrade in safety norms and system in line with updated technological developments i.e., I4.0. Safety in I4.0 aims to develop digital safety representation of an activity, process, or system. Challenges in its implementation are yet to be identified. Primarily, industrial safety systems safeguard men and machines. Large investments are involved to procure, install, and for maintenance of safety components. Also, it has to comply with safety regulatory standards. Technological advancements require updating of the safety standards and to upgrade the safety systems. Industrial IOT (IIoT) has created new opportunities for integration of safety with compliance. The data collected through sensors is transmitted to the cloud via internet connectivity with required security. Real-time analysis is then conducted, using third-party software to generate alerts and evacuation plans in the case of any accident or emergency, providing real-time monitoring capabilities to safety systems. The demand for industrial safety is expected to grow with the adoption of IIoT in industrial safety compliances/standards. In addition to finance, manpower, society, environment, and technology, sustainability has evolved as a new pillar of development [2]. Sustainability, safety, agility, and high efficiency are considered four primary objectives of I4.0 [3]. Health and safety are the prime foci of sustainability and I4.0, with a focus on hazard, safety, accident reduction, environment, and spreading safety knowledge. I4.0 is still striving with emerging risks and safety requirements [4].

Though workplaces conditions have improved a lot with the application of labor regulations, even then, integration of safety in working has shown better outcomes. Hence, labor laws and regulations need to be updated for I4.0 technologies.

Revolutionary developments in safety are the need of the hour. Safety guidelines for changed technological processes and work methods need to be rolled out. Safety should shift focus to prevent accidents, rather than precautions after accidents. Any business remains healthy until it is safe to work in that business environment. Safety needs to be rewritten for I4.0 with decentralized information and decision-making systems. Intelligent robots are emerging as big safety tools [5].

It's time to analyze existing safety norms. A safety consequence of the fourth-generation industrial revolution needs to be revisited. Optimistic replies to things like proactivity gain, moderating influence, and required apprehension are necessary for integration of safety into I4.0.

IOT, AI, and ML have emerged as major tools and technologies for today's complex systems with a wide range of applications in manufacturing, planning, supply chain management, vendor management, marketing, finance, design, automation, operations management, safety management, disaster management, resource management, and all fields of sciences, technology, management, governance, social sciences, and humanities, with a focus on data science for prediction of events and activities and their management. Safety management with an application of IoT, AI, and ML for the I4.0 environment is the focus in the present article. The objective of this research article is to first analyze the literature published on industrial safety in the context to I4.0, to outline the major findings in context to safety and technology, and the need to revisit the safety standards under I4.0 illumination.

17.2 BACKGROUND

The fourth-generation industrial revolution is rewriting history with internet, IoT, AI, and ML, but the picture is still not very clear. Literature availability for I4.0 safety is limited to a very few articles and applications. Safety or health with logical combinations in digital manufacturing, smart manufacturing, smart production, factory of the future, smart industry, smart factory, advanced manufacturing, etc. are some buzzwords for I4.0. Table 17.1 presents some definitions associated with I4.0.

Complexity of new generation systems has increased to the next higher level [6,7]. I4.0 was the most discussed term in technology and business circles in 2011 [8,9]. From several different publications in the field of I4.0, only a few are found to have focused their interest on safety. An effort of IoT and big data in I4.0 needs to be described deeply with a focus on health and safety through intelligent sensor cyber-physical systems and advanced computation capabilities [10]. A wide range of personal protective devices need to be updated to meet the compliances of I4.0 [11]. Dynamic risk management systems need to be developed. The alarm of new safety risks with advanced manufacturing technologies are required to be installed on the shop floors [12]. The conventional tools are no longer capable for risk assessment and safety with changed technology and manufacturing environments. There is a need to implement new risk analysis models with the capability of monitoring and controlling all safety requirements. Cyber-physical systems are found to aid safety [13]. There is a need of significant input for ergonomics in newer industrial revolutions with the latest technologies [14]. Safety concerns with robots also need to be focused on [5]. A number of safety issues were found associated with big data [15] and wireless communication [16,17]. The use of historical data with real-time data was found to be useful for industrial safety [18]. Optimization

TABLE 17.1
Definitions Associated with I4.0

Category	Definition
Machine Learning (ML)	It is the study of computer algorithms to improve itself automatically through experience.
Internet of Things (IoT)	Inter-connecting machines to transfer data over a network with a unique identity and without human intervention.
Internet of Everything	Bringing people, process, data, and things together to make network connections more valuable and relevant.
Cloud of Things (CoT)	Connecting everything to the internet for the highest performance through a cloud computing based platform.
Big Data Analysis	Large amount of data that needs to be examined to uncover hidden patterns and correlations.
Artificial Intelligence (AI)	The process of self-learning through computer programs and algorithms.
Artificial Intelligence IoT (AIIoT)	The combined platform of AI and IoT for redefinition of industries, businesses, and economics functions to simulate smart behavior and decision making with a full automated approach.
Cyber Physical System (CPS)	Real-time integration of computing, networks, and physical processes (real with virtual) to optimize performance of a system.

for productivity with reduced manufacturing costs and lead times were found critical in safety concerns [19–21]. Several new tools and technologies were found unsafe for workers' health and safety (Uhlmann et al, 2017).

Present safety laws and regulations are also obliging for industries. Safety is favored by a I4.0 framework [22]. Even though the safety practices are guided by the safety management standards (SAFETYAS 18001, CSA Z1000-06, Z1002-12, etc.), even then the safety status is deeply influenced by the corporate goodwill [23]. Safety is usually overseen over production cost and productivity [24]. It is easy to reduce risks in full automation. Researchers and experts are continuously warning about the risks with newer technologies [25]. An industrial hazard needs to be identified through information systems (Ross et al., 2005). Real-time risk management may soon become practical [11,26,27] with the application of AI and ML in decision making [8]. Several countries and unions have produced safety guidelines for newer industrial revolutions. The European Commission (2013) have proposed the following guidelines for integration of safety with I4.0:

- Research is required for integration of human and intelligent equipment.
- Risks associated at all levels of production should be identified.
- Psychosocial risk analysis needs to be done.
- Need to rewrite safety in accordance with I4.0.
- Business models need to be re-examined for safety.
- Need to consider physical, adaptive, and cognitive factors.

- Focus should be on human safety and comfort.
- Modeling of human behavior with stress, reactions, and uncertainty.
- Safety information needs to be updated and protected.

With SAE J4000 standards, I4.0 is useful in lean manufacturing [28]. Zhen et al. integrated intelligence with computing simulation in robotics and demonstrated the social and industrial endeavor of I4.0 in a diversified manner [29]. The cyber-physical system (CPS) was found to be a core element for transformation and acceptable implementation of I4.0 in pre-existing industrial setups. A theoretical analysis of such a system was presented with prompting benefits by Sander et al. [30]. CPS with operation management has proved to be the most successful transformative tool in this successful transformation for manufacturing industries [31]. Data science with prior knowledge of handling and managing disaster situations has proved to be successful in hazard management at the present time [32]. Leadership capabilities like interpersonal, cognitive, business, and strategic skills are found to be another set of parameters for acceptable implementation of I4.0 in present-day industrial setups [33].

Some prompting results by IoT-based safety devices are shared by Akram et al. for women's safety [34]. Niemeyer et al. proposed the idea of a Digital Capability Centre (DCC) to take advantage of IoT-based digital transformation to revolutionary SMEs on the I4.0 platform [35]. A sociotechnical perspective of I4.0 is found to be necessary for sustainable business in the future [36]. Readiness of the employees in adoption of I4.0 is found to be another quantifying factor [37]. The mobile technology is found to have a great impact on performance and sustainability of I4.0 technologies in the industrial sector [38]. Human reliability is another factor in successful implementation of I4.0. Simulated framework for assessment of human reliability was presented by Anastasia et al. [39]. The relationship between I4.0 technology sets and their assessing parameters like productivity have proved to be useful to remain competitive in the market for manufacturing firms [40]. For smart and safe homes, IoT-enabled devices are found useful [41]. Industrial firms can develop robust business models by describing, analyzing, and classifying their business with I4.0 standards [42]. Implementation of I4.0 in several stages reduces the risk and challenges to be successful [43]. I4.0 is applicable to conventional centralized applications also [44]. The impact of I4.0 innovations needs to be studied before their implementation [45]. The industrial safety market is estimated to reach a figure of about 8 billion USD by 2025 and will be driven by reliable safety systems for people [46]. I4.0 is emerging with complex risks over traditional risks. It is now well established that I4.0 will revolutionize industrial safety and its management [47]. Though sufficient papers are published on the I4.0 domain, only a few address safety or risk. The width of the I4.0 technology set is very wide [48]. Academic contribution is desirable in enriching cyber-physical systems, data handling, digital transformation, IT infrastructure, human machine interaction, IoT, cloud computing, and decision making for acceptance and implementation of industry [49].

To take advantage of the I4.0 technologies, evaluation of safety-related issues, their impact on safety management, and implementation guidelines/compliances with proactive strategies and standards need to be examined for a smooth transition towards I4.0.

17.3 SAFETY STANDARDS/REGULATIONS

Standards bring state-of-the-art information and technology to applications. Accidents bring new perspectives and chemistry with the latest technologically upgraded machinery in the market with different safety requirements. Rapid technology changes give new types of stresses/hazards. There always exists a necessary need to rewrite the regulatory standards on safety and health for different countries and evaluate their extent of usage in the industries in the I4.0 era. New advancement also brings better control techniques and safety features. The statutory provisions ensure a minimum standard to guarantee workplace safety. Process of its amendment is time consuming and lengthy to meet the state-of-the-art requirements.

Some non-regulatory standards by national and various different international countries, organizations, and agencies can serve as working and dynamic guidelines. Many times, standards are voluntary and are left open to the organization to ensure product quality and reputation. Several organizations are reluctant in implementation of standards, until they are compulsory or are related with trade. Safety-related ISO Standards are compulsory at some agencies controlling international trade. Several agencies regulate and provide these safety standards or guidelines. Factories Act-1948; Manufacture, Storage & Import of Hazardous Chemicals Rules-1989 (Under EP Act 1986), Indian Explosive Act-1984; Indian Petroleum Act -1934; Building & Other Construction Workers Act-1996, Boiler Act, and Indian Electricity Act-1910 etc. are some regulatory standards in India.

The Factories Act-1948 protects the workforce from industrial accidents and occupational diseases. Disposal, ventilation, pressure, temperature, human density, visibility, working area, material handling, climate, environment, gases, and processes of metal, wood, chemical, rubber, service or any other type of industry is the primary focus of the act.

Some Indian standards include IS 18001-2000 for safety management, IS 14489-1998 for safety audit, IS 3786-1983 for frequency of severity rates, IS 8091-1976 for plant layout, IS 9474-1980 for mechanical guards on machine, IS 8235-1976 & SP 53–1992 for hand operated tools, IS 11016-1984 for machine tools, IS1991:1987 for abrasive grinding wheels, IS:13367:1992 for cranes, IS 181–1968 for welding and thermal cutting, IS 2825-1969 for unfired pressure vessels, IS 5903-1970 for gas cylinders, IS 6044-2000 for LPG storage, IS 2379-1990 for pipelines color codes, IS 4209-1987 for chemical laboratories, IS 11457-1985 for rubber and plastic – fire safety, IS 8964:1978 for wood working machine, IS 1446-2002 for classification dangerous goods, IS 1260–1973 for pictorial marking dangerous goods, IS 11451-1986 for asbestos, IS:4015 Pt I :1967 first aid measures, IS:4015 Pt II :1967 for symptoms, diagnosis & treatment, IS 3043-1987 for earthing, IS:4691:1985 for rotating electrical machinery, IS 5424-1969 for rubber mats, IS 5571-2000 for equipment for hazardous areas, IS 5572-1994 for hazardous area classification, IS 5780-2002 for intrinsically safe equipment, IS 6381-2004 for increased safety equipment, IS 15451-2004 for encapsulated equipment, IS 6539-1972 for telephone for hazardous area, IS 8607-1978 for electrical measuring equipment for explosive areas, IS 11005-1984 for dust proof equipment, IS: 11006: 1984 for flash back arrestor, IS 10300-1982 for road tanker for LPG, IS 4657-1978 for fork lift stability test, IS 6305-1980 for powered industrial trucks, IS 7155 for conveyors, IS 8216-1976 for lifting wire rope inspection, 12735-1994 for wire rope slings, IS 8324-1988 for lifting chain & chain in slings, IS 13367-1992 & IS 13583-1993 for cranes, IS 7194-1994 for noise measurement, IS 3483-1965 for noise reduction, IS 14817-2004 for steam turbine assessment, IS 14817-2004 for aircraft assessment, IS 3646-1992; 3646-1996, 3646-1998, 3787-1983 for interior, IS 875:1987 for structural safety, IS 7969:1975 for building material, IS 4130:1976 for demolition, IS 2190-1992 for first aid fire extinguishers, IS 2189-1999 for fire detection and alarm, IS 2726-1988 for cotton ginning and pressing industrial buildings, IS 3034-1993 for electrical generating & distribution industrial buildings, IS 3594-1991 for general storage and warehouses industrial buildings, IS 6382-1984 for fixed CO2 fire extinguisher, IS 9668-1990 for provision and maintenance of water supplies and fire-fighting, IS 3521-1999 for safety belt and harness, IS 4501-1981 for aprons, IS 6153-1971 for leather clothing, X-ray IS 7352-1974 for lead rubber, IS 8519 1977 for guide for selection, IS 8990-1978 for maintenance and care of industrial clothing, IS 9167-1979 for specifications, IS 8520-1977 for selection, IS 6229-1980 for measurement of attenuation, eye and face protectors IS 1179-1967 for eye and face protectors in welding, IS 5983-1980 for eye protectors, IS 7524-1980 for non-optical test for eye protectors, IS 8521-1994 for face shield – plastic visors, IS 8521-1994 for face shield – wire mesh visors, IS 10667-1983 for selection guide, IS 1989-1986 for leather safety boots and shoes for heavy metal industries, IS 10348-1982 for safety shoes for steel plant, IS 3738-1998 and 5557-1999 for rubber boots, and many other types. There is an urgent need of standard draft with uniform pattern to bring uniformity in approach. Guidelines are also issued by national industry associations, such as MSDS international agencies like ISO, ILO, WHO, ACGIH, British standards, HSE guidelines, and American Society for Testing and Materials (ASTM) codes of practices. Safe and healthy working conditions provide a sense of security among the employees of the industry/organization that in turn increases the productivity and profitability. The sense of belongingness increases with a safe and healthy working environment.

17.4 EMERGING RISKS

Numerous definitions and models of risk management are available in the open-source literature. The sum of consequences, damage, severity, and uncertainty is the most appropriate claimed model of risk [50]. According to ISO 31000:2018, risk can be expressed in terms of its sources, potential events, their consequences, and likelihood [51]. ISO 45001:2018 (OH&S) has defined risk as a combination of the occurrence possibility and the injury severity at the workplace [52]. Risk when viewed from a systemic perspective includes emerging and new conditions, with uncertainty as the main characteristic. ISO 31050 is guidance for managing these risks [53]. I4.0 also generates emerging risks [54]. Decision making needs to be decentralized in integration with artificial intelligence (AI) [55] according to system hierarchy. ISO 31000:2018 & ISO 45001:2018 standards provide guidelines and a generalized approach for risk management. ISO 31050 and CWA 16649: 2013 are standards for managing emerging risks [53]. One of the most common necessities for updating the safety standards is the need of a common technological model around which the safety guidelines for I4.0 can be re-framed. We have proposed such an open-ended model that can be used to re-phrase the existing safety standards for various different application sectors.

17.5 FRAMEWORK FOR AI SAFETY MANAGEMENT SYSTEM (AISMS)

A model as like a computer-aided design and manufacturing system that is proposed in the article. It is similar to computer assistance in design, manufacturing, planning, maintenance, and other associated activities in the production system. A framework with sensors, actuators, IoT gateway, communication protocols, and cloud service is proposed. An IoT gateway communicates the sensor data safely and securely to the publisher. Usually, mqtt protocols are used to secure the data communication. An IoT gateway shares the data with the cloud safely by generating a SAS token for SSL security. The data generated is of the order of several GB to TB per unit time; thus, we need the cloud to store such data, called big data. Big data is stored in cloud services like Azure, IBM, AWS, etc. Different application software packages are available for data analysis. The data stored in the cloud can be accessed using the SSL certifications. After data cleaning and checking the data for all errors and irregularities, the data is fed to ML algorithms. Part of the data is used to train the algorithms and part of the data is used to test the algorithms. Once the algorithm is trained and tested, it leads to the development of artificial intelligence in the machines. The amount of data fed to the developed algorithms leads to increased levels of artificial intelligence in the system. Figure 17.1 presents a generalized Safety Management System (SMS) with I4.0 technologies at its core.

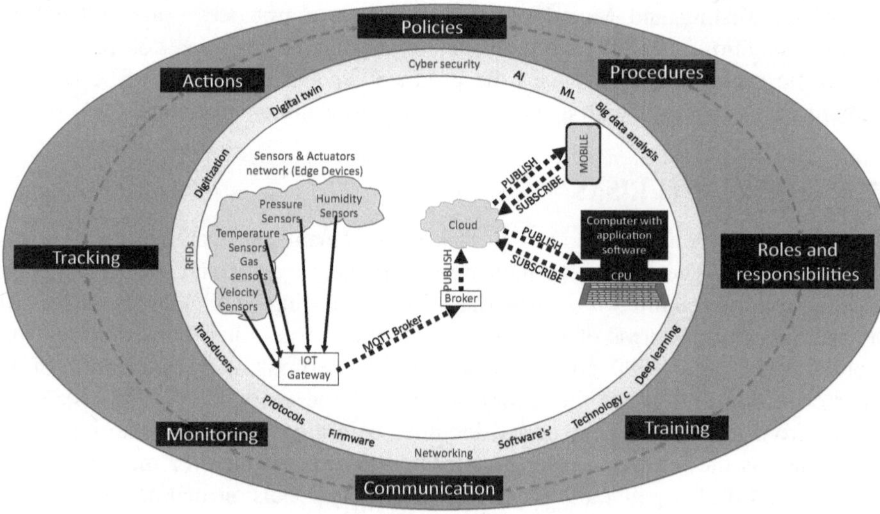

FIGURE 17.1 Safety Management Cycle (SMC) with I4.0 technologies at its core.

Policies, procedures, roles and responsibilities, training, communication, monitoring, tracking, and actions are tasks of the safety management cycle, derived from the standard and well-known Federal Motor Carrier Safety Administration's (FMCSA) Safety Management Cycle (SMC). The data collection, data security, extracting information from the data, decision making, implementation of decision, and actual action on the ground are under the influence of AI and ML devices in I4.0 enabled SMS, which can be called AISMS. The AISMS is claimed to overpower human monitored, controlled, and supervised safety systems. AISMS will have zero tolerance in terms of manipulation when empowered with distributed authorities and blockchain technology. Authority distribution on the one hand fixes roles and responsibilities; blockchain on other hand eliminates the possibilities of any manipulation in the system.

17.6 DISCUSSION

New risks will be generated by advanced manufacturing systems/processes, where conventional risk analysis tools will be incapable of identifying emerging risks [12]. Safety in I4.0 needs significant input for risk management [14]. Work conditions can be significantly improved by adopting proper communication systems [16] and specifically wireless communication is capable of detecting risks in the workplace [17]. Historical data in combination with real-time data improves performance, safety, reliability, and sustainability of industrial systems [18]. Integration of safety with technology positively affects cost, quality, and productivity of the production houses [56]. Businesses houses are focusing to develop production facilities with accident prevention strategies [54]. Though complexity of the systems are increasing with adoption of I4.0 [6], its adoption reduces risks along with optimization of resource, wastage, performance, and flexibility [57] reducing lead time (time to market) and planning costs by 30 to 40 percent respectively [58] improving interaction among different hierarchy levels. Such analyses are yet to be done to prove its worth for industrial safety [19–21]. Overlooked risks might become serious at times [54]. Accidents and hazards can be management by using the right information at the right time. Real-time management of risk is possible with present technologies [11]. AI is likely to be a central player in decision making for better safety and reduced risks [59]. Long-term data observation and its management need to be the core activity for proper safety. Industry needs to enhance I4.0 linked organizational performance along with risk management. To enhance performance, challenges need to be studied first. Deviation from safety and failure are two major risk factors. There is a need of a human-focused dynamic risk management system for the I4.0 environment [60]. Only a few articles focusing on safety and risk in the I4.0 context are available in open-source literature [54]. Personal protective devices use these I4.0 technologies [11].

I4.0 integrates safety systems enabling digital management of operations with a safe working environment. To name, virtual reality (VR) and RFID technology help in workforce management and safety implementation. Safety and risk challenges can be solved by static and dynamic management and by performance analysis of organizations and humans in complex systems. Understanding risks and their complexity and interdependence, gap analysis, modeling and measuring human and organizational performance, adapting the concept of a complex governance system, and using relevant experience can be the potential tools for improving human and organizational performance along with safety in I4.0. The proposed AISMS will revolutionize the safety management cycles in the near future. More such systems like the proposed AISMS need to be proposed for comparative analysis and optimization of the application processes. Blockchain technology needs to be a mandate in safety management to make it fool-proof and manipulation-proof.

17.7 RECOMMENDATIONS

A potential positive and negative concern of the I4.0 technologies needs to be analyzed first. A collaborative effort by academicians, industrialists, professionals, and experts is needed for safety and risk management in I4.0. In this context, recommendations by European Commission (2013) can be used as compliance. To list their recommendations, integration among man and intelligent machines, need of focused research with simulation and analysis on emerging risks, upgradation of existing or development and adoption of new safety, risk and hazard management standards, re-examination of all business models with safety and risk management at its core, and developing safety and emotion sensors need to study and analyze human behavior, their intentions, and stress reactions under difficulty and uncertainty from several different prospectives. The proposed AISMS needs to be implemented in real time to analyze its capabilities.

17.8 CONCLUSION

In 2011, Germany introduced the concept of I4.0 in its growth stage. In addition to advantages, I4.0 also carries safety-related risks. Robust approaches need to be developed to deal with emerging risks. Safety should be incorporated with the devices and their usage with AI and IoT. Safety must be an in-built accessary of the system rather than a mounting. More research articles need to be published on safety in the I4.0 era. Academicians and industrialists should collaborate to develop best safety guidelines, rules, and devices to meet I4.0 challenges. Safety guidelines, rules, laws, and regulations need to be rewritten for the fourth-generation industrial revolution. There is a need of exhaustive research to identify strategic, philosophical, economic, and social outcomes with advantages, disadvantages, and limitations. A multi-objective formulation of the system with probable risk analysis needs to be carried out. Risk zones and risk types need to be redefined. Integration of intelligences and computing simulation; cyber-physical system (CPS); operation management; data science with prior knowledge of handling and managing disaster situations; leadership capabilities like cognitive, interpersonal, business, and strategic skills; sociotechnical perspective; readiness of employees; mobile technology; human reliability; relationship between I4.0 set of technologies and their assessing parameters like productivity; IoT-enabled devices; description; analysis; and classification of business with I4.0 standards needs to be done. Implementing I4.0 in several stages, impact analysis of I4.0 innovations, modeling, physical prototyping, and testing are necessary for successful implementation of the industry. AISMS will surely be the future of safety management systems. Sooner or later, the safety tools, devices, and systems will be updated to new technological requirements.

REFERENCES

[1] H. Kagermann, W. Wahlster and J. Helbig, "Recommendations for Implementing the Strategic Iniciative Industrie 4.0," *Final Report of the Industrie 4.0 Working Group, Acien-National Acadamy of Science and Engineering*, 2013.

[2] K. Jilcha and D. Kitaw, "Industrial occupational safety and health innovation for sustainable development," *Engineering Science and Technology, an International Journal*, vol. 20, no. 1, pp. 372–380, 2017.

[3] F. Qian, W. Zhong and W. Du, "Smart process manufacturing—perspective fundamental theories and key technologies for smart and optimal manufacturing in the process industry," *Engineering Journal*, vol. 3, no. 2, pp. 154–160, 2017.

[4] F. Brocal, C. Gonzalez, D. Komljenovic, P. F. Katina and M. A. Sebastian, "Emerging Risk Management in Industry 4.0: An Approach to Improve Organizational and Human Performance in the Complex Systems," *Complexity*, 2019.

[5] M. Beetz, G. Bartels, A. Albu-Schaffer, F. Balint-Benczedi, R. Belder and D. Bebler, "Robotic agents capable of natural and safe physical interaction with human co-workers," in *IEEE International Conference on Intelligent Robots and Systems*, Hamburg, Germany, 2015.

[6] B. Waschneck, T. Altenmuller, T. T. Bauernhansl and A. Kyek, "Production scheduling in complex job shops from an industrie 4.0 perspective: a review and challenges in the semiconductor industry," in *Centre for European Union Resarch Workshop Proceeding*, 2017.

[7] C. Block, S. Freith, N. Kreggenfeld, F. Morlock, C. Prinz, D. Kriemeier and B. B. Kuhlenkhotter, "Industry 4.0 as a Socio-technical Area of Tension Holistic View of Technology, Organization and Personnel," *ZWF Journal of Economic Factory Management*, vol. 110, no. 10, pp. 657–660, 2015.

[8] A. Rojko, "Industry 4.0 concept: background and overview," *International Journal of Interactive Mobile Technologies*, vol. 11, no. 5, pp. 77–90, 2017.

[9] J. Qin, Y. Liu and R. Grosvenor, "A categorical framework of manufacturing for industry 4.0 and beyond," *Proceedia CIRP*, vol. 52, no. 1, pp. 173–178, 2016.

[10] J. Kaivo-Oja, P. Virtanen, H. Jalonen and J. Stenvall, "The effects of the internet of things and big data to organizations and their knowledge management practices in Knowlegde Management in Organizations," in *Lecture Notes in Business Information Processing*, vol. 224, Slovenia, Ed., Springer, 2015, pp. 495–513.

[11] D. D. Podgorski, K. Majchrzycka, A. Dąbrowska, G. Gralewicz and M. Okrasa, "Towards a conceptual framework of OSH risk management in smart working environments based on smart PPE, ambient intelligence and the Internet of Things technologies," *International Journal of Occupational Safety and Ergonomics*, vol. 23, no. 1, pp. 1–20, 2017.

[12] F. Fernandez and M. Perez, "Analysis and Modeling of New and Emerging Occupational Risks in the Context of Advanced Manufacturing Processes," *Procedia Engineering*, vol. 100, no. 1, pp. 1050–1059, 2015.

[13] D. Kuschnerus, A. Bilgic, F. Bruns and T. Musch, "A hierarchical domain model for safety-critical cyber-physical systems in process automation," in *IEEE International Conference on Industrial Informatics*, Cambridge, 2015.

[14] C. E. Siemieniuch, M. A. Sinclair and M. J. C. Henshaw, "Global drivers, sustainable manufacturing and systems ergonomics," *Applied Ergonomics*, vol. 51, no. 1, pp. 104–119, 2015.

[15] S. Mattson, J. Partini and A. Berglund, "Evaluating Four Devices that Present Operator Emotions in Real-time," *Procedia CIRP*, vol. 50, no. 1, pp. 524–528, 2016.

[16] J. A. Palazon, J. Gozalvez, J. L. Maestre and J. R. Gisbert, "Wireless solutions for improving health and safety working conditions in industrial environments," in IEEE 15th International Conference on eHealth Networking, Applications and Services Healthcom," Boston, 2013.

[17] J. R. Gisbert, C. Palau, M. Uriarte, G. Prieto, J. A. Palazon, M. Esteve, O. Lopez, J. Correas, M. C. Lucas Estan, P. Gimenez, A. Moyano, L. Collantes, J. Gozalvez, B. Molina, O. Lazaro and A. Gonzalez, "Integrated system for control and monitoring industrial wireless networks for labor risk prevention," *Journal of Network and Computer Applications*, vol. 39, no. 1, pp. 233–252, 2014.

[18] G. W. Vogl, B. A. Weiss and M. Helu, "A review of diagnostic and prognostic capabilities and best practices for manufacturing," *Journal of Intelligent Manufacturing*, vol. 25, no. 1, pp. 1–17, 2016.

[19] I. Bucker, M. Hermann, T. Pentek and B. Otto, "Towards a methodology for Industries 4.0 transformation," *Lecture Notes Business Information Processing*, vol. 255, pp. 209–221, 2017.

[20] D. Kiel, C. Arnold, M. Collisi and K. Voigt, "The impact of the industriala internet of things on established business models," *25th International Assocaition for Management of Technology Conference*, 2016.

[21] M. Reuter, Q. Henning, M. Wannoffel, D. Kreimeier, J. Klippert, P. Pawlicki and B. Kuhlenkotter, "Learning Factories' Trainings as an Enabler of Proactive Workers' Participation Regarding Industrie 4.0," *Procedia Manufacturing*, vol. 9, no. 1, pp. 354–360, 2017.

[22] E. MacEachen, A. Kosny, C. Stahl, F. O'Hagen, L. Redgrift, S. Sanford, C. Carrasco, E. Tompa and Q. Mahood, "Systematic review of qualitative literature on occupational health and safety legislation and regulatory enforcement planning and implementation," *Jorunal of Work, Environment & Health*, vol. 42, no. 1, pp. 3–16, 2016.

[23] R. Burke, C. Cooper and S. Clarke, "Occupational Health and Safety," *Gower*, p. 392, 2011.

[24] U. T. Schwarz, H. Hasson and S. Tafvelin, "Leadership training as an occupational health intervention: Improved safety and sustained productivity," *Safety Science*, vol. 81, no. 1, pp. 35–45, 2016.

[25] F. Brocal and M. Sebastain, "Identification and Analysis of Advanced Manufacturing Processes Susceptible of Generating New and Emerging Occupational Risks," *Procedia Engineering*, vol. 132, no. 1, pp. 887–894, 2015.

[26] T. Niesen, C. Houy, P. Fettke and P. Loss, "Towards an integratuve big data analysis framework for data-driven risk management in industry4.0," in *Proceedings of the Annual Hawaii International Conference on System Sciences*, Koloa, Hawaii, 2016.

[27] M. Malinowski, P. Beling, Y. Haimes, A. Laviers, J. Marvel and B. Weiss, "System iSystem interdependency modeling in the design of prognostic and health management systems in smart manufatcuring," in *Proceedings of the Annual Conference of the Prognostics and Health Management Society*, San Diego, 2015.

[28] A. P. Pacchini, W. C. Lucato, F. Facchini and G. Mummolo, "The degree of readiness for the implementation of Industry 4.0," *Computers in Industry*, vol. 113, no. 1, pp. 103–125, 2019.

[29] G. Zhen, T. Wanyama, I. Singh, A. Gadhrri and R. Schmidt, "From Industry 4.0 to Robotics 4.0- A Conceptual Framework for Collaborative and Intelligent Robotic System," in *International Conference Interdisciplinarity in Engineering*, Romania, 2020.

[30] S. Lass and N. Gronau, "A factory operating system for extending existing factories to Industry 4.0," *Computers in Industry*, vol. 115, no. 1, pp. 103–128, 2020.

[31] A. Napolene, M. Macchi and A. Pozzetti, "A review on the characteristics of cyber-physical systems for the future smart factories," *Journal of Manufacturing Systems*, vol. 54, no. 1, pp. 305–335, 2020.

[32] K. Adikaram and C. Nawarathna, "Businees sector preparedness in disaster management: Case study with businesses in southern Sri Lanka in both aspects of natural and technological disasters," in *International Confernence on Building Resilience; Using scientific knowledge to inform policy and practice in disaster risk reduction*, Bangkok, 2017.

[33] V. E. Guaman, B. Muschard, M. Gerolamo, H. Khol and H. Rozenfeld, "Characteristics and Skills of Leaderships in Context of Industry 4.0," in *Global Conference on Sustainable Manufacturing*, Shanghai, China, 2020.

[34] W. Akram, M. Jain and C. Hemalatha, "Design of a Smart Safety Device for Women using IoT," in *International Conference on Recent Trends in Advanced Computing*, chennai, India, 2019.

[35] C. L. Niemeyer, I. Gehrke, K. Muller, D. Kusters and T. Gries, "Getting Small Medium Enterprsies started on Industry 4.0 using retrofitting solutions," in *Conference on learning Factories*, Garz, Austria, 2020.

[36] G. Beier, A. Ullrich, S. Niehoff, M. ReiBig and M. Habich, "Industry4.0: How it is defined from a sociotechnical prespective and how much sustainability it includes- A literature Review," *Journal of Cleaner Production*, vol. 259, no. 1, pp. 120–156, 2020.

[37] S. Fareri, G. Fantoni, F. Chiarello, E. Coli and A. Binda, "Estimating Industry 4.0 impact on job profiles and skills using text mining," *Computer in Industry*, vol. 118, no. 1, pp. 103–222, 2020.

[38] B. Chunguang, P. Dallasega, G. Orzes and J. Sarkis, "Industry 4.0 technologies assessment: A sustainability prespective," *International Journal of Production Economics*, vol. 229, no. 1, 2020.

[39] A. Angelopoulou, K. Mykoniatis and N. Boyapati, "Industry 4.0: The use of simulation for human reliabilty assessment," in *International Conference on Industry 4.0 and Smart Manufacturing*, Rende, Italy, 2019.

[40] S. K. Backhaus and D. Nadarajah, "Investigating the Realtionship between Industry 4.0 and Productivity: A conceptual Framework for Malaysian Manufatcuring Firms," in *Information Systems International Conference*, Hyderabad, India, 2019.

[41] S. Aheleroff, X. Xun, L. Yaqian, M. Aristizabal, J. Velasquez, J. Benjamin and Y. Valencia, "IoT-enabled smart appliances under industry 4.0: A case study," *Advanced Engineering Informatics*, vol. 43, no. 1, pp. 1010–1043, 2020.

[42] J. Weking, M. Stocker, M. Kowalkiewicz, M. Boham and H. Krcmar, "Leveraging Industry 4.0- A business model pattern framework," *International Journal of Production Economics*, vol. 225, no. 1, pp. 1075–1088, 2020.

[43] M. Hirman, A. Benesova, F. Steiner and J. Tupa, "Project Management during the Industry 4.0 Implementation with Risk Factor Analysis," in *International Conference in Flexible Automation and Intelligent Manufacturing*, Limerick, Ireland, 2019.

[44] V. Alcacer and V. Machado, "Scanning the Industry 4.0: A Literature Review on Technologies for Manufacturing Systems," *Engineering Science and Technology an International Journal*, vol. 22, no. 1, pp. 899–919, 2019.

[45] J. Y. Won and M. Park, "Smart fatcory adoption in small and medium-sized enterprises: Empirical evidence of manufatcuring industry in korea," *Technological Forecasting & Social Change*, vol. 157, no. 1, pp. 1201–1217, 2020.

[46] S. S. Kamble, A. Gunasekaran and S. A. Gawankar, "Sustainable Industry 4.0 framework: a systematic literature review identifying the current trends and future perspectives," *Process Safety and Environmental Protection*, vol. 117, pp. 408–425, 2018.

[47] G. Buchi, M. Cugno and R. Castagnoli, "Smart factory performance and Industry 4.0," *Technological Forecasting & Social Change*, vol. 150, no. 1, 2020.

[48] P. Osterrieder, L. Budde and T. Friedli, "The smart factory as a key construct of industry 4.0: A systematic literature," *International Journal of Production Economics*, vol. 122, no. 1, pp. 219–222, 2020.

[49] T. Aven, "The risk concept-historical and recent development trends," *Reliability Engineering & System Safety*, vol. 99, pp. 33–44, 2012.

[50] ISO-31000:2018, "Risk management–guidelines," in *International Organization for Standardization (ISO)*, Geneva, 2018.

[51] ISO-45001:2018, "Occupational health and safety management systems-Requirements with guidance for use," in *International Organization for Standardization (ISO)*, Geneva, 2018.

[52] ISO/NP-31050, "Guidance for managing emerging risks to enhance resilience," in *International Organization for Standardization (ISO)*, Geneva, 2019.

[53] A. Badri, B. B. Trudel and A. S. Souissi, "Occupational health and safety in the industry 4.0 era: A cause for major concern?," *Safety Science*, vol. 109, pp. 403–411, 2018.

[54] T. Stock and G. Seliger, "Opportunities of sustainable manufacturing in industry 4.0," *Procedia CIRP*, vol. 40, pp. 536–541, 2016.

[55] B. J. V. Holland, R. Soer, M. R. D. Boer, M. F. Reneman and S. Brouwer, "Preventive occupational health interventions in the meat processing industry in upper-middle and high-income countries: a systematic review on their effectiveness," *International Journal of Architecture, Occupatational & Environmental Health*, vol. 88, no. 4, pp. 389–402, 2015.

[56] S. Simons, P. Abe and S. Neser, "Learning in the AutFab – The fully automated Industrie 4.0. Learning factory of the University of Applied Sciences Darmstadt," *Procedia Manufacturing*, vol. 9, pp. 81–88, 2017.

[57] Kip Hanson/ Protolabs, "Data, Digital Threads, and Industry 4.0," Manufaqcturing tommorrow, an *online Trade Magazine - Industry 4.0 Advanced Manufacturing and Factory Automation*, 18/08/2016.

[58] S. Percy, "Artificial Intelligence: The Role of Evolution in Decision-Making," 7wdata, 23 MARCH 2017.

[59] D. Komljenovic, G. Loiselle and M. Kumral, "Organization: a new focus on mine safety improvement in a complex operational and business environment," *International Journal of Mining Science and Technology*, vol. 27, no. 4, pp. 617–625, 2017.

[60] G. K. Badhotiya, V. P. Sharma, S. Prakash, V. Kalluri and R. Singh, "Investigation and assessment of blockchain technology adoption in the pharmaceutical supply chain", *Materials Today: Proceedings*, vol. 46, pp. 10776–10780, 2021.

18 CPM/PERT-Based Smart Project Management

A Case Study

Fatih Erbahar
Sakarya University, Natural Sciences Institute, Engineering Management Program, Türkiye

Halil Ibrahim Demir
Sakarya University/Industrial Engineering Department, Sakarya, Turkey

Rakesh Kumar Phanden
Department of Mechanical Engineering, Amity School of Engineering & Technology, Amity University Uttar Pradesh, Noida, India

Abdullah Hulusi Kökcam
Sakarya University/Industrial Engineering Department, Sakarya, Turkey

CONTENTS

DOI: 10.1201/9781003333760-18

18.1 INTRODUCTION

As long as humanity exists, human beings have made many projects, big and small. The Great Wall of China and the Egyptian pyramids are examples of great projects made in the past. Although projects are as old as human history, CPM and PERT, which are common techniques used in project management, are techniques that emerged much later, that is, recently. Before the CPM and PERT techniques, the Gantt chart began to be used. Henry Gantt first used the Gantt chart when World War I was declared in 1914. Petri nets (Kumar et al., 2021, 2022), Critical Chain Method, Earned Value Management, and many others are the useful tools and techniques for project management. However, CPM and PERT were developed in the late 1950s. CPM/PERT has an advantage over the Gantt chart because it utilizes a network to display the order in which activities must occur. In contrast, the Gantt chart does not clearly depict the connection among activities, particularly in complex networks. Therefore, a CPM/PERT system offers a clearer and visually more accessible representation, which makes it a widely used method among project planners and managers (Taylor, 2013).

It is obvious that our need for energy in the world will never end under today's conditions. In this context, companies are making serious investments in the energy sector. Each region has its own cheaper energy potential. Since the Caribbean Region is close to Venezuela and this country is one of the leading countries in terms of oil resources, generators operating with fuel oil are preferred (Al Mustanyir, 2023). These systems are like ship machinery systems.

Our aim is to optimize the design and installation time of the 8-megawatt power plant to be built for a cement factory in Barbados. Here, the project management will be executed separately using both CPM and PERT methodologies. Furthermore, the PERT technique will be used to estimate the likelihood of completing the project prior to the 395-day mark. Then the time cost analysis was performed, and the project was solved with Lingo optimization software; first, the project was shortened to a maximum, and then shortened with a budget of 50,000 dollars, and finally, the project was shortened as long as it was profitable.

The next section of the chapter discusses related literature and section three presents the materials and method used. Section four presents the definition of the problem and section five discuss the results. The conclusion has been presented in the last section of the chapter.

18.2 LITERATURE SURVEY

A project is *"a temporary endeavor undertaken to create a unique product, service, or result"* (Pasian & Williams, 2023). After the definition of the project, it would be appropriate to define project management. Project management is defined according to the British Standard as *"The planning, monitoring and control of all aspects of a project and the motivation of all those involved in it to achieve the project objectives on time and to the specified cost, quality and performance"*. Projects can be classified into various types, including civil or chemical engineering and construction projects, manufacturing projects, IT projects and management projects, and pure scientific research projects (Lock, 2007).

Pinto (2002) examined the concept of project management under many titles, such as the planning stage, risk management, project team coordination, and project budget, and discussed all stages. Critical activities and responsibility matrices were also mentioned in the project.

Ballard & Howell (2003) compared traditional project management studies with lean methods and determined that the project could be managed by minimizing costs with certain interventions. Schieg (2006) worked on risk management in project management and determined that businesses should pay attention to risk management, which consists of a six-step process, in order to be successful in the market. In addition, it has been studied that risk management systems and project management should be integrated. Morris (2010) researched the development of project management in the last 50 years in many subjects, especially in design, methodology, and approach, with certain findings and arguments.

To obtain detailed information about project management, the books of Lock (Lock, 2007) and Schwalbe (Schwalbe, 2015) can be examined. Jiang & Klein (2014) emphasized the importance of project management in the field of informatics. It has been determined that the project management understanding of the project managers and the project team will lead the projects to success. Seymour & Hussein (2014) conducted a study on the historical development of the project management approach. The importance of project management is emphasized with examples from both the first time and the recent past time.

Ömürbek et al. (2015) investigated the selection of the program that can be used with AHP and TOPSIS, which are multi-criteria decision-making techniques, and evaluated the programs of many companies by determining the weights of the criteria. Demirkesen & Ozorhon (2017) emphasized the importance of integration management in construction projects. They investigated the effects on construction project management and determined that the systems had a strong effect on the project by establishing various integration systems in all project management processes.

Gencer & Kayacan (2017) conducted studies on project management with the waterfall method applied to software projects. They also discussed the agile methods approach, compared the two approaches, and presented a project management methodology. Atkinson (1999) focused on the trio of time, cost, and quality, as well as the success of project management and success criteria. The author studied how projects can be developed by making certain estimates and the benefits/costs of stakeholders.

18.3 MATERIALS AND METHOD

The critical path method (CPM) and program evaluation and review technique (PERT) were both created in the late 1950s. In 1956, Morgan R. Walker and a computer scientist named James E. Kelley, Jr. initiated a project aimed at enhancing their company's engineering programs' planning, scheduling, and reporting through a computerized system. The outcome of this initiative was the network approach known as the critical path method (CPM). Around the same time as the development of the CPM method, the PERT method emerged in 1958. The U.S. Navy established a research team led by D.G. Malcolm, consisting of Navy Special

Projects Office members, to create a management control system for the Polaris Missile Project, resulting in the development of PERT. In this project, the PERT method and networks were used.

The primary difference between CPM and PERT lies in their respective use of either fixed duration estimates or probabilistic time estimates. PERT projects utilize three time estimates for each activity, allowing for the determination of the mean and variance through a beta distribution of activity durations, which is shaped by the three time estimates. These three time estimates consist of a = *optimistic time* (when things are going well), m = *most likely time* (when things are going normal), and b = *pessimistic time* (when things are going bad). Based on these times, the meantime and variance values are calculated in the PERT method. Although these two methods were developed by different people, they serve the same purpose. For this reason, these two methods are sometimes called the CPM/PERT method (Pasian & Williams, 2023).

The pioneer of this method is the Gantt diagram found by Henry Gantt in 1910. This diagram is used to create project planning and scheduling. Based on this diagram, planning and scheduling have been made easier by software such as MS Project and Primavera. Microsoft's Excel application is also available to work with a visual in this function. Both PERT and CPM rely heavily on networks to aid in the planning and presentation of activity coordination. In addition, they typically employ a software package to create program information and manage all necessary data related to tracking project progress. Project management software, such as MS Project, is now extensively utilized for these functions.

There are two different representations in the CPM/PERT method solution. These are the AOA (activity on arc) and AON (activity on node) notations. In AOA notation, activities are indicated by arrows, while nodes indicate the beginning and end of the activity. In AON notation, on the other hand, activities are represented by nodes, and arrows show the order of precedence among activities.

Identifying critical activities and project completion time is very important. The project manager or management will understand the importance of these activities while planning the project and will realize the importance of these activities in cost reduction. Many project managers did not adopt the project management approach, and many managed projects ended in failure.

18.4 PROBLEM DEFINITION

Since the Caribbean region is close to Venezuela and this country is one of the leading countries in terms of oil resources, generators operating with fuel oil are preferred. These systems are similar to ship machinery systems. The engine turns the alternator, and the motion energy is converted into electrical energy. Fuel oil used to turn the engine is a cheap and efficient fuel since it is unprocessed; that is, the distilled form of crude oil. Diesel is expensive because it is processed. The engine is capable of using both fuels.

Generally, when the engine is cold, the engine is started with diesel fuel and then the system switches to fuel oil without interruption. When maneuvering on ships, diesel fuel is also preferred because it responds to speed changes in a short time and safely. When the system is stable on long journeys or for electricity generation, it is switched to the fuel oil system. Since cheap fuel is suitable for cheap energy needs, this power plant in Barbados will be established with an installed power of 8 megawatts to meet the electricity needs of the cement factory. In this case, we need project planning considering the design and assembly stages.

Our aim in the project is to examine how we can optimize the design and installation time of the 8-megawatt power plant to be established for the cement factory in the country of Barbados. In short, CPM and PERT methods have been applied to project management. In addition, a necessary time cost analysis was made.

18.5 DISCUSSION AND RESULTS

In this section, the civil, mechanical, and electrical works requested by the main company, namely the customer, are examined in terms of administrative and technical aspects. Project contract scopes are included as a guide for the project in this regard. Our project was in the city of St. Lucy. An 8-megawatt power plant planned to be built for a cement factory in this city. The activities to be carried out during the site period are, respectively, construction activities, mechanical works, electrical works, and commissioning works.

The materials used in the project first reach Panama and then Barbados by sea transportation from Turkey. During this route, customs and transportation costs belong to the contractor company. A 33-day trip was planned for the materials, as shipping by sea is slow. The time difference between the two countries is 7 hours. The difficulties to be experienced due to the time difference were taken into account before the project.

The number of days for project design is limited to 180 days. For this reason, the work to be done in the field will be divided into branches and the design team will share the work among themselves according to their areas of expertise and get approval from the customer. However, if the customer does not comment within 1 week from the date of the presentation, that design will be considered accepted. All design works will be progressed through the transmittal documents, with the approval of the customer through the SharePoint application. For this reason, besides the design team, there will also be a DCC employee; that is, the document controller.

The project language is English as the client is from Barbados. For this reason, those who work in positions such as the project manager and the field staff who will subsequently contact the customer or the design manager must have a command of the English language. The project target delivery time is 13 months, with 6 months for design and 7 months for field assembly. Since the contract was signed on 01.01.2022, 8-megawatt production should be realized by 01.02.2023. If the time is over, the contractor company will pay a fine of 5,000 USD per day to the main company.

The personnel will stay in a hotel close to the construction site and personnel needs such as three meals a day will be provided by the customer. Within the scope of the project, cars, services, etc. vehicles will be provided by the contractor company. If there are visa costs, the costs will be borne by the contractor company, and the work permit costs will be borne by the customer. The contractor firm may choose its own subcontractors. In this context, the workers of the subcontractor company will work under the responsibility of the contractor company and will work with the rights of the main company.

In order to shorten the project period in case this subcontractor is insufficient, details such as additional subcontractor recruitment and personnel increase will be examined, and the project will be examined with three different crashing methods. The three methods entail maximum crashing of the project, project crashing with an additional budget of $50,000, and crashing the project as long as it remains profitable.

First of all, if the project activities and durations are determined, a list will be created, as in Table 18.1 below. Critical paths are activities A, B, F, G, H, I, L, M, N, O, P, Q, R, S, T, Z4, Z7, Z8, Z9, and Z10, painted in yellow.

If we solve the problem with Lingo software, we get the following result. In order to write the Lingo code, we will need a correctly drawn AOA-type representation of the project, as shown in Figure 18.1.

To solve the problem with Lingo software according to the CPM method, it will be necessary to write a model like the one below. Here, for example, x2E represents the earliest event time of the second node, and x2L represents the latest event time of the second node. x2E-x1E>=180; constraint tells us that the earliest event time of the second node must be at least 180 days after the earliest event time of the first node. The same is true for the relationship between the latest event times. x34L=x34E at the last node represents the earliest event time equals latest event time, and this gives us the project completion time. In the objective function, the sign of the early times is positive, and these times are penalized, and these times are ensured to be the earliest. The sign of the late times is negative, which rewards the late times and ensures that these times are the latest. Because it is long, only a part of the program is given below.

```
Min = (x1E+ x2E+ x3E+ x4E+ x5E+ x6E+ x7E+ x8E+ x9E+ x10E+ x11E+ x12E+
x13E+ x14E+ x15E+ x16E+ x17E+ x18E+ x19E+ x20E+ x21E+ x22E+ x23E+ x24E+
x25E+ x26E+ x27E+ x28E+ x29E+ x30E+ x31E+ x32E+ x33E+ 100*x34e) - (x1L+
x2L+ x3L+ x4L+ x5L+ x6L+ x7L+ x8L+ x9L+ x10L+ x11L+ x12L+ x13L+ x14L+
x15L+ x16L+ x17L+ x18L+ x19L+ x20L+ x21L+ x22L+ x23L+ x24L+ x25L+ x26L+
x27L+ x28L+ x29L+ x30L+ x31L+ x32L+ x33L+ x34L); .......

x2E-x1E>=180;      x3E-x2E>=33;       x4E-x3E>=0;        x4E-x2E>=2; .......
x32E-x30E>=6;      x33E-x32E>=6;      x34E-x33E>=9;      x2L-x1L>=180; ......
x3L-x2L>=33;       x4L-x3L>=0;        x4L-x2L>=2;        x5L-x4L>=12; .....
x33L-x32L>=6;      x34L-x33L>=9;      x34L=x34E;         end
```

The result found as a result of this coding is as follows: X34E and X34L values are 364 days.

Variable	*Value*	*Reduced Cost*	*Variable*	*Value*	*Reduced Cost*
X1E	0.000000	99.00000	X1L	0.000000	0.000000
....					
X34E	364.0000	0.000000	X34L	364.0000	0.000000

TABLE 18.1
CPM and PERT Table

Activity Type	Activity Name	Activity Description	Predecessors	CPM Data	PERT DATA		
				Activity Time (day)	a= Optimisitc Time (day)	m = Most Likely Time (day)	b = Pessimistic Time (day)
DESIGN	A	PROJECT DESIGN PROCESS	–	180	150	180	210
LOGISTIC	B	TRANSPORTATION OF MATERIALS TO BARBADOS AND CUSTOMS PROCESS	A	33	30	33	36
MOBILIZATION	C	ACTIVITIES AFTER THE TEAM REACHES BARBADOS (OHS TRAINING, TIME DIFFERENCE ADAPTATION, ETC)	A	2	1	2	3
CONSTRUCTİON	D	MACHINE ROOM CONCRETE AND DUCT WORKS	B,C	12	8	12	18
CONSTRUCTİON	E	PREPARATION OF ENGINE BASE	D	3	2	3	4
CONSTRUCTİON	F	TANK AREA CONCRETE WORKS	B,C	10	8	10	12
CONSTRUCTİON	G	TANK SITE INSTALLATION	F	9	7	9	11
CONSTRUCTİON	H	STEEL WORKS	G	12	10	12	16
MECHANICAL	I	COOLING SYSTEM INSTALLATION (RADIATORS)	H	6	5	6	8
MECHANICAL	J	GENERATOR ASSEMBLY (MACHINE ROOM)	E	7	5	7	9
MECHANICAL	K	FLUE INSTALLATION	J	9	8	9	11
MECHANICAL	L	MACHINE ROOM DIESEL LINE PIPING	I,K	10	9	10	12
MECHANICAL	M	MACHINE ROOM FUEL OIL LINE PIPING	L	12	11	12	14
MECHANICAL	N	TANK AREA DIESEL LINE PIPING	M	10	9	10	12
MECHANICAL	O	TANK AREA FUEL OIL LINE PIPING	N	14	12	14	16
MECHANICAL	P	ENGINE AIR LINE PIPING	O	7	5	7	9
MECHANICAL	Q	ENGINE OIL LINE PIPING	P	7	5	7	9
MECHANICAL	R	ENGINE WATER LINE PIPING	Q	12	10	12	14

(Continued)

TABLE 18.1 *(Continued)*

CPM and PERT Table

Activity Type	Activity Name	Activity Description	Predecessors	CPM Data	PERT DATA		
				Activity Time (day)	a= Optimisitc Time (day)	m = Most Likely Time (day)	b = Pessimistic Time (day)
MECHANICAL	S	ALL FIELD PIPE TESTS	R	5	4	5	8
MECHANICAL	T	MACHINE ROOM VENTILATION SYSTEM INSTALLATION	S	7	6	7	9
ELECTRİCAL	U	INSTALLATION OF ELECTRICAL EQUIPMENT SUCH AS TRANSFORMER AND PANELS	J	3	2	3	4
ELECTRİCAL	V	CABLE CONDUCT INSTALLATION	U	21	18	21	25
ELECTRİCAL	W	PIPE HEATING SYSTEM INSTALLATION	V	5	4	5	6
ELECTRİCAL	Y	HIGH VOLTAGE WIRING	V	6	5	6	7
ELECTRİCAL	Z	LOW VOLTAGE WIRING	W,Y	7	5	7	8
ELECTRİCAL	Z1	FIRE ALARM SYSTEM INSTALLATION	Z	4	3	4	6
ELECTRİCAL	Z2	CABLE STRENGTH AND ACCURACY TESTS	Z1	3	2	3	5
ELECTRİCAL	Z3	WIRING AND PAN COVER ASSEMBLY	Z2	7	6	7	9
PUT INTO USE	Z4	1. ENGINE PROTECTION TESTS	T,Z3	3	2	3	5
PUT INTO USE	Z5	2. ENGINE PROTECTION TESTS	Z4	3	2	3	5
PUT INTO USE	Z6	3. ENGINE PROTECTION TESTS	Z5,Z7	3	2	3	5
PUT INTO USE	Z7	1.ENGINE COMMISSIONING	Z4	6	5	6	8
PUT INTO USE	Z8	2.ENGINE COMMISSIONING	Z5,Z7	6	5	6	8
PUT INTO USE	Z9	3.ENGINE COMMISSIONING	Z6,Z8	6	5	6	8
PUT INTO USE	Z10	72 HOURS PERFORMANCE TESTS AND TRAINING	Z9	9	8	9	10

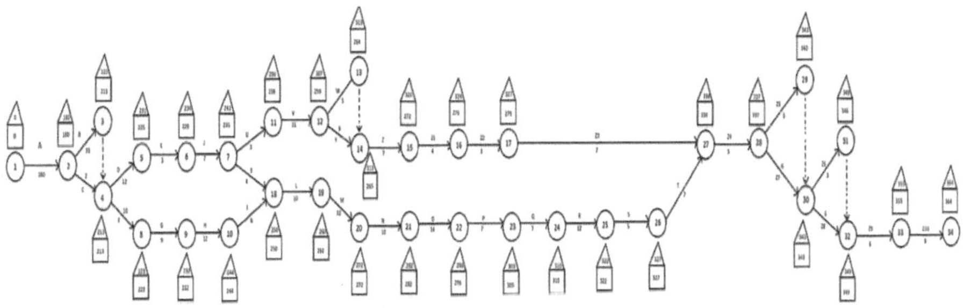

FIGURE 18.1 AOA representation of normal times (project completion time is 364 days).

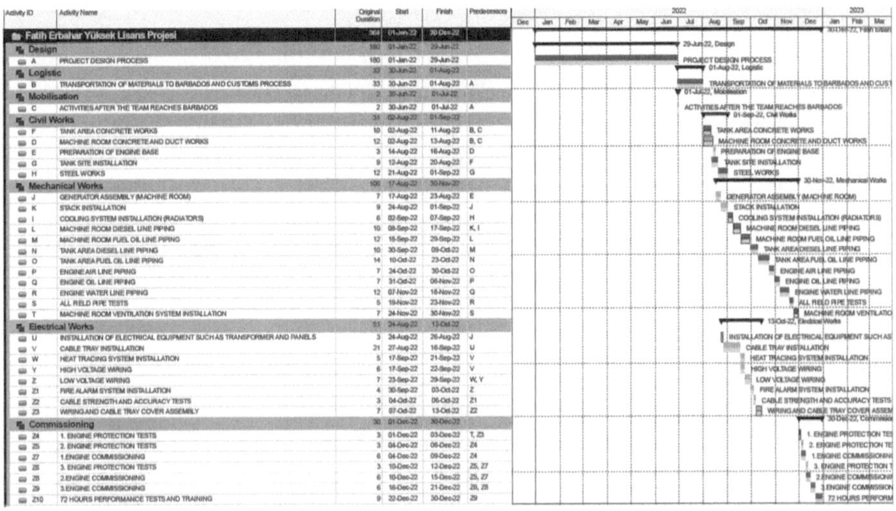

FIGURE 18.2 Screenshot of Project Gantt Chart (Primavera example).

The timeline created in the Primavera program of the project is also shown in Figure 18.2. Thus, the total duration of the project, which started on 1 January 2022 and will end on 30 December 2022, was found to be 364 days. After the table is created, critical activities are created by finding the earliest and latest times with the AOA notation method. As seen in the Primavera program, the project completion time is again 364 days.

18.5.1 Possibility of the Project to Be Finished Before 395 Days

Probabilistic calculations can be made in the PERT method. Optimistic durations, most likely durations, and pessimistic durations of each activity are given in Table 18.1. First, mean activity times and variance of each activity are also calculated according to three-time estimates. The total variance is obtained as 107.47 by summing all variances of critical activities. Since the standard deviation is the square root of this sum, it is found to be 10.36. After these solutions are made, it is desired to calculate the probability of the project that will be finished before 395 days (13 months project duration agreement). According to the formula, the difference in days is divided by the deviation. Therefore, the value (395-365)/10.36= 2.990 is obtained. From the Standard Normal Z table, the value 0.9986 is obtained for 2.99. This means that the project will end earlier than 395 days with a 99.86 percent probability, according to the PERT method.

18.5.2 Maximum Project Crashing

Activities can be shortened by applying methods such as arranging additional subcontractors, personnel bonuses, and increasing working hours. In this case, a table should be created again by predicting how much each activity of the project can be shortened (refer Table 18.2).

It is also desirable to reduce the budget to a certain extent by shortening the project duration, but this is not always the case. In fact, there is an inverse relationship between time and cost. The shorter the time, the higher the costs, but up to a point the benefit of completing the project early offsets the increase in costs. By reducing the duration of the project up to a certain point, there is a decrease in costs, but after a point, shortening the project extra increases the cost of the project. In the creation of the project budget, the shortening cost of each activity should be determined. In order to shorten the normal durations, that is, to shorten the project duration, some costs must be taken into consideration. After determining the crashed days and crashing costs of each activity, Table 18.2 is created.

According to the crashed times given in Table 18.2, the earliest and latest times are found by drawing AOA Project network again. According to crashed activities, critical activities remained the same as activities A, B, F, G, H, I, L, M, N, O, P, Q, R, S, T, Z4, Z7, Z8, Z9, and Z10. The project time decreased to 308 days. Similar values are obtained when Lingo codes were generated according to the maximum shortening.

The Lingo code objective function to be created in the maximum shortening of the project is as follows. Whereas CRASH = Total crash cost, x1E = Earliest event time for node 1, x1L = Latest event time for node 1, 5000*x34E = Punishment cost if project is delayed one more day (or reward if project is finished one more day earlier).

```
Min=CRASH+ (x1E+ x2E+ x3E+ x4E+ x5E+ x6E+ x7E+ x8E+ x9E+ x10E+ x11E+
x12E+ x13E+ x14E+ x15E+ x16E+ x17E+ x18E+ x19E+ x20E+ x21E+ x22E+ x23E+
x24E+ x25E+ x26E+ x27E+ x28E+ x29E+ x30E+ x31E+ x32E+ x33E+ 5000*x34E) –
(x1L+ x2L+ x3L+ x4L+ x5L+ x6L+ x7L+ x8L+ x9L+ x10L+ x11L+ x12L+ x13L+
x14L+ x15L+ x16L+ x17L+ x18L+ x19L+ x20L+ x21L+ x22L+ x23L+ x24L+ x25L+
x26L+ x27L+ x28L+ x29L+ x30L+ x31L+ x32L+ x33L+ x34L);
```

In the Lingo solution, the CRASH value gives the budget to be spent if the project is shortened to the maximum. This value was found to be $75,250. The duration of the project decreased to 308 days. The critical path is the same.

TABLE 18.2
Time-Cost Trade-Off

Activity	Predecessors	Normal Time (day)	Crashed Time (day)	Normal Cost ($)	Crashed Cost ($)	Cost Per Day ($)	Max. Days to Crash
A	–	180	150	150.000	180.000	1,000	30
B	A	33	32	30.000	32.000	2,000	1
C	A	2	2	5.000	5.000	0	0
D	B,C	12	8	55.000	58.000	750	4
E	D	3	2	12.000	13.000	1,000	1
F	B,C	10	8	48.000	51.000	1,500	2
G	F	9	7	76.000	80.000	2,000	2
H	G	12	10	85.000	90.000	2,500	2
I	H	6	5	24.000	26.000	2,000	1
J	E	7	7	6.000.000	6.000.000	0	0
K	J	9	8	26.000	28.000	2,000	1
L	I,K	10	9	16.000	19.000	3,000	1
M	L	12	11	15.000	17.000	2,000	1
N	M	10	9	16.000	18.000	2,000	1
O	N	14	12	15.000	18.000	1,500	2
P	O	7	5	5.000	8.000	1,500	2
Q	P	7	5	7.000	9.000	1,000	2
R	Q	12	10	5.000	8.000	1,500	2
S	R	5	4	2.000	3.500	1,500	1
T	S	7	6	24.000	26.000	2,000	1
U	J	3	2	42.000	44.000	2,000	1
V	U	21	19	50.000	55.000	2,500	2
W	V	5	4	7.000	8.000	1,000	1
Y	V	6	5	15.000	16.000	1,000	1
Z	W,Y	7	5	130.000	145.000	7,500	2
Z1	Z	4	3	12.000	14.000	2,000	1
Z2	Z1	3	2	2.000	3.500	1,500	1
Z3	Z2	7	6	2.000	3.500	1,500	1
Z4	T,Z3	3	2	2.000	3.000	1,000	1
Z5	Z4	3	2	2.000	2.500	500	1
Z6	Z5,Z7	3	2	2.000	2.500	500	1
Z7	Z4	6	5	3.000	4.000	1,000	1
Z8	Z5,Z7	6	5	3.000	4.000	1,000	1
Z9	Z6,Z8	6	5	3.000	4.000	1,000	1
Z10	Z9	9	8	9.000	12.000	3,000	1

18.5.3 Crashing the Project with a $50,000 Additional Budget

The following crash budget constraint is added to the Lingo codes to be created when the project is shortened with an additional budget of $50,000: CRASH<= 50000.

The crash value in the Lingo solution indicates that the project duration will be completed in 319 days if an expenditure of $49,500 is made. At the maximum shortening, it was possible to shorten by $75,250, which is why it was 308 days lower. With a budget of $49,500, it can only be shortened to 319 days. The $500 deviation from the $50,000 crash budget is because only integer values are allowed in activity crashing. After shortening the project at $49,500, the extra shortening will be over $500, and this is where the shortening ends.

18.5.4 Crashing the Project as Long as Profitable

Lingo codes can be created in shortening the project as long as it is profitable and are as follows, whereas 3000*x34E tells us delaying the project one more day costs $3,000. CRASH is the total crashing cost. 1000*y_1_2 tells us that the one-day crashing cost of activity from node 1 to node 2 (Activity A) is $1,000.

```
Min=CRASH+(x1E+x2E+x3E+x4E+x5E+x6E+x7E+x8E+x9E+x10E+x11E+x12E
+x13E+x14E+x15E+x16E+x17E+x18E+x19E+x20E+x21E+x22E+x23E+x24E
+x25E+x26E+x27E+x28E+x29E+x30E+x31E+x32E+x33E)+3000*x34E; ............
CRASH-(1000*y_1_2+2000*y_2_3+750*y_4_5+1000*y_5_6+1500*y_4_
8+2000*y_8_9+2500*y_9_10
+2000*y_10_18+2000*y_7_18+3000*y_18_19+2000*y_19_20+2000*y_20_
21+1500*y_21_22
+1500*y_22_23+1000*y_23_24+1500*y_24_25+1500*y_25_26+2000*y_26_
27+2000*y_7_11
+2500*y_11_12+1000*y_12_13+1000*y_12_14+7500*y_14_15+2000*y_15_
16+1500*y_16_17
+1500*y_17_27+1000*y_27_28+500*y_28_29+500*y_30_31+1000*y_28_
30+1000*y_30_32
+1000*y_32_33+3000*y_33_34)=0; ...........
y_33_34+x34E-x33E>=9; end
```

In the Lingo solution, the CRASH value indicates that $69,000 should be spent if the project is shortened as long as it is profitable. The X34E value indicates that the project will be completed in 310 days.

Variable	*Value*	*Reduced Cost*	*Variable*	*Value*	*Reduced Cost*
CRASH	69000.00	0.000000			
X1E	0.000000	3033.000	X18E	213.0000	0.000000
X2E	150.0000	0.000000	X19E	222.0000	0.000000
....					
X16E	245.0000	0.000000	X33E	301.0000	0.000000
X17E	248.0000	0.000000	X34E	310.0000	0.000000

18.6 CONCLUSION

Each project may not be completed within the time frame as planned. Its lengthening or shortening depends on different conditions. The important thing is to be planned and programmed. As the best words to describe this, *"A project without a Critical Path is like a ship without a rudder"*; starting an unplanned project is not knowing where the end is. At this level, factors such as the experience of the project manager and the quality of the team they work with are very important. All the inspections done during the project design period mean that the personnel in the field experience fewer problems. No matter how small the problem that arises, the work done in the field will be completed in a short time and in this context, the project will continue to be profitable.

In the budgeting part of this project, the crashing was applied in three different ways. If we look at the results in order, of the maximum shortening solution, it can be thought that it will be to the advantage of both the customer and the contractor company that the project period ends in 308 days instead of 364 days. In total, 56 days of earnings mean a daily profit of $168,000 over $3,000. However, when the maximum shortening is made, our shortening cost is $75,250, so it will be beneficial for the project to go for this type of shortening. This benefit is $92,750. Maximum shortening is the preferred method for this project.

If an additional budget of $50,000 is allocated, which is the second solution we implemented to crash the project, the duration of the project decreases to 319 days when an additional budget of $49,500 is spent. With a similar calculation, the cost of the 45-day shortening will be $135,000 as a result of calculating the daily income of $3,000; $49,500 will be spent from the project budget and $135,000 will have profited. The benefit is $85,500.

The third solution we applied was to shorten the project; in case the project is shortened as long as it is profitable, the project duration decreased from 364 days to 310 days and the crash cost is calculated as $69,000. While it earns $162,000 over 54 days of earnings and $3,000 per day, the benefit of the project, which is calculated to be completed in 310 days, is $93,000, unlike the maximum shortening. The most lucrative approach among the three options is to earn $93,000 by shortening the project duration as long as it remains profitable. As a result, the project will be completed within 310 days.

ACKNOWLEDGMENTS

We would like to acknowledge the efforts of reviewers and editors to help us in improving the content of the chapter. Also, we are grateful to the publisher to provide us with a platform to showcase our research work.

REFERENCES

Al Mustanyir, S. (2023). Government healthcare financing and dwindling oil prices: Any alternatives for OPEC countries?. *Cogent Economics & Finance*, 11(1), 2166733.

Atkinson, R. (1999). Project management: Cost, time and quality, two best guesses and a phenomenon, its time to accept other success criteria. *International Journal of Project Management*, 17(6), 337–342. 10.1016/S0263-7863(98)00069-6

Ballard, G., & Howell, G. (2003). Lean project management. *Building Research & Information*, 31(2), 119–133. 10.1080/09613210301997

Demirkesen, S., & Ozorhon, B. (2017). Impact of integration management on construction project management performance. *International Journal of Project Management*, 35(8), 1639–1654. 10.1016/j.ijproman.2017.09.008

Gencer, C., & Kayacan, A. (2017). Yazılım Proje Yönetimi: Şelale Modeli ve Çevik Yöntemlerin Karşılaştırılması. 335–352. 10.17671/gazibtd.331054

Jiang, J. J., & Klein, G. (2014). Special Section: IT Project Management. *Journal of Management Information Systems*, 31(1), 13–16. 10.2753/MIS0742-1222310101

Kumar, A., Kumar, V., Modgil, V., Kumar, A., & Sharma, A. (2021). Performance analysis of complex manufacturing system using Petri nets modeling method. In Journal of Physics: Conference Series (Vol. 1950, No. 1, p. 012061). IOP Publishing.

Kumar, A., Kumar, V., Modgil, V., & Kumar, A. (2022). Stochastic Petri nets modelling for performance assessment of a manufacturing unit. *Materials Today: Proceedings*, 56, 215–219.

Lock, D. (2007). *Project management* (9th ed). Gower.

Morris, P. W. G. (2010). Research and the future of project management. *International Journal of Managing Projects in Business*, 3(1), 139–146. 10.1108/17538371011014080

Ömürbek, N., Makas, Y., & Ömürbek, V. (2015). AHP Ve TOPSIS Yöntemleri İle Kurumsal Proje Yönetim Yazılımı Seçimi. 21, 59–83.

Pasian, B. L., & Williams, N. L. (Eds.). (2023). *De Gruyter Handbook of Responsible Project Management*. Walter de Gruyter GmbH & Co KG.

Pinto, J. K. (2002). Project Management 2002. *Research-Technology Management*, 45(2), 22–37. 10.1080/08956308.2002.11671489

Schieg, M. (2006). Risk management in construction project management. *Journal of Business Economics and Management*, 7(2), 77–83. 10.3846/16111699.2006.9636126

Schwalbe, K. (2015). *Introduction to project management (Fifth edition)*. Schwalbe Publishing.

Seymour, T., & Hussein, S. (2014). The History Of Project Management. *International Journal of Management & Information Systems (IJMIS)*, 18(4), 233–240. 10.19030/ijmis.v18i4.8820

Taylor, B. W. (2013). *Introduction to management science* (11th ed). Pearson.

Index

Note: Page numbers in italics indicate a figure and page numbers in bold indicate a table on the corresponding page.